W9-AVC-679

1000 WONDERS of NATURE

1000 WONDERS of NATURE

Published by

THE READER'S DIGEST ASSOCIATION LIMITED

LONDON • NEW YORK • SYDNEY • MONTREAL

CONTENTS

INTRODUCTION

MOST PEOPLE HAVE HEARD of the Seven Wonders of the World, even though far fewer can name them. But the world is also home to wonders of a quite different kind – ones created not by humans, but by the forces of nature. For more than 4 billion years, these forces have been shaping our planet, and its precious cargo of living things. The result is a bewildering array of natural marvels and spectacles, some of which scientists fully understand, and others which they are only just beginning to explain.

This book focuses on a thousand of these natural wonders, and reveals what lies behind them. Beginning with living things, it looks at some of the strangest inhabitants of our planet, including animals that are even smaller than some bacteria, fungi that digest themselves, and plants that live and flower underground. Here you can also discover wonders of animal behaviour: fish that spend the night in slimy 'sleeping bags' and birds that feed on blood. The list of these living wonders also includes nature's fastest movers, its greatest travellers, and its finest architects – animals such as beavers and termites, which create fantastically elaborate homes without needing to learn how to build.

1000 Wonders of Nature also looks at our planet as a whole, beginning with its setting in the Solar System. Dip into this section of the book and you will learn about phenomena that have puzzled people through the ages, and others that are still new to science. What are comets, and why do they have glowing tails? What is a quasar, and where would you look for one? What makes the Sun shine, and how much longer will it burn before it finally goes out?

Closer to home, you can discover nature at its most violent and destructive. Here, the list of wonders includes volcanic eruptions and earthquakes, as well as extreme weather events such as thunderstorms, hailstorms, and tornadoes. This is also the place to find out about the most spectacular features of the Earth's surface, from its highest mountains to its deepest oceans and caves, and to discover what lies behind some of nature's rarest spectacles, such as the legendary 'green flash' and St Elmo's Fire.

Centuries ago, people explained nature's wonders with myths and legends, rather than by facts. Today, thanks to science, we know much more about how nature works. But as this book shows, the stories behind nature's wonders often makes them seem even more wonderful still.

NATURE'S GREAT EVENTS

1

NATURE'S GREAT EVENTS

Whenever and wherever huge numbers of animals gather together, it is sure to be spectacular. They congregate to meet, court, and mate; to feed when supplies are plentiful; to rest safely in a group; and to prepare for migration.

WALRUSES THROW A BACHELOR PARTY

Walruses love a crowded beach. Each summer more than 12,000 males congregate along the shore and rocky shelves of Round Island, on Alaska's south-west coast.

They lie in rows, eyes closed, their brown and pink bodies packed so tightly together that the pairs of 50 cm (20 in) long ivory tusks are the only indication of which end is which.

During the preceding winter mating season, they will have fought fiercely with the other males with whom they now share the beach, battling for the right to mate with the females.

The bulls arrive at Round Island in shifts during June, usually about 3,000 at a time. They spend a couple of days ashore, then go to sea to feed for a week before returning to the island for a two-day rest.

Meanwhile, the females are feeding and raising their young hundreds of kilometres north among the ice floes of the Bering and Chukchi seas.

HOT SPOT *Large male walruses force smaller ones to give way at the warm centre of the melee.*

RED CRABS' LONG TREK TO THE COAST

A 100-million-strong army of red land crabs embarks on a hazardous journey every November. The onset of the rains signals the start of the exodus from their native rain forest on Christmas Island in the Indian Ocean to the sea. Marching in the cooler parts of the day, the crabs clamber across anything in their path. A million perish on the roads during the two week journey.

Mature males arrive at the coast first and fight for the best mating burrows before the females join them. After mating, the males head home while the females wait for their eggs to develop. Two weeks later, at the water's edge, they shake out thousands of eggs from their brood pouches and set off home themselves, followed after 25 days by an army of miniature crabs, each no more than 5 mm (1/4 in) wide.

WINTER WARMTH *American bison once roamed from Texas to the Yukon, and adapt well to extremes of temperature and climate.*

BISON FIND HOT SPOTS IN THE SNOW

North American bison know just how to keep warm. Each winter in Yellowstone National Park, USA, hundreds of them gather around the geyser fields and hot springs to survive the worst of the icy weather. The snow cover here is not as deep as in the rest of the park, so the grasses and sedges are more accessible for the animals to feed on, and it is more comfortable for them to stay close to the warmth of the hot springs.

As the bison stand next to the geysers, steam and spray freeze on their fur and faces, covering them with a thin layer of hoarfrost. Should they leave the immediate area, the snow can be so deep that they have to shovel through 1.8 m (6 ft) drifts with their broad noses to reach the plants below. In blizzards they stand facing the wind, protected by their thick, shaggy coats.

RELAXING WITH THE BELUGA WHALES

July is holiday time for the thousands of belugas (white whales) that gather in the shallow inlets of Canada's Arctic. They come into the relatively warm fresh water – where they are safe from their main enemy, the killer whale – to give birth, moult, and play. The congregations are noisy, for the whales constantly chatter with chirps, clangs, screams, grunts, and whistles – behaviour that has earned them the nickname 'sea canaries'.

In the estuaries, they play with stones and fronds of seaweed. Large stones are held in the mouth or balanced on the top of the head, and seaweed is draped over the body. As soon as the carrier is spotted, other whales bump and jostle it until the stone is dislodged. One game has tragic undertones. Females have been seen swimming with planks of wood on their backs. They are thought to be whales that have lost a calf and have adopted the wood as surrogate babies.

TIGER MOTHS SEEK AN AEGEAN RETREAT

In the searing heat of July and August clouds of Jersey tiger moths descend on the Greek island of Rhodes. For a month or more, among the rocks and trees of the 'Valley of the Butterflies', they remain immobile by day to conserve energy. At night they go in search of a mate.

The moths are attracted to the valley by the aromatic resin that exudes from the bark of oriental sweet gum trees. Water evaporating from the River Pelekanos below keeps them cool and hydrated. When the worst of the heat has passed, the moths fly off to settle in the surrounding countryside.

TIGER SWARM *A sign in the 'Valley of the Butterflies' on the Greek island of Rhodes warns visitors not to remove any of the resting moths.*

WHY BUTTERFLIES GATHER ON TROPICAL SANDS

Millions of dazzling yellow male heliconia butterflies are drawn to South America's tropical rivers, such as the Amazon and Orinoco. Attracted by the mineral-rich wet sand and puddles, they crowd together on the shore, each one drawing up a solution of minerals and salts through its long, curling proboscis.

During mating, the male butterfly transfers much of his sodium to the female in a sperm package, and he comes to the riverbanks to replenish his supplies. Sodium is essential for the working of their nerves and muscles.

MARINE IGUANAS WORSHIP THE SUN

Like prehistoric monsters from the age of the dinosaurs, rows of 20 or 30 marine iguanas line up on roosting rocks to bask in the sun. These 1.2 m (4 ft) long creatures live exclusively on the Galapagos islands off the coast of Ecuador. They are one of the few sea-going lizards in the world, making brief excursions to feed on the green sea lettuce that grows on submerged rocks 5 m (16 ft) or more below the water's surface.

The time they spend sunbathing raises their body temperature sufficiently to enable them to swim in the cold waters of the Humboldt Current that sweeps up from the Antarctic to bathe the Galapagos. The iguanas are attended by rock crabs, which gather round them to pluck ticks and dead skin from their motionless bodies. Occasionally, a basking iguana bursts into life, sneezing a shower of salty vapour to expel excess salt from glands in its nose.

HAMMERHEAD SHARKS FIND SAFETY IN NUMBERS

For protection, hammerhead sharks congregate in vast schools. These are found close to the islands of the eastern Pacific Ocean, such as the Cocos and Galapagos, where the hammerheads are safe from killer whales and other sharks.

The school consists mainly of females. They swim up and down without feeding and every so often the oldest sharks twist and turn, using their body language to keep the younger ones in their place. The school is also a rendezvous for males and females. Occasionally, a male will

COLONISTS *The marine iguanas of the Galapagos evolved from a land-living species, which arrived by sea, probably from South America.*

DAZZLING ARRAY *Butterfly mass gatherings are often a response to seasonally available resources – in this case minerals and salts.*

grab a female by the pectoral fin. They mate in the depths below. In the evening the school breaks up and the sharks go their separate ways to feed.

MASSED RANKS OF THE HIGH-FLYING QUELEA

Flocks of red-billed queleas can blacken the sky. These small weaverbirds, each weighing no more than 28 g (1 oz), gather in large numbers. Foraging flocks can contain a few thousand birds – one night roost in the Sudan was said to contain 32 million birds. Each red-billed quelea eats about one-sixth of its body weight in wild grass seeds a day. At the end of the dry season, when food is in short supply, the flocks descend on crop-growing areas in river valleys with devastating results for farmers. With more than 1,500 million of them in Africa, red-billed queleas are the most abundant birds in the world.

SPAWNING CORALS MAKE UNDERWATER FIREWORKS

On a December night each year, all the corals on the 2,000 km (1,250 mile) long Great Barrier Reef begin to spawn. Eggs and sperm float to the surface in what looks like an upside-down snowstorm.

Stony coral reefs are made up of soft-bodied coral polyps that secrete calcium carbonate (chalk). When the sea warms up, the polyps' sex organs produce coloured packages of eggs or sperm. By spawning simultaneously, sometimes within a single hour, the coral polyps maximize their chances of fertilization. Within a day the packages burst, eggs are fertilized by the sperm, and the larvae float away.

MASS REPRODUCTION *Corals release huge numbers of eggs and sperm in order to produce enough larvae to recolonize their reef habitat.*

MONKEYS FIND SPRING WARMTH IN WINTER

Japanese macaques, which live the farthest north of any monkey or ape, like to take a hot bath during a winter snowstorm. Although their thick coats keep out the worst of the cold winter weather of northern Japan, they bathe in steaming spring water to warm up to a comfortable temperature.

One mountain population of macaques on Honshu island even relax in hot water while a blizzard rages around their heads. They feed on bark in the depths of winter when most of the vegetation is under snow and so avoid having to move to the lowlands in search of food. In summer they feast on fruits, flowers, and leaves.

MONKEY BATH *Warm-water springs provide Japanese macaques with a comfortable place to relax in winter, when most other animals move down to the valleys.*

FEATHERED FEAST FOR OCEAN TIGERS

Tiger sharks appear with unfailing regularity every July along the shores of the westerly islands of Hawaii. At this time, the local nesting population of Laysan albatross chicks is about to take its first flight. The young birds practise their take-offs on the beach, but inevitably they are blown out to sea by the trade winds. If the winds drop,

many ditch in the water where the sharks are waiting.

How the sharks know when and where to take advantage of this seasonal bonanza is a mystery. They rush up from below, pushing their noses out of the water. The chicks have a good chance of escaping, because they are forced away by the pressure wave in front of the sharks' snouts. Some sharks are wise to this and speed across the sea's surface to land open-mouthed on their victims.

MANTA RAY PERFORMS WATER GYMNASTICS

The manta ray 'flies' through the tropical waters with great sweeps of its large, wing-like pectoral fins. Sometimes it leaps clear of the water and cartwheels with one fin emerging as the other descends into the sea.

The manta, or devil, ray is the world's biggest ray. It has a wingspan of 6 m (20 ft) and can weigh in excess of 1,360 kg (3,000 lb). Unlike other rays that remain close to the seabed, mantas live near the surface, where they feed on the plankton that swarms mainly at night. They can be seen swimming in formation like a squadron of aeroplanes.

Mantas are recognized instantly by the 'horns' on either side of their broad head, a feature that led ancient sailors to call them 'devil fish'. They are, in fact, quite harmless and use their horns to guide their prey into their broad rectangular maw.

CAPE BUFFALO CHARGE

Cape buffalo have an aggressive way of deciding who is top bull. These powerfully built East African creatures clash head-on, crashing their awesome curved horns together with extraordinary force.

The buffalo's only natural enemy is a pride of lions, which targets the cows and calves. In defence, the herd lines up, each animal standing and staring, nose held high. A few steps forward, accompanied by loud snorts and a tossing of the head, is often enough to deter a predator, but should it attack it could be in for a shock. A mighty buffalo bears down on the predator at 56 km/h (35 mph), head held high so it can focus on its adversary right up until the last moment, when it lowers its horns and butts its opponent with a force that can kill outright.

SEA EAGLES ADD A SEASONAL SPLASH OF COLOUR

The spectacular bright yellow bills and white-on-black markings of the Steller's sea eagles cut a dash on the ice floes of the Japanese island of Hokkaido. The eagles arrive in winter from their summer breeding grounds in north-east Siberia. They roost in the coastal forests during the night. By dawn, 40 or more Steller's eagles can be seen following local fishing fleets for discarded fish. The birds swoop down with feet lowered and tail fanned. If the fishing is poor, they scavenge on the carcasses of dead seals.

PARROTS SEEK NATURAL REMEDY

Parrots and macaws in Peru's Manu National Park go to great lengths to obtain their daily medicine. This is found in the grey-pink cliff-side clay of the park's high riverbanks: the clay is a natural kaolin mixture thought to help to neutralize plant poisons. This enables birds like macaws to consume young, unripe fruits that are high in toxins before other creatures can feed on them.

The clay-licks are a dangerous place for the birds. They not only attract clay-eaters, but also their predators, such as jaguars and pumas. Following the principle of safety in numbers, the parrots and macaws gather every morning in the surrounding trees. They chatter animatedly for a couple of hours. If no more than 20 birds turn up, the group will abandon its visit and disperse. If there is a quorum, the birds will drop down for a dose of kaolin.

HANGING TOGETHER *Green-winged macaws gather at a clay-lick. By visiting in numbers, they reduce the risk of being surprised by a predator.*

JELLYFISH TRACK THE SUN

Jellyfish in lakes on the islands of Palau in the western Pacific Ocean perform a daily ritual. They pulsate their bells to move them to the sunniest stretches of water. The jellyfish were trapped here millions of years ago when land levels rose, creating saltwater lakes.

Deprived of their natural food, such as small fish, they took to a symbiotic way of life with tiny green algae that now live in their tissues. The algae synthesize food from sunlight, giving some to their jellyfish hosts. In return the jellyfish keep the algae in sunlight by following the sun around the lakes.

BALD EAGLES FLY IN FOR FOOD CONVENTION

About 3,000 bald eagles gather in Alaska's Chilkat Valley each November. This is just when a late run of chum salmon swims upstream towards its spawning ground. A geothermal upwelling of warm water keeps stretches of the river ice-free, allowing the birds to feed even in winter. The eagles swoop down from the trees, talons extended, and snatch the fish. They are not averse to acts of piracy, bullying ospreys, gulls, river otters, and immature bald eagles and stealing their hard-won food.

2

FAMILY LIFE OF ANIMALS

ANIMAL FAMILIES

In the animal world, there are advantages to be gained from being part of an extended family. Under the leadership of the most experienced member, the group cooperates to raise and protect its young, to spot danger, and to hunt.

SILVER MASTER KEEPS ORDER IN THE JUNGLE

Petty jealousies and wayward offspring are part of daily life in the mountain gorilla family. It is the dominant male, known as the silverback, who settles these conflicts. He achieves this with nothing more than a penetrating stare or a cuff to the back of the offenders' heads. As his aggression rises, that of his females diminishes and peace is restored once more among the group.

Mountain gorillas, from the forests of Central Africa, are the world's largest living primates. Though large and powerfully built, these creatures are gentle and peace-loving. The family is central to gorilla life and the large and dominant male is the linchpin. It is this powerful patriarch who controls the group, defending it against predators, such as leopards, and chasing away rival males.

Adult mountain gorillas may live together for a lifetime, making the gorilla family one of the most stable social groups of all the great apes, but not necessarily the most harmonious.

While many females in a silverback's harem remain with him for most of their lives, not all stay at home. Young females can have their heads turned by another handsome silverback and switch allegiance. As a result, many of the females in a gorilla group are unrelated and so have little interest in each other. This is one of the main causes of family squabbles, particularly at mealtimes.

STABLE FAMILY *A mountain gorilla family group contains five to ten individuals: a mature male, younger males and females, and their offspring.*

KILLER WHALES OPT FOR LIFE IN THE POD

Killer whales live together in groups known as pods, and go to great lengths to protect each other. In one instance off Canada's west coast two whales were seen supporting a companion injured by the propeller of a boat.

The whales, also known as orcas, usually remain with the same pod for life. Their leader is the oldest female. A large pod may split into smaller groups consisting of a 'granny' and two generations of offspring. When food is plentiful, or at breeding time, pods often join forces, forming a 100-strong super-pod.

MATRIARCH'S IRON RULE IN THE ELEPHANT HERD

Females stick together in the elephant world, living in herds under a matriarch. The experience and knowledge of this elder stateswoman, who may be 60 years old, are of great benefit to the group. She will remember, for example, the location of water holes and seasonal

GRANDMOTHER'S FOOTSTEPS *A herd of African elephants in the Amboseli National Park, Kenya, is led by the oldest and wisest female.*

food supplies. In times of danger, such as when threatened by a pride of lions, the group bunches around the young and it is the matriarch who decides whether to flee or confront the threat.

A family herd comprises of up to 20 adult females and their young, including the matriarch's grown-up daughters and her sisters with their offspring. If the matriarch should die, her eldest daughter will often take over.

Adolescent males leave at puberty to form their own 'bull bands'. When they mature, they become solitary nomads, attracted to a family group only when one of its mature females, or 'cows', is ready to mate.

GUARDING THE FUTURE *The large male lions in a pride protect the youngsters, which will carry their genes on to the next generation.*

SOCIABLE LIONS FIND SAFETY IN NUMBERS

Lions are the only cats to live in a large family group, known as a pride. Native to Africa and India, they live in groups which can contain up to 40 animals, but usually have about 15: five related females and their offspring, guarded by a pair of powerful males.

Family life is relatively harmonious with the exception of mealtimes, when tempers fray. Cubs may suckle from any mother with milk, so orphaned cubs do not starve. Cast out at puberty, males live in the wilderness for some years before seeking their own pride by challenging resident males in a bloody battle. The male lion's role is to protect his pride from other males, which would kill his young, and predators, such as hyenas.

FEMALE HORNBILL'S GIFTS OF COURTSHIP

Southern ground hornbills present their partners with a nuptial gift. Anything from insects, frogs, and snakes to small mammals will suffice.

Only one pair in a group of eight birds breed; the rest act as helpers. Courtship is noisy and active: the birds emit loud booming calls at dawn and the enlarged section at the top of the beak, known as the 'casque', helps the sound resonate. The female beats her beak on the ground, flashes her wing feathers, and presents the male with gifts, usually insects. When the female is brooding her pair of eggs, members of the group bring her food.

Reluctant to fly, ground hornbills roam their 100 km^2 (40 sq mile) territory in the forests and grasslands of Central and East Africa on foot, travelling up to 11 km (7 miles) a day in search of prey. They eat fruit occasionally, but otherwise are mainly carnivorous, sometimes scavenging on the carcasses of antelopes and zebras that have been killed by lions or hyenas.

HOW ROYAL SCENTS KEEP HONEYBEES IN ORDER

The queen bee keeps her colony in check by producing special scents called pheromones. These scents prevent sexual development in the workers, which are all females, and therefore repress any sex drive.

There may be 50,000 honeybees in a single colony. Of these, 1,000 or so are idle male bees (drones), and the rest are workers that do just about everything: they build wax combs; collect pollen, nectar, and water; attend eggs; feed larval bees; cool or defend the colony; and carry out dead bodies.

The only individual to produce eggs is the queen, who lays up to 2,000 a day. She leaves the colony on just two occasions. At the beginning of her reign she takes a nuptial flight during which she mates on the wing with up to ten drones. Towards her reign's end, when the scent she produces is too dilute to maintain social order, she takes 1,500 to 30,000 workers and

A TASTE FOR FLESH *Unlike most kinds of hornbill, which search for fruit in the trees, the southern ground hornbill prefers to eat meat. This frog will be swallowed whole.*

drones with her to pastures new. Several possible new queens will have been reared by the workers to take her place. The first to emerge stings her sisters to death to become monarch.

THE TOP DOGS IN A WOLF PACK

Mating among wolves is strictly limited to the alpha male and female of the pack. The rest of the pack, which might have up to 30 or 40 members in exceptional cases, but more usually consists of six or seven wolves, supports the alpha pair, helping to raise and protect the cubs.

The pack travels and hunts together over a territory whose size depends on the availability of food: a pack of ten wolves in Alaska, for example, requires a home range of 12,000 km^2 (4,600 sq miles) to find enough food.

SHOWING RESPECT *With tails lowered and ears held back, two submissive wolves greet a more dominant member of the pack.*

Each wolf is aware of its status, but all wolves defer to the dominant male and female. Body postures, such as the position of the tail and ears, reinforce the hierarchy. Older members of the pack will fight for higher status, and even challenge the alpha pair.

YOUTHFUL BEE-EATERS HELP OUT IN THE NEST

White-fronted bee-eaters differ from most other birds in that the young do not all leave after fledging. They stay with their families to become helpers, a behaviour known as 'cooperative breeding'. They might dig nesting chambers, catch food, and feed the chicks of their parents, uncles, and aunts. Females take on incubation duties, while males guard against predators, such as snakes.

Found in the savannah of Central and East Africa, there may be up to 17 white-fronted bee-eaters in a group, with four generations present. About 15 to 25 of these large families live in a single colony.

COMMUNAL NEST *A female ostrich visits the nest of the 'major hen' – a simple depression in the sand. Although laid by different females, the eggs will all hatch together.*

OSTRICHES PUT ALL THEIR EGGS IN ONE BASKET

The partners of the polygamous male ostrich all deposit their eggs in the same nest. A primary female will then join the male in incubation duties, looking after all the eggs. Should there be too many eggs, she will push away those on the outside, while ensuring her own are in the middle. When the eggs hatch, the chicks are well camouflaged in the undergrowth. If a predator, such as a lion or warthog, approaches, one of the parents will perform a 'distraction display' during which it feigns an injury and tries to entice the threat away from their offspring.

Ostriches are at home in the dry savannah and semi-deserts of Africa and south-west Asia. They travel in bands of 10–50 individuals, the most conspicuous birds being the 2.5 m (8 ft) tall males, with their striking black-and-white plumage.

The largest birds in the world, ostriches cannot fly: they run, reaching speeds of up to 70 km/h (44 mph), making them the fastest-running animals on two legs. Should a predator come too close, their powerful limbs can deliver a kick that is strong enough to kill a lion.

ROLL CALL FOR HAWAII'S SPINNER DOLPHINS

Late afternoon in the bays of Hawaii sees the spinner dolphins living up to their name. One by one they leap clear of the water, spinning on their long axis for as many as seven turns before crashing back into the water. This enthusiastic behaviour is thought to be a kind of roll call, during which each spinner announces that it is ready to depart for a night's hunting and feeding.

With the register completed, the dolphins head for the ocean where they join other groups to form hunting parties of 100 animals or more. In the morning they return to their bays, where they are less likely to be surprised by predators while they rest.

LITTLE MONKEYS HELP OUT WITH THE CHILDREN

Assisting at birth and carrying the young comes naturally to cotton-top tamarins. This behaviour is not seen in any other primates except human beings.

These squirrel-sized primates live in groups of up to 19 individuals in the forests of northern Colombia in South America. Each group consists of a pair of mature adults that mate for life, the babies of the year, and other young and subordinate animals of both sexes. Only the dominant female mates and gives birth, usually to twins, once a year. The father, brothers, sisters, sons, and daughters all help her look after the infants, and sometimes carry youngsters on their backs.

MEERKAT PATROLS KEEP PREDATORS AT BAY

The sight of a meerkat standing sentinel is one of the strangest in the animal world. While on guard duty, this apparently selfless individual holds itself bolt upright for an all-round view of potential predators. If danger approaches, the meerkat emits an alarm call to warn its colleagues, urging them all to scuttle quickly into their communal burrow.

Members of the mongoose family, meerkats are found in the Kalahari Desert of Southern Africa. They are burrowing animals that live in groups of up to 20. To feed in safety, each group keeps alert to predators such as hawks and eagles by posting a guard. The sentinel might appear vulnerable to attack, but it takes up position close to a tunnel entrance and after sounding the alarm is the first to reach the burrow.

ON GUARD *Meerkats frequently stand on their hind legs to look for danger. One will always keep watch while the rest of the group forage.*

THE BIGGEST YAWN IN THE ANIMAL KINGDOM

The hippopotamus is one of the most dangerous animals in Africa. Females are especially aggressive when they have young. As well as seeing off predators, females have to protect their young from being bullied by grumpy males. Mothers teach their young to stay close, giving them a head butt or a nip if they wander off track, and if a mother needs to leave the crèche for any reason she places her calf with a female baby sitter.

Hippopotamuses live in groups of 15 or more females and subordinate males, governed by a dominant male. He will scare off any challengers by opening his mouth very wide and showing his enormous 50 cm (20 in) long canine teeth. It is an awesome sight, since the hippopotamus has by far the largest mouth of any animal outside the oceans. Hippopotamuses are highly territorial and it pays not to stray into their domain: there are many reports of small boats being capsized by hippopotamuses and their occupants being gored to death.

PROTECTING THE UNBORN

Some animals go to great lengths to ensure their offspring have the best chance of survival. They build safe nests with disguised entrances, lay eggs in toughened cases, bury eggs, carry, guard, and even hide them in unexpected places.

THE FISH, THE MUSSEL, AND THE TEST-TUBE BABIES

Bitterling have a curious way of protecting their eggs and young. During the breeding season from May to July, the female bitterling, a finger-sized fish found in European lakes, canals, ponds, and slow-moving rivers, grows a thin 5 cm (2 in) long egg-laying tube. She then seeks out a living freshwater swan mussel and pushes the yellow or red coloured tube into its inhalent siphon (which sucks water in), without disturbing the shellfish's sensitive shell-closing muscle. Having deposited her eggs, she withdraws the tube gently and swims away to lay more eggs in another mussel nearby.

Meanwhile, the male bitterling – decked out in his pinkish courtship colours with a bright blue streak near the tail – positions himself over the mussel's inhalant siphon and sheds his sperm. This is drawn into the cavity, where it fertilizes the eggs. He follows the female to the next mussel, and the next, until the female has completed egg-laying. She departs, but he lingers, guarding the patch of mussels containing his offspring.

The eggs and the newly hatched fish remain protected inside the mussels until they have used up their yolk sacs. They are ejected through the mussel's exhalant siphon, and then the young bitterling are on their own. The relationship is not always so one-sided, however: the adult bitterling often becomes the unwitting host to the mussel's parasitic larvae.

EGG SHELL *The female bitterling inserts its egg-laying tube into the siphon of a freshwater swan mussel, while the male waits to fertilize the eggs once they are inside the shellfish.*

MOTHER TOAD KEEPS HER EGGS IN A BACKPACK

The pancake-shaped Surinam toad, or pipa, has developed a novel way of protecting its unborn. The female toad carries her eggs in tiny pockets in the skin on her back.

Surinam toads live in the coffee-coloured waters of the Amazon and Orinoco rivers of South America.

The male and female mate by performing a series of somersaults at the surface of the water, during which the male not only fertilizes the eggs, but also places them on the female's back. A patch of spongy tissue grows over them, hiding the eggs from view.

During the next three months, each egg develops into a tadpole. It absorbs nutrients from its mother through the skin of its protective pocket. The offspring remain inside until they are ready to break out, either as well-developed tadpoles or as tiny toads.

FISH OUT OF WATER

The female grunion ensures that her eggs are not washed out to sea to be gobbled up by other fish. When the spring tides are at their highest, thousands of silvery, 18 cm (7 in) long female grunion ride in with the waves to deposit their eggs along the tideline of Cabrillo Beach near Los

Angeles in California. The female drills her tail into the wet sand and lays about 3,000 eggs, while the male curves around her and releases his sperm, which trickles down through the sand to fertilize them. The male returns to the sea on the next wave, but the female remains for about 20 minutes before leaving. As the tide ebbs, the eggs are left behind, buried in the sand.

After about eight days, the embryos are ready to hatch, but they remain another six days until the next very high spring tide. When the water reaches them, they burst out and are washed into the ocean.

EARLY START *A green turtle prepares to leave her nest at dawn, on Bartolome Island in the Galapagos.*

THE EPIC JOURNEY OF THE SEA TURTLE

In a demonstration of extreme endurance sea turtle mothers go to great lengths to protect their eggs. Green turtles make long journeys to isolated islands, such as the Galapagos. One population travels more than 2,000 km (1,250 miles) from feeding sites on the Brazilian coast to Ascension Island, a tiny speck in the middle of the Atlantic Ocean. Why the turtles travel so far and how they find their way are mysteries, but when they arrive they undertake another gruelling journey, this time on land.

During the night, each female hauls herself out of the surf and onto the beach. Without the support of the water, the weight of her body presses down on her lungs, making breathing difficult. She drags herself laboriously to the top of the beach where the sand is at just the right level of dampness for incubating eggs but safe from flooding by the sea. Here she digs a nest hole, using her flippers, and deposits about 140 spherical eggs. She then fills the hole, smoothes the sand, and crawls slowly back to the sea, her duty completed. The entire process takes a couple of hours, and from this moment the developing turtles are left to their own devices.

The eggs are protected in their underground bunker for nine to ten weeks, at which point the hatchlings emerge simultaneously, dig their way to the surface, and head for the sea.

BABY DOGFISH GROWS IN A MERMAID'S PURSE

While many female sharks give birth to live young, others deposit eggs. The common, or lesser spotted, dogfish, a small shark found in British waters, produces 20 to 25 eggs every spring, each encased in a translucent, leathery capsule known as a 'mermaid's purse'. This oblong case has a horn at each corner, ending in curly tendrils. When the time comes to lay her eggs, the dogfish swims among seaweed and sponges to anchor her capsule firmly.

Inside, the embryo is attached to a yolk sac on which it depends for its food. It makes swimming movements to circulate the water in the capsule, helping the exchange of oxygen and carbon dioxide across the capsule wall. The embryo develops in its protective case for nine months, finally emerging as a baby fish with dark stripes. As it grows, the stripes break up into black and brown dots, the distinctive livery of the dogfish.

PROTECTIVE CAPSULE *Inside its egg case, the dogfish embryo is relatively safe from predators during its nine month development period. Tendrils anchor the case to seaweed.*

THE MALE WATER BUG'S OXYGEN DANCE

To survive in streams in the deserts of Arizona, the giant water bug copes with extremes. One moment the streams form raging torrents, the next a string of stagnant pools. The water bug breeds during the dry season, when aquatic insect prey is concentrated in the pools and the female bugs can find enough food to produce eggs.

The female lays about 100–150 eggs, which she places in batches on the back of her mate. He ensures that all the eggs have been fertilized by his own sperm by mating frequently. The female then leaves, for it is up to the male to ensure the eggs' survival.

The stagnant pools contain little of the oxygen that the embryos need, so the father stands on a twig or rock just below the water's surface, where most oxygen is concentrated, and performs a little dance. He bends and extends his legs continuously, so that oxygenated water passes over the eggs. He also exposes the eggs to air from time to time, to prevent mould. After three weeks the nymphs emerge and the egg cases fall off the male's back.

THE FEMALE CROCODILE'S CAMOUFLAGE NEST

The saltwater crocodile is the world's largest reptile, yet its offspring still need protection. Monitor lizards and rats eat crocodile eggs, and birds of prey take hatchlings.

The female crocodile from northern Australia protects her brood by depositing her eggs in a large nest mound, up to 90 cm (3 ft) high and 2.5 m (8 ft) across, composed of plant material mixed with soil. It can take her seven or more days to build. In swampy areas the nest may be built on a floating raft.

About a month after mating she deposits between 20 and 90 white,

BACK PACK *The male giant water bug undertakes egg-guarding duties by carrying them on his back until the larvae hatch.*

EXTREME EGGS FROM THE WORLD OF BIRDS

The largest egg laid by a bird is that of the North African ostrich. It is about 15 cm (6 in) long and weighs nearly 1.8 kg (4 lb), the equivalent of two dozen hen's eggs. Even this monster pales in significance alongside the egg of the extinct elephant bird of Madagascar. This 3 m (10 ft) tall bird laid an egg 76 cm (30 in) long with a capacity of 9 litres (2 gallons), enough to make 60 three-egg omelettes. This is the limit of a viable egg size: any larger and the shell would have to be so thick to hold the contents that the chick inside would be unable to break out.

A living bird that sometimes has problems with egg-laying is the brown kiwi of New Zealand. It produces the biggest egg relative to its body size, a 1.7 kg (3 lb 12 oz) hen laying a 400 g (14 oz) egg – one-quarter of the hen's body weight. The smallest bird's egg is that of the Vervain hummingbird, measuring just 1 cm (3/8 in) across.

RELATIVE SIZE *The eggshell of the ostrich (left) and the hen (right) is made of calcium absorbed from the mother's food and bones.*

leathery eggs in the nest. She then completes the structure with a layer of vegetation and stands guard in a nearby wallow (muddy resting place) throughout the incubation period. The eggs are relatively safe in the nest, and the rotting vegetation produces heat to help them to develop.

FATHER MALLEE FOWL'S TEMPERATURE CONTROL

The role played by male mallee fowl to protect the female's eggs is that of heating technician. His task is to maintain the temperature of the mound in which the eggs have been laid at a constant 33–34°C (91–93°F) throughout the long period of incubation lasting 62–64 days, regardless of the outside temperature.

Mallee fowl live in southern Australia. To create the nesting mound, a pair of birds use their big feet to dig a pit about 90 cm (3 ft) deep and 5 m (16 ft) across before the winter rains arrive. Again using their feet, they fill it with dead vegetation from the surrounding forest. When rain falls, the rotting of this vegetation accelerates and the temperature in the pit can rise to more than 60°C (140°F). The fowls cover most of the vegetation with a 50 cm (20 in) layer of sand, forming a substantial mound with a depression in the centre. The female lays her eggs in the depression and the pair then cover them.

The male mallee fowl checks the temperature by inserting his tongue into the mound, and then goes to work making any adjustments needed. In spring, he opens the mound in the early morning to let heat escape; in summer, he scrapes sand onto the mound to protect the eggs from the midday sun; and in autumn, when the sun's rays are weakening and the rotting process is declining, he spreads out the sand to warm it up before scraping it back onto the heap.

THE PYTHON THAT KEEPS ITS EGGS WRAPPED UP

Most snakes bury eggs in a safe place and then abandon them, but not the green tree python. The snake, from New Guinea and north-east Australia, lays a clutch of about 100 eggs and then coils her muscular body around them. If the outside temperature is too cold, she can incubate the eggs by shivering. She makes rhythmical contractions of her powerful muscles, and can raise her body temperature by about 7°C (13°F).

GULLS BEAT THE HEAT TO NEST IN SAFETY

The grey gull, or garuma, of Chile nests in one of the driest places on Earth – the Atacama Desert. The nest is just a scrape in the ground, and one parent always remains on it to shade the eggs and chicks from the sun. Here, daytime temperatures can reach 50°C (122°F), but for the grey gulls it is worth the discomfort. Few creatures, except the odd snake or scorpion, would venture into such an inhospitable place, so the nesting birds and their offspring are relatively safe.

Although the gulls feed and mate on the fish-rich Pacific coast of South America, they fly 80 km (50 miles) inland every day to nest in the desert, possibly because there were once lagoons there.

DESERT NEST *A grey gull shields its egg from the desert sun. In the evening it will be relieved of incubating duties by its partner.*

EARLY DAYS

On hatching, or after birth, some baby animals must rely on their instincts to survive. Others enjoy a further period of parental care during which they gain nourishment, learn the ways of their world, and are protected from predators.

EMPEROR PENGUIN COSSETS HIS OFFSPRING

In the Antarctic winter, emperor penguin chicks spend their first days warming on their fathers' feet. Emperor penguins breed far inland on the Antarctic continent, and lay their eggs not in spring, but at the onset of winter. There is no nest: the female holds the egg on her feet to keep it away from the ice, but she does not incubate it. She shuffles over to the male and passes him the egg. He balances it on his feet and covers it with a thick fold of skin hanging from his lower belly. Here it remains throughout winter, centrally heated by a bare patch of skin within the male's 'pouch' that is well supplied with blood vessels.

The female departs for the Southern Ocean to feed, leaving the male to fend for himself alongside 6,000 other male emperors who huddle together for warmth. The bird endures temperatures that plummet to –60°C (–76°F) and strong and bitter winds.

Two months later the egg hatches and the chick perches precariously on its father's feet. At the same time the females return and take over parental duties. If the partner should be late, the male feeds its chick a special 'milk' produced by the lining of his gullet.

COLD COMFORT *Two-week-old penguin chicks perch on their fathers' feet, where they will stay for a further four weeks before being left to cope on the ice on their own.*

THE GENTLE JAWS OF THE MOTHER CROCODILE

Mother Nile crocodiles take their newborn young into their jaws – but without harming them. Having waited 90 days for her eggs to hatch, the mother crocodile is alert to the first persistent chirping calls from her emerging youngsters. She digs into the sandy nesting mound to release around 45 young.

Gently, she picks them up one by one in her formidable jaws and transports them to the water in a pouch in the bottom of her mouth. She also takes any unhatched eggs and rolls them on her tongue to help each youngster to break out. The newly hatched reptiles are just 28 cm (11 in) long and are easy prey for fish eagles, marabou storks, and other crocodiles. So she guards them in shallow-water nursery sites while they learn to hunt.

HOW THE RED KANGAROO POCKETS ITS YOUNG

After only five weeks in the womb, a red kangaroo's baby emerges from its mother's birth canal. Peanut-sized, undeveloped, and blind, the little 'joey' struggles through the fur of her underbelly towards the pouch on her abdomen, and clambers in. It then attaches itself to one of four available nipples and hangs on for all it is worth until the teat swells and

NURSE FOR A MONTH *The newborn red kangaroo 'joey', from Australia, attaches itself to a nipple and does not let go for 28 days.*

'locks' the youngster in place. Kangaroos are marsupials. Unlike other mammals, the female does not have a placenta, but gives birth to an immature baby that completes its development in her pouch.

The joey stays fixed to the teat for a month, until its jaws are developed enough for it to let go, but mother's milk remains its staple diet until it emerges at about seven months old and begins to explore its surroundings. At 11 months the offspring vacates the pouch.

PIGGY-BACK RIDE FOR YOUNG SCORPIONS

Scorpions are unusual among arthropods – spiders, centipedes, wood lice, crabs, and insects. They do not lay their eggs. Instead, the female stores her fertilized eggs in chambers within her body, and nutrients are transferred from her own digestive system to her offspring throughout their development.

It can take as long as a year before the tiny white baby scorpions are 'born'. They are caught in a 'birth basket' formed from the mother's pincers and they clamber up her legs and onto her back.

If any baby should fall off en route, the mother recognizes it by its smell and will scoop it up. After their first moult at 7–12 days, when they shed their hard outer skin and grow while the new skin is still soft, the youngsters drop off her back intentionally to fend for themselves.

THE FROG WITH TADPOLES IN HER STOMACH

The most bizarre means by which a frog or a toad protects its young must be that of Australia's gastric brooding frog. After mating, the female stops normal feeding and swallows 18–25 of her fertilized, cream-coloured eggs. The eggs, and later the tadpoles, develop in her stomach. Chemicals in the jelly surrounding the egg and those secreted by the tadpoles switch off the production of gastric juices from the stomach wall and the entire digestive system shuts down, preventing the absorption of her young.

When a batch of froglets is ready to emerge, the mother opens her mouth wide and expands her throat. The young are pushed from the stomach into her open mouth, from which they hop away.

BABIES ON BOARD *Protected by the mother's sting, young scorpions crowd onto her back for up to 12 days before setting off alone.*

BABIES FIND REFUGE IN THEIR FATHER'S MOUTH

The male aruana is a fish with a cavernous mouth in which his offspring hatch. The father gulps in the fertilized eggs, and when the young fish hatch they survive the first three weeks inside his mouth. Every so often he opens wide and a cloud of offspring swims out for a 10–15 second exercise period to feed on algae, aquatic larvae, and water fleas. If danger threatens, they crowd back in, using his mouth as a refuge.

This flat-sided freshwater fish from the Amazon is about the length of a human arm. It has a habit of waiting expectantly beneath a tree like a coiled spring before leaping clear of the water to pluck insects from the branches.

LIVING ON THE EDGE *On the precipitous cliffs of Bear Island in the Norwegian Arctic, thick-billed murres are tightly packed on their nesting ledges.*

LITTLE MURRE CHICKS TAKE THE PLUNGE

During the Arctic summer, the cliffs are packed with thick-billed murres and their chicks. On Canada's Digges Island and Norway's Bear Island, the parents go fishing and return regularly to narrow ledges on the cliff face to feed their growing young. One false move and the chick will fall onto the rocks below, and any unattended youngster becomes prey to voracious gulls.

When the time comes to leave, the chicks are only a third grown. They hurl themselves unsteadily from their nesting ledges and in a short, undignified flight using rudimentary wings they reach the water, shadowed by their father. If gulls should swoop down, the fledglings dive below the surface to escape.

Then, under cover of darkness, father and offspring swim many kilometres to the feeding grounds. Here, for the next couple of months, he continues to provide food for his youngsters until they are ready to fish for themselves.

SIBLING SHARKS' RIVALRY ENDS IN DIGESTION

Unborn sand tiger sharks eat their brothers and sisters while still in the womb. The sharks hatch out inside their mother's body, and just two dominant embryos survive – one in each of her two uteruses.

The two babies first eat all of the other eggs and embryos present, then are fed on unfertilized eggs released regularly by their mother. After nearly a year, they are born, headfirst, and, within seconds, swim away into shallow water to avoid the predatory attentions of larger sharks.

TURTLE HATCHLINGS' TERRIFYING DASH TO THE SEA

Green sea turtle hatchlings emerge from their underground nests on a Costa Rican beach as if going up in a lift. The youngsters on top scrape away the ceiling while the turtles below undercut the walls and those at the bottom compact the falling sand. The floor rises, the ceiling collapses and the hatchlings pour out towards the sea.

Lying in wait are the predators: black vultures, iguanas, ghost crabs, and frigate birds. A few baby turtles make it to the ocean, where sharks are lurking. Of the survivors only one in 5,000 reaches maturity.

SEAL PUPS' SNOWY DEN

The ringed seal of the Arctic Ocean is the only seal that builds a den. At the onset of winter, a mother seal excavates a lair under the snow. Snug inside during the worst of the weather, she gives birth. She suckles her pup with fat-rich milk for one to two months until it has put on enough blubber to keep it warm in the icy water. A female ringed seal might make several snow caves, and should a hunting polar bear intrude, mother and pup can dive through an escape hole and flee to a safer den.

WHITE-TOOTHED SHREWS HANG ON FOR SAFETY

When a white-toothed shrew mother goes foraging, she takes her youngsters with her. Should danger threaten, such as a small cat or bird of prey, they quickly line up behind her, one baby grasping her tail and the other five or six holding on tightly to the tail of the baby in front. The train of shrews hurries to safety.

The mother white-toothed shrew, from Europe, Africa, and Asia, burns up calories rapidly and needs to eat her own weight in insects every day. So, as soon as the babies can see, she takes them out with her and they supplement their diet of mother's milk with insects.

TALLEST BABY CAN RUN AN HOUR AFTER BIRTH

The giraffe calf's first traumatic journey is a 2 m (7 ft) drop headfirst to the ground. Mother giraffes stand 4.6 m (15 ft) high, and give birth standing up so that they can keep a close watch for lions, hyenas, and wild dogs, which could easily snatch a vulnerable youngster.

Within an hour, when the 1.8 m (6 ft) tall calf has mastered its long legs and is able to run, its mother can lead it away from danger, the pair fleeing across the plains of East Africa at speeds up to 52 km/h (32 mph).

FAST MOVERS *The baby giraffe stays with its mother for 10–17 months, until it is weaned. If danger threatens, the pair can flee at speed together.*

GROWTH AND DEVELOPMENT

Seals are weaned within two weeks, elephants take two years. Environmental conditions, parental attentiveness, and the threat from predators all dictate just how quickly an animal develops before it is able to survive on its own.

SEAL PUPS FATTEN UP FAST FOR SAFETY'S SAKE

Harp seals are weaned within two weeks, one of the fastest weaning times of any mammal. Only the hooded seal, another Arctic wonder, does it quicker – birth to weaning in just seven days. The reason for this very rapid development is the need to beat the predators, such as polar bears and Arctic foxes. The less time the pups spend on the ice, the less likely it is that a predator will chance upon them.

Every March, off the coast of Newfoundland, the sea ice is dotted with hundreds of fluffy, white harp seal pups, which are suckled by their mothers 10–12 times a day for ten minutes. Harp seal milk has a 43 per cent fat content, so the pups grow fast, gaining 1.4 kg (3 lb) a day. When they reach 32 kg (70 lb), the pups moult their camouflage white coats, revealing a grey pelt.

At this point they are abandoned by their mothers and must enter the water to fish for themselves. At first they can catch only surface-living shrimps and small fish. Sometimes they lose up to 6 kg (14 lb) in weight before they learn to dive deeper in search of herring and cod.

MURDER IN THE NEST

Masked boobies produce just two eggs, but these Pacific sea birds brood each one as it is laid. The chicks therefore hatch several days apart, giving the first-born a considerable head start, which it quickly turns to its advantage. The older bird, although still wobbly, attacks or drives out its sibling, ensuring that it receives most of the food the parents bring back. When times are hard and food is scarce, the older bird will even kill the younger one. By not intervening the parents encourage the conflict, ensuring that the stronger chick survives.

SPEEDY GROWTH *A harp seal pup, known as a 'white-coat', trebles its weight in 12 days, half of which is blubber for insulation.*

BLUE CRAB BREAKS OUT IN ORDER TO GROW

A young crab gets through up to 12 shells in its first year. All hard-bodied creatures moult their rigid external skeleton in order to develop and grow. And the blue crabs of Chesapeake Bay on the eastern coast of the United States are no exception. First the crab absorbs calcium and any organic materials from the shell it is about to discard. Then its body swells by absorbing water, causing the shell to split along a line of weakness. It takes the crab about three days to ease slowly out of its old shell.

Young crabs moult frequently during their first year until they reach sexual maturity; thereafter they shed their shell just once or twice a year. Some crabs eat their old shell so nothing is wasted. At moulting time, the crab is vulnerable not only to predators, but also to others of its own kind. A soft-shelled crab will hide away in seaweed or beneath a stone until its new shell is hard enough to protect it.

A ROYAL FEAST FOR TOMORROW'S QUEEN BEES

Only a few female honeybee larvae receive 'royal jelly' or 'bee milk' as they develop. These are the ones destined to become queens. The rest of the colony's thousands of new females feast on the same royal jelly for just three days after hatching, and thereafter are given pollen and honey.

Royal jelly is a thick, milky food produced by glands in the mouths of

LIVING COMB *The hexagonal cells of a nursery comb in a honeybee colony contain well-developed larvae and pupae, cared for by workers.*

young nursing worker bees. It consists of a mixture of proteins, sugars, fats, vitamins, and a special fatty acid.

Deprived of this rich foodstuff, the ovaries of female honeybees fail to develop, rendering them incapable of producing offspring. Their lot is to spend the rest of their days as workers, serving the needs of their monarch and her larvae.

HOW THE TSETSE FLY SUCKLES HER YOUNG

Few insects care for their young so diligently as the tsetse fly. Despite many attempts to eradicate them, tsetse flies are still a scourge, their continued survival due in part to their unusually attentive mothers.

Unlike most insects, which give rise to myriad offspring each time they breed, then leave their larvae to fend for themselves, the female tsetse fly raises only one larva at a time. Her single fertilized egg is retained in a pouch in her body, and when the larva hatches it feeds on milk secreted from glands in the pouch wall. When almost as big as its mother, the giant larva is 'born'. Within an hour it burrows into the ground and pupates.

The female tsetse fly produces only six offspring during her three-month life, but so much care is lavished on each individual that most survive. This poses a serious health problem for the human population of Africa, where tsetse flies are common, because they transmit a particularly debilitating disease known as sleeping sickness.

ELEPHANT BABIES THRIVE ON LONG-TERM CARE

Gestation and weaning periods among African elephants are the longest of all mammals. A mother gives birth to her calf about two years after fertilization, then suckles it for a further two years or more. During this lengthy period of dependency, the youngster not only feeds well on its mother's milk, but also learns the finer points of elephant society, and picks up vital information about how to find the best places to eat and drink.

The baby grows rapidly from about 120 kg (265 lb) at birth to 1,000 kg (2,200 lb) at six years old – the weight of a black rhinoceros – by which time it has switched from drinking milk to munching vegetation. It continues to grow at this rate until it is about 15 years old. Bull elephants put on a spurt of growth at about 20 years, which accounts for their larger size.

Elephants can live for 70 years, although about half of all elephants born and bred in the wild do not reach their 15th birthday. They succumb to diseases, poachers, and predators.

SUCKLING TIME *An African elephant calf drinks its milk from paired breasts between its mother's forelegs, using its mouth not its trunk.*

THE CARNIVOROUS CATERPILLAR

When it hatches, the caterpillar of the European large blue butterfly eats thyme, but it soon turns carnivorous. After it has grown and changed its coat for the third time, it takes to eating ants: the same ants which are also its protectors.

On leaving its first food plant, the caterpillar wanders about on the ground. If it comes across a particular species of red ant (*Myrmica sabuleti*) it rears up and produces a drop of 'honey dew' from a gland on its abdomen. The ant finds the sweet liquid attractive, so it grabs hold of the caterpillar and hauls it back to its underground nest.

Rather than attacking it as they would any other intruder, the ants crowd around the caterpillar and encourage it to make more honey. The caterpillar ensures it is not torn to pieces by the ants by producing a pheromone that suppresses any attack.

Safe in the subterranean nest from other predators, such as birds, the caterpillar repays its hosts by eating their larvae and eggs. After about six weeks of fattening up, it develops into a pupa and continues to placate the ants by producing sugar and pheromones. It also makes scratching noises on the pupal case that mimic the ants' communication signals.

When the adult butterfly emerges, it too produces a sugar solution and the ants escort it from the colony. As the wings of the butterfly expand and dry, the ants guard it from predators, such as ground beetles, until it is ready to fly away.

METAMORPHOSIS

Many insects and amphibians follow a life cycle that sees a complete change of body form, appearance, and behaviour in recognizable stages. This process is known as metamorphosis and is controlled by a balance of hormones.

The monarch butterfly, for example, lays her eggs on the leaves of milkweed, and tiny caterpillars hatch out. These represent the feeding and growing stage of the butterfly. At first the caterpillars eat their egg cases and then they set about demolishing the leaves, storing the poisons they contain for later use when they will help to protect the adult butterfly from bird attacks.

After several weeks, each caterpillar changes into a packet-like pupa or chrysalis which is attached to the underside of a leaf. Inside the chrysalis a remarkable transformation takes place, during which the caterpillar's cells and tissues rearrange themselves to form the adult butterfly – the reproductive stage. Fully formed, the butterfly emerges from the chrysalis, dries its wings, then flies away to find a mate, and so the life cycle continues.

Amphibians, such as frogs, newts, and toads, are the only vertebrates to undergo a metamorphosis, apart from some fish. The common frog tadpole that emerges from an egg is completely different from the adult. It has a tail and external gills. After ten days its muscles develop, enabling it to swim. At four weeks its external gills are replaced by internal ones. Between six and nine weeks its hind legs appear, its body elongates and the head becomes more distinct. Finally, all four limbs are formed and the tail is absorbed. The tadpole has metamorphosed into a froglet.

1 *Monarch butterfly eggs laid underneath milkweed leaves.*

2 *Caterpillar feeds on leaves and grows in a series of moults.*

3 *Chrysalis suspended from the stem of a milkweed leaf.*

4 *Adult monarch butterfly emerges from its chrysalis.*

5 *Butterfly dries and expands its brightly coloured wings that warn birds it is unpalatable, having acquired milkweed poisons when a caterpillar.*

LEARNING AND PLAY

Many animals acquire their survival skills by watching their parents or other adults. Some practise these new techniques by fighting without causing injury or jumping about without apparent reason, a behaviour known as 'play'.

EDUCATING CHEETAH TO CATCH ITS PREY

A cheetah mother will catch a live gazelle calf so that her cubs can practise hunting. She releases it in front of them and when it makes a dart for freedom, the cubs try to chase it and bring it down. If it escapes, the mother will retrieve it and they start all over again.

During these deadly games the cubs learn how to stalk silently, bring down prey by cuffing its legs with their forepaws, and how to kill with a single bite to the throat. They also learn not to be afraid, for even a gazelle can be intimidating to a cheetah cub.

YOUNG IBEX LEARN TO STAND THEIR GROUND

A young ibex will jump onto a rock or tree stump and defend it vigorously against challengers. This game is a serious practice session for confrontations later in adult life. Young male Alpine ibex have mock battles during which they clash heads, and push and shove their opponents until one backs down or a stalemate is called. These head-banging skills come in useful later when, armed with large scimitar-like horns, the ibex faces an equally skilled fighter in a battle for a place in the hierarchy and the right to a harem of females.

HOW POLAR BEARS HONE THEIR ICE WARRIOR SKILLS

The polar bears' annual autumn gathering in Churchill, northern Canada, provides an ideal opportunity for play-fighting. After spending summer in the cool of the forest, the bears in this region head for the sea, arriving hungry and tetchy on the shores of Hudson Bay around Churchill.

While they wait for the ice to form so they can go hunting for seals, the young males bide their time with mock fights. These practice bouts are a useful rehearsal for the more bloody battles that occur during the breeding season. As cubs, play-fighting in the

HUNTING SCHOOL
1 *A cheetah mother has caught a young gazelle, but not killed it.*

2 *The mother releases the gazelle to teach her cub how to hunt.*

3 *The young cheetah practises bringing down prey. Often the cubs are so inept that the gazelle escapes.*

Arctic snow will have helped to develop the rapid reflexes they need to capture their prey.

MONKEYING AROUND WITH SERIOUS INTENTIONS

Young rhesus macaques play for most of their waking hours. To warn an opponent that an impending fight is only in play and is not an act of aggression, a young macaque will first peer at its chosen adversary upside down through its legs.

The chasing and wrestling that follow can help to control aggression between members of the macaque troop because the youngsters find out just how far they can go before causing injury to their opponent.

Play also decides pecking orders, the strongest males taking higher social positions than the less able members. A male automatically loses his rank when he leaves a troop. Often he will move on to join the same troop as his brother. Then he must use the skills learned as a juvenile to establish his place.

A juvenile female macaque ranks just below her mother and, unlike in human society, above her older sisters.

ROUGH AND TUMBLE GAMES OF THE LION CUBS

Lion cubs goad each other into mock fights from an early age. They even target their parents, attacking their black-tipped tails. These play-fights are vital rehearsals for adult life as a top predator.

Mock wrestling matches prepare male cubs for the time when they will have to fight other lions to win control of a pride. Cubs practise catching prey by stalking and chasing each other, swatting at legs to bring an adversary down, or grabbing its rump.

NATURE'S MASTERPIECES

Learning from mother

The skills needed to catch a slippery salmon do not come naturally to North America's grizzly bear cubs. With their eyes fixed on their mother's every move, they watch her find a quiet spot and hook salmon out of the water with her claws. They see her take her catch to the shore and carefully strip away the flesh from either side of the bone. If the cubs are lucky they might pick up some leftovers. Then it's their turn. They fling themselves in erratic belly flops into shallow pools where the salmon have gathered. Inevitably the fish escape, but the young bears will spend two years with their mother perfecting these skills.

Grizzlies fatten up prior to winter hibernation by feasting on the glut of salmon that in autumn fills rivers along the Pacific coast of North America. The bears intercept the fish at rapids and waterfalls, taking full advantage of the annual bonanza.

CHIMPANZEES PASS ON TOOL MAKING TIPS

Chimpanzees have developed ingenious ways of obtaining food by using tools. They learn by watching their mothers closely. In Gombe, Tanzania, a mother will show her youngster how to catch termites using a stick or stem. First she helps the juvenile to peel the bark off a twig and fray the end. Then she inserts the implement into a termite mound to extract a wriggling mass of termite soldiers. She draws the twig between her lips and swallows the termites.

Similarly, chimpanzee females from certain West African populations will help their offspring to find a suitable hammer, and then direct its hands to help it to position the tool properly to crack a nut successfully. Chimpanzees in the Ivory Coast have discovered how to use large stones or branches to crack particularly hard coula nuts without crushing the kernel inside.

THE WATERY ROLL CALL OF THE HUMPBACK WHALES

One piece of dramatic behaviour a humpback whale calf must master is 'breaching'. This is when the whale leaps almost clear of the water and plunges back down in a fountain of spray. Baby humpbacks mimic their mothers, but without their parents' finesse. Time and time again, they will lift their chins above the surface and bring them down with a crash. Humpbacks use this spectacular performance as a way of keeping in touch with other whales that are some distance off, a kind of roll call to check that everyone feeding or travelling together is present.

Humpback whales are found in inshore waters worldwide. A calf is suckled by its mother for almost a year after it is born and accompanies her everywhere, learning about whale life. It will also learn to slap its tail against the surface to attract attention – a behaviour known as 'tail lobbing' – and to splash with its long, white pectoral flippers, a defensive manoeuvre for use against marauding killer whales.

FRISKY DOLPHINS PLAY GAMES UNDER WATER

Wild dolphins, like those in captivity, play spontaneously. An Atlantic spotted dolphin may frolic for hours with a strand of seaweed, balancing it on its nose and rushing through the water with it draped over a flipper, like a child with a streamer.

Many species of dolphin appear to revel in riding the bow waves of boats and ships, and mischievous white-sided dolphins may splash the human occupants as they peer over the side.

TERMITE FEAST *Using a grass stem or twig, a young chimpanzee extracts juicy termites from a colony.*

IN-BUILT NAVIGATION SYSTEM

Immediately an Arctic tern chick hatches from its egg it is alert to the positions of the Sun, Moon and stars in the sky. Within six weeks of hatching it is ready to leave its nesting site in the Arctic Circle and migrate to the Southern Hemisphere. It uses the Sun, Moon and stars to navigate, as well as the Earth's geomagnetic field, as it flies to the opposite end of the world and then home again to the same spot. Before the bird migrates for the first time it explores the countryside surrounding its nest site to memorize landmarks.

When a juvenile bird reaches Antarctica it may spend up to three years circling the Southern Ocean and its islands before returning north.

GLOBAL TRAVELLER *No bigger than a pigeon, the Arctic tern makes one of the most staggering journeys in the animal kingdom.*

PLAYFUL PARROTS APE MAMMAL BEHAVIOUR

The mountain-dwelling New Zealand kea is a parrot that enjoys rolling and frolicking in the snow just like a monkey. They even turn onto their backs, primate-style, with their feet waving in the air.

The kea is an exceptionally inquisitive and cheeky bird. Around ski resorts in the Southern Alps, these parrots break off car windscreen wipers, strip away rubber mouldings and tear at seats and electrical wiring. Their behaviour is very similar to that of free-ranging monkeys in drive-through safari parks.

While parrots are sometimes called 'honorary mammals', the kea is considered an 'honorary primate'.

BAD BEHAVIOUR *The instinctively curious kea likes to explore and investigate novel objects, whether natural or man-made.*

BEACH SCHOOL FOR KILLER WHALES

Along South America's Patagonian coast killer whales ride in on the surf and pluck sea lions right off the beach. They then struggle back out to sea on the next wave. Youngsters learn the behaviour from older killer whales, perfecting their skills at special practice beaches.

Having caught their prey, the adults 'play' with it, like a cat playing with a mouse, propelling it high into the air. Elsewhere, killer whales play with objects in the water. Off the Namibian coast of south-west Africa, they harass jackass penguins and Cape cormorants, grabbing birds in their mouths and then releasing them.

The Patagonian killer whales tend to beach where the sand or shingle shelves steeply and there is less chance of ending up high and dry.

On the Crozet Islands in the southern Indian Ocean, where the beaches have a lower gradient, a row of adults and youngsters will race forwards and beach together to catch elephant seals. Parents and 'aunties' help any young whale that should become stuck by nudging it back into deeper water.

SERIOUS SIDE TO THE JOYS OF SPRING

Foals, fawns, lambs, calves, and kids frequently explode into life, prancing, leaping, and running about for no apparent reason. These exaggerated movements are not simply for fun. They are thought to be a way of exercising growing muscles and ensuring that nerves and muscles are working in harmony.

Although they are no longer common prey, horses, deer, sheep, cattle, and goats still practise these antipredator manoeuvres. Gambolling and frisking are forms of play that help to tone up their bodies.

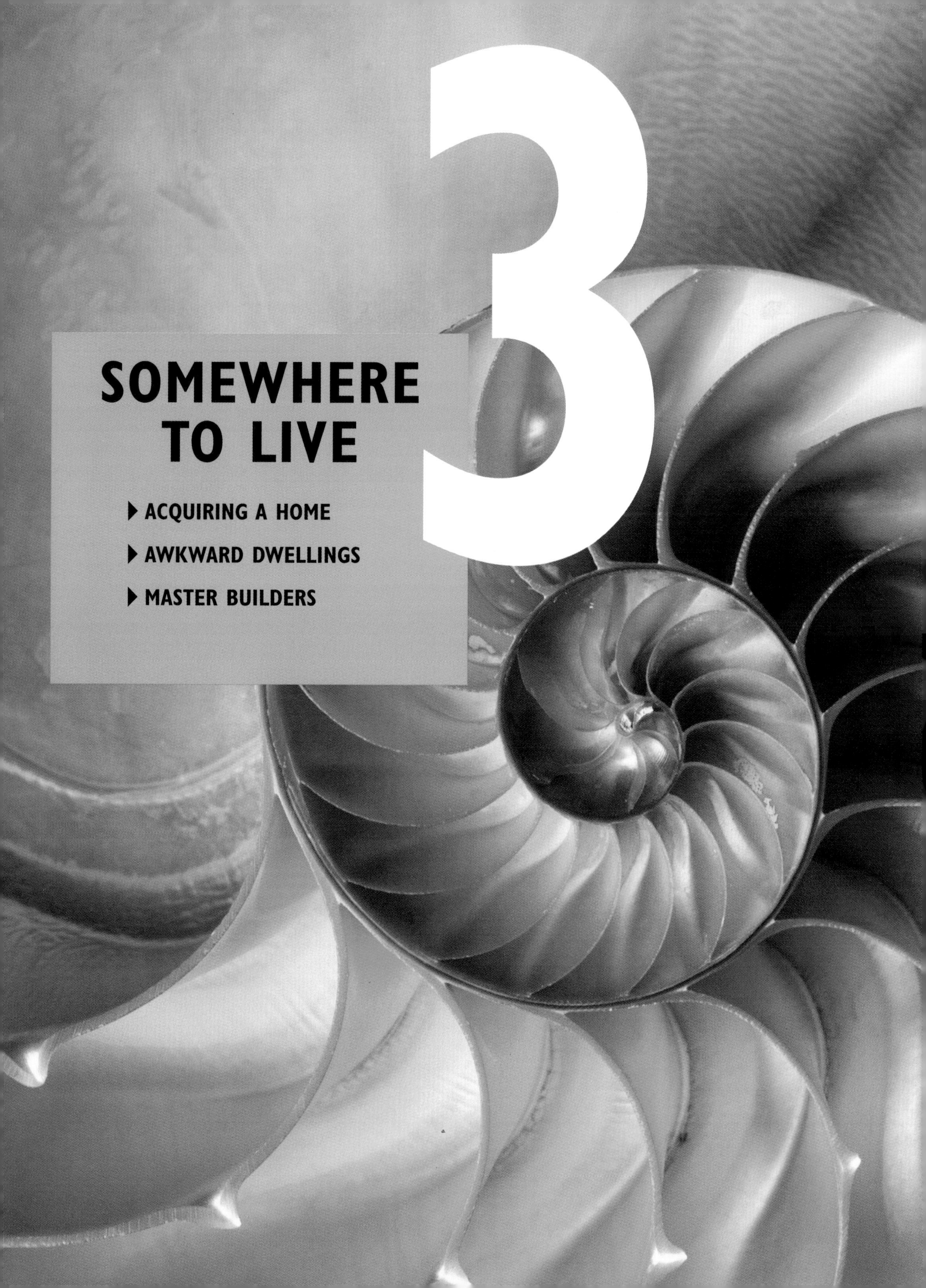

3

SOMEWHERE TO LIVE

- ACQUIRING A HOME
- AWKWARD DWELLINGS
- MASTER BUILDERS

ACQUIRING A HOME

From mobile homes to squats, solid foundations to living walls of flesh, when it comes to putting a roof over their heads, animals will explore every possibility in the search for a comfortable, safe residence in which to settle down.

THE PARROTFISH'S SLIMY SLEEPING-BAG

The parrotfish, found in tropical and subtropical oceans, makes a nightly home. It camps inside a sleeping-bag made of mucus. While other fish doze off on their sides or lean against rocks, or seek out temporary shelters by slipping into crevices or wriggling into soft mud, the parrotfish beds down in a cocoon of its own making.

To form the 'sleeping-bag', the parrotfish secretes a transparent jelly-like slime that slowly envelops its body. It can take about an hour to make this, and just as long to get out of it the next day. The cocoon also serves to mask the fish's smell from predators such as moray eels.

By day, the parrotfish uses its heavy-duty, parrot-like 'beak', formed by fused front teeth, to scrape off the organisms from rocks and corals.

SLEEPING IN SLIME *A bullethead parrotfish from Hawaii rests in a bag of its own ooze. The mucus helps to protect the fish from predators.*

TREE-CREEPER TURNS A TINY NICHE INTO A HOME

There is no space too small for the tree-creeper. This quiet, unobtrusive and well-camouflaged Eurasian bird has mastered the art of nesting where no nooks or crannies appear to exist. As well as being superbly concealed, its nest is also very narrow, barely 3.5 cm (1 1/3 in) in diameter. The bird designs the nest to exploit whatever space is available, moulding it into an oval or a crescent shape, for example, in order to squeeze it into a crack in the tree trunk or tuck it under a flake of bark or behind a curtain of ivy.

The result is surprisingly robust. A foundation of twigs and stems supports a cup of rootlets, grass, moss, and strips of bark, with soft furnishings of hair, feather, and wool.

LIVING WITH THE ANEMONE'S LETHAL STING

The clownfish, or anemone fish, is not bothered by the stinging tentacles of the sea anemone. Unlike most fish, which avoid touching the venomous sea anemone, clownfish even go so far as to seek out one to live with. Exactly how they remain unaffected is a mystery. Perhaps their mucus coat fails to set off the anemone's arrows, or perhaps it is thick enough to protect them.

Gingerly at first, then more confidently, the clownfish slips in between the tentacles of the anemone to set up home. Thereafter the fish seldom strays far other than to feed. If the anemone itself senses danger, and

SAFE FROM HARM *The tiny orange clownfish from Micronesia seeks a living shelter among the stinging tentacles of an anemone.*

pulls its tentacles in, the clownfish is drawn deep within the anemone's soft body for safety.

Anemones obtain their food by stinging. When a creature brushes against its long wavy tentacles, it triggers miniature coiled darts called 'nematocysts', which shoot out, injecting the victim with venom.

In return for its living safe house, the clownfish cleans debris from the anemone's tentacles, blows sediment away with its darting swimming motion, and chases off creatures that might want to eat its host.

GOING UNDERGROUND WITH THE SQUATTING OWL

Most birds have to put some work into building their nests, but not America's burrowing owl. This most terrestrial of all owls moves into underground tunnels made by other animals in the grasslands and deserts of the Americas. Ready-made, out of sight, and sheltered, these subterranean holes are prime nesting spots. The owl makes its grass-lined nest down a hole 1 m (3 ft 3 in) underground and at the end of a meandering tunnel 3 m (10 ft) long.

In Florida, the burrowing owl can, if it must, scrape out its own tunnel in the sandy soil by using its legs. Elsewhere, it kills the owner, such as a ground squirrel or prairie dog, and then moves in.

Sometimes the owl takes up residence with another animal. Gopher tortoise tunnels, which can be up to 12 m (40 ft) long, are favourites of the owls. They may be used by several generations of tortoises, and are often still in use when the bird moves in. The burrowing owl, standing by the entrance of its new squat, will screech to ward off intruders. If the tortoise lumbers home, the owl will fix it with an indignant steely glare from its big yellow eyes, but will not attack it. Burrowing owls also scare predators away by making a noise like a rattlesnake.

HOME SHARE *A burrowing owl emerges at dusk to stand guard over the entrance to the borrowed hole where it builds its nest.*

REPTILIAN HOUSE SITTER KEEPS THE PLACE CLEAN

On islands and rock-stacks off the New Zealand coast, a rare reptile makes a bird's home its own. Birds such as shearwaters, fairy prions, and petrels spend months feeding at sea. When they return to the nests they have tunnelled into the clifftops, they may find that the lizard-like tuatara has moved in.

Living to the age of 120 years or more, and measuring about 45–65 cm (18–26 in) in length, the tuatara becomes the hole's permanent year-round caretaker. While the birds are at sea, the reptile keeps the tunnel clean for their return. The tuatara will eat eggs and chicks, but usually from other nests in the colony. It rarely raids its own landlords' nest chamber.

The tuatara, meaning 'old spiny back' in Maori, is the sole living member of an entire reptile order, the Rhynchocephalia, which came into being about 220 million years ago – well before the dinosaurs. All except the tuatara became extinct 65 million years ago.

SAFE HOUSE *Living in an empty shell gives both protection and mobility to the hairy-legged hermit crab of south-east Australia.*

THE HERMIT CRAB'S SEARCH FOR A VACANT HOME

With its soft body, a hermit crab needs shelter, and it finds it in the shell of another animal. Instead of hiding in, say, rock or coral, it slips into an abandoned whelk, winkle, or other similar shell.

Unlike true crabs, the hermit's abdomen curls to one side to fit the spirals of gastropod shells. Once the creature is safely inside a shell, it seals the entrance with its claw. Some species of hermit crab will coax a sea anemone onto their shell to keep predators at bay.

As the crab grows and the fit gets tighter, it must move house. When it finds another empty shell of about the right size, the crab surveys it with great care. It holds the shell with its front walking legs, climbs on it, and rolls it over, running its opened claws across the surface to check its texture and shape. Then the crab removes any debris with its claws.

The next step is to see how the shell fits. The crab flexes its abdomen and reverses into the shell, going in and out several times to see how the new accommodation feels from the inside. If it is snug, the crab will move in, but it will not always leave straightaway. If the old shell had an anemone, the crab taps and massages it until it releases its grip. Once the anemone is safely stuck to its new roof, the move is complete.

HOW BLISTER BEETLE LARVAE HITCH A LIFT

There is only one known example of parasites acting cooperatively to fool their host. A group of blister beetle larvae on a plant stem not only looks like a female bee, it emits a chemical that makes it smell like one, too. When a male bee tries to mate with the fake 'bee', the larvae climb aboard. The next time he meets a genuine female, the larvae – up to

2,000 individuals – disembark. They stay with the female as she digs her burrow, infesting her nest and feeding on the pollen she gathers.

HOW THE PARASITIC FLUKE MOVES HOUSE

There is one species of parasitic fluke, or flatworm, that actually changes the behaviour of the snail it infests when searching for a new host. If, for instance, a grazing snail is unlucky enough to eat any of the flatworm eggs that land on the ground in droppings from small birds such as flycatchers and thrushes, it is doomed.

Inside the snail's body, the eggs hatch into larvae which tunnel their way into the liver. There they reproduce and form mobile cysts. Every morning, these cysts travel into the snail's tentacles.

Once the parasite has forced its way inside, the tentacle swells into a thick club, which is stretched so tight that it becomes transparent.

The parasite, striped bright yellow, orange, and dark brown, is easy to see inside, and it pulses. To a bird, it probably looks like a living, juicy caterpillar. Somehow, the parasite, which is found in northern Europe, makes the snail change its behaviour. Instead of seeking cover at dawn, it stays out in the open and becomes prey to a bird. As the snail enters the bird's digestive system, the parasite successfully moves to its next home.

SHELLFISH LARVAE THAT STICK TO THEIR POSITIONS

Unlike other crustaceans such as crabs and lobsters, barnacles are not mobile and stay stuck hard to a surface for ever. Their larvae, though, are free-swimming and feed for several weeks before looking for somewhere to settle.

Barnacle larvae often attach themselves to rocks, shells, coral, boats, and floating objects such as timber, coconuts, and bottles. Some end up on whales, turtles, and other slow-moving creatures.

Once the larva has found a good home – preferably one with plenty of other barnacles already – it cements itself on, using a substance from glands at the base of its antennae, and begins to change into an adult.

RARE PARROT SHARES ITS HOME WITH INSECTS

The rare golden-shouldered parrot is a fussy tenant. It lives in a tiny area of dry savannah in northern Queensland, Australia, where it nests in termite mounds, and nowhere else. Between April and August, when the earth in the mounds is still damp from the wet season, the parrot excavates a nest chamber in the crumbly earth of the termites' galleries. The nest's entrance hole is about 4 cm (1 1/2 in) wide. Inside, a tunnel leads to a roomy, rounded chamber about 25 cm (10 in) long.

Cone-shaped mounds made by the termite *Amitermes scopulus* are the best, because the temperature inside remains more constant. Meridian, or magnetic, mounds made by other termites have broad, flattened flanks that are often too thin and a burrowing parrot may break right through to the other side.

UNUSUAL SITE *The golden-shouldered parrot is nicknamed the 'ant hill parrot' because of its preference for nesting in termite mounds.*

AWKWARD DWELLINGS

The biggest concern for any house-hunter is security: where you live must be safe from predators. Instead of locks, keys, or burglar alarms, some animals gain security from living in the strangest, most inaccessible places.

DESIGNED TO SAVE CHICKS FROM DISASTER

How do kittiwakes, which nest in huge colonies on sea-cliffs, prevent their chicks from falling? The answer lies in the design and location of the nest itself, a neat little construction of compacted mud, grass, and seaweed. This is attached to a tiny ledge or bracket that juts out from an otherwise sheer cliff face in Britain and other northern countries. The nest has one refinement that offers the chick some security: it features a nest cup deeper than that of any other sea bird. Eggs cannot accidentally roll out, and chicks cannot easily climb out. In any case, kittiwake offspring by nature tend to stay quite still in the nest.

HANGING SAFELY *Patches of white droppings might make the black-legged kittiwake's nest conspicuous, but it does not matter since few predators can reach these sheer cliffs.*

NO WAY IN FOR THE OROPENDOLA'S ENEMIES

When making her nest, the female oropendola of Central and South America takes no chances. While the male bird spends his days singing and performing to attract other females, she works at making her nest as inaccessible as possible to predators. She tears strips of leaves from banana trees and weaves them so tightly that the nest is waterproof.

The top of the nest is bound to a branch so thin that no spectacled bears, monkeys, or snakes would risk testing their weight on it. Its long, narrow neck dangles down for as much as 2 m (7 ft), then bulges out into a bulb-shaped, padded nest chamber. This neck is much too tight a squeeze for aerial hunters such as hawks.

Female oropendolas also seek safety in numbers: up to 100 will nest in the same tree, guarded by a single male. Some birds choose trees near wasps nests. The wasps leave the oropendolas alone, but will chase away other animals that come too close.

NESTING EAGLE LOOKS DOWN ON ITS NEIGHBOURS

The eyrie of the golden eagle seems vulnerable to predators and the elements. Scorning the protection of overhangs or vegetation, eagles build conspicuous, messy-looking nests, either on rocky outcrops or treetops.

A newly built eyrie on a stony ledge may be no more than a few sturdy sticks and branches arranged into a ring. Golden eagle pairs return year after year to the same spot and over time the eyrie may build into a huge basket, up to 5 m (16 ft) deep. The interior is lined with grasses, ferns, and greenery. Yet there is room for only one chick. So, though golden eagles lay two eggs, the older chick usually kills the younger in order to rid itself of competition for food.

AIR POCKET *The European water spider drags bubbles of air under water and traps them beneath a domed sheet of silk to form a kind of diving bell.*

WEATHERPROOF BUNKERS ON THE CLIFFTOP

Buffeted by sea winds, clifftops may appear undesirable places for puffins to raise a family. And so, without a shrub or tree in sight, the birds find shelter underground. Here, in their isolated location, they are safe from land-based predators such as foxes and stoats.

In the Northern Hemisphere, when a puffin returns from sea to nest in May and June, it takes over abandoned rabbit holes or Manx shearwater burrows. If there are no ready-made tunnels, the puffin excavates its own weatherproof bunker, scratching away with its webbed feet. The tunnel may be up to 1 m (3 ft 3 in) long, with several exit holes. Sometimes, the nest at the end is lined with grass and feathers.

Puffins nest in colonies of tens of thousands, and return to the same cliffs year after year. On the island of St Kilda in the Scottish Hebrides, up to a million arrive within three days. The grass-covered ground is so riddled with tunnels that, occasionally, an area caves in and the puffins have to find another clifftop.

CLIFFTOP HAVEN *An Atlantic puffin loiters near the entrance to its burrow, taking advantage of the wind-proof shelter it offers.*

THE UNDERWATER SPIDER'S BUBBLE DIVING BELL

Many spiders are to be found near water, but only the European water spider lives, hunts, and breeds under water. Like human divers, it takes its own air supply down by building a diving bell.

The female water spider weaves an underwater silken sheet, which she anchors to water plants. Swimming to the surface, she uses her rear abdomen and hind legs to capture a bubble of air and drags it under water, trapping it beneath her sheet, which arches to form a bell. She makes this trip several times until she has enough air.

Safe inside her bell she lies in wait for passing prey. In summer she will lay her eggs inside the dome.

AT HOME ON THE WATER IN MORE WAYS THAN ONE

In North America, pairs of Clark's grebes hitch their nests to weeds, rushes, and other vegetation. No land predators can cross water to reach the middle of a lake, so many water birds lay their eggs on floating nests like these.

The grebes' nest is a sturdy piece of architecture made from sticks, mud, reeds, and weeds. The platform above the surface is firmly lashed to an underwater pyramid that may be as much as three times broader than the nest itself.

Working together to keep their nest in good condition, the grebes plug leaks, plaster down stray strands and tuck fresh weed around the edges.

ISLAND LIFE *Clark's grebe has feet so far back on its body that it is almost unable to walk. It only leaves the water to fly or sit on its floating nest.*

THE BURROWING FROG'S WATERTIGHT SUIT

Australia's central desert, where rain may not fall for years, would seem hardly the best place for a moist-skinned frog. But the burrowing frog manages to make a home in this inhospitable landscape by burying itself in a skintight wet suit.

When the last pools of water dry to small puddles, the frog goes to bed. It wriggles backwards into a muddy pool, burying itself in the slime until it disappears completely from view. The frog continues to burrow until it has dug down to about 30 cm (12 in). It excavates a hole about twice its own size, using its feet to tamp down the soil along the walls. Then it goes to sleep. Its heartbeat and breathing slow down to the minimum necessary to maintain life.

After about two weeks, the top layers of the frog's skin loosen and mesh together to form an all-over body suit that is completely watertight. Two tiny tubes link the frog's nostrils to the outside. The frog, well protected against drying out, stays there until the rains come.

Sensing moisture trickling down through the sand, it wakes up, breaks out of the skin-sack and clambers up to the surface. With only a few days before the puddles dry out again, the frog gorges on insect prey, drinks its fill, mates, and spawns.

WHY HORNBILL FATHER IMPRISONS HIS MATE IN MUD

The female hornbill is content to be walled up in jail as she incubates her eggs. After finding a suitable nest hole in a tree, she settles on her eggs, while the male brings a mixture of mud and saliva to the nest site. The two of them use this to build a wall in front of the hole, the mud hardening as it dries, cementing the female inside. By the time they have

finished building, the only opening left is a narrow slit in the mud wall.

She stays inside, sitting on her eggs, safe from predators, while the male feeds fruit and insects to her through the slit. After about 84 days, when the chicks are half-grown, the female breaks out of her prison. She and her partner seal the nest again, leaving a slit through which the parents feed the chicks. Two weeks later the chicks break out, too.

SNUG IN THE SNOW *A polar bear cub will not leave its secure snow den in Churchill, Manitoba, until it is about three months old.*

HOW POLAR BEARS KEEP THEIR CUBS WARM

Lacking any ready-made shelter in the snowdrifts, the female polar bear constructs her own. Using her massive paws, she scrapes a 1.8 m (6 ft) tunnel into a bank of dry, compacted snow – one bear is known to have built a den 12 m (40 ft) long – to protect herself and her cubs from the Arctic chill. Alternatively, she might allow herself to get snowed in, later enlarging the hole that forms around her body. At the end of this tunnel she digs out a vaulted chamber with a raised platform.

Outside, the temperature can be as low as –30°C (–22°F). Inside, the den is warmed by the polar bear's body heat and can be as much as 21°C (38°F) higher. She may add a ventilation shaft in the roof, though such holes could be made accidentally when the polar bear scrapes ice from the surface. The cubs, born in December and January, spend the first three months of their lives in the den.

HOLLOW HOME *The gilded flicker likes to drill holes in cacti where it can nest, safe from snakes, among the spiky ribs.*

PRICKLY SITUATION FOR FLICKER BIRDS

A cactus's needle-sharp spines are enough to deter most predators. These make it a prime location for gilded flicker birds which, in the deserts of south-western USA and Mexico, nest only among the spiky ribs of the giant saguaro cactus.

In the rainy season the cactus absorbs so much water that it sometimes swells to twice its original size. Its flesh becomes soft and spongy, so the flickers can easily peck a hole. Sap oozes from the damaged spot and hardens, forming a smooth nest floor for the birds' eggs and chicks.

After one nesting season, the pair of flickers move out. The nest has become too full of parasites, feathers, and food scraps for them. New tenants, such as the tiny elf owl, soon take up residence.

NATURE'S MASTERPIECES

Stacks of sticks

The untidy heaps of sticks that are home to large tree-nesting birds, such as these painted storks in India's Keoladeo National Park, may at first appear less sophisticated than the neat, tightly woven basketwork of other birds. But piling sticks together in such a way that they actually interlock and hold tight is not as easy as it might look. Often the sticks have been deliberately criss-crossed or laid down according to some underlying symmetry. So, when the tall, long-legged, large-winged storks come flapping and crashing down on their nests, they do not knock them to pieces.

The nests of some species, such as the white stork from Europe, are added to year by year. On returning to its nest, a stork spends time pecking at the floor and 'airing' it out. It will then smarten it up, adding twigs until after several years the nest forms a tall stack, sometimes weighing hundreds of kilograms. Smaller birds, such as sparrows, occasionally nest in the base of a tall stork nest. Eventually a storm will bring the nest down whereupon the owner has to start building from scratch.

MASTER BUILDERS

Many animals move into ready-made shelters: others build their accommodation themselves. Nature's potters, weavers, paper-makers, construction engineers, and tailors create some of the most extraordinary homes.

POTTER WASP SCULPTS A NEST OF MUD

The female potter wasp is fussy about her materials. If the mud is too soggy, she will wait for it to dry. If it is too dry, she will regurgitate water from her stomach to soften it, kneading it in with her jaws and legs. When the consistency is just right, she works off a small pellet and flies with it to her nest, which might be on a leaf, under a piece of bark, or among the undergrowth. There, she starts to craft her first pot.

Moulding the pellet into a long strip, she lays it into a ring. Returning to the mud patch for another pellet, she builds the next layer on the hardening rim of the first. Before long, she has sculpted a shapely vase, complete with bulbous base and slender neck. Then she catches a caterpillar or spider and pushes it into the pot, into which she lays a single egg and seals the neck with clay.

When the grub hatches a few days later it eats the imprisoned insect, pupates, and breaks out of the pot as a new adult potter.

STOCKING THE LARDER *A potter wasp stuffs food for her grub into a clay-pot nursery which she has crafted.*

SPINNING A NEST OF PURE SALIVA

Swifts and swiftlets do not need to go in search of nest-building materials, they use their own spit. The edible-nest swiftlet of Malaysia and Borneo needs nothing but saliva to spin a translucent 'white' nest. In the breeding season, glands in the cave swiftlet's throat become enlarged, enabling the bird to produce copious amounts of glutinous saliva.

Once it has chosen a nesting spot – a small ledge or overhang – it flies towards the spot, dabbing it with its tongue. The saliva hardens quickly. The bird repeats this and within a few days it has built a cup-shaped lattice of threads, big enough to hold two eggs. These nests are much sought after for bird's-nest soup, considered a delicacy in South-east Asia.

HOW THE CHIMPANZEE WEAVES ITS TREE HOUSE

It takes years of observation and practice for a chimpanzee to learn how to make a nightly nest. Having mastered the trick, the ape from West and Central Africa chooses a tree with a stable bough 9–12 m (30–40 ft) up for a base. It squats on the bough and pulls three thick branches towards itself, pressing these under its feet and then weaving them into a disc 60–80 cm (24–32 in) around the base. Then it works thinner branches into a wreath and lines it with plucked leaves and twigs. A skilled chimpanzee can make a night's nest in just a minute.

A SPIDER'S STONE CIRCLES

In the middle of the Namib Desert, a stonemason is at work. The so-called 'mathematical' spider builds a near-perfect circle around its burrow using seven small pebbles. The stones are roughly the same size, shape, and colour, and the spider shows a preference for quartz. Exactly why the spider counts and arranges the stones is a mystery, but it may be that vibrations from the pebbles help it to detect the approach of prey.

SPIDER WEB STYLES

Orb webs
Spiders build an enormous variety of webs, all designed to ensnare their prey. The best known and most complex are the symmetrical orb webs, as built by the female garden spider of northern Europe (see right). A typical web may have 20 m (65 ft) of silk, meeting at 1,000 junctions. It weighs less than 0.5 mg, yet the spider it supports is 4,000 times as heavy.

Hammock and sheet webs
Other spiders build webs made of sheets of silk. The tiny money spider of northern Europe builds this type of web and 50 or more can cover a gorse bush with their miniature silk sheets.

Hammock webs can reach up to 30 cm (12 in) across. Any insect that blunders into the hammock or flies just above is tripped up by a maze of guy-line threads. The spider, which has been waiting underneath, grabs the insect from below and pulls it through the hammock, wrapping it in silk. It repairs the hole later.

Scaffold webs
A web that looks like an untidy mass of scaffolding on tall herbs, bushes, and small trees is likely to belong to a spider of the Theridiidae family. The web is studded with blobs of glue that hold its victims tight. A European species, *Achaearanea riparia*, lightly anchors a long, taut thread coated with sticky blobs to the ground. When an insect walks into the snare and begins to struggle, the thread breaks anchor and hauls the insect up to the scaffold web above like a bungee rope.

Web-casting spiders
The web-casting spider, also known as the ogre-faced spider because of its huge eyes, spins a net the size of a postage stamp. The spider holds this cat's cradle between its front legs. It then hangs upside down from a leaf and when a moth flutters by, the spider lurches forward and stretches the web taut, netting its victim.

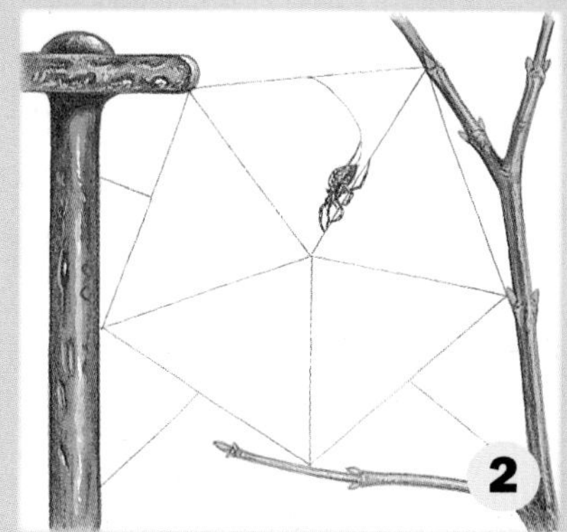

1 *From a horizontal bridge-line, the orb spider drops down on a vertical thread until it reaches a fixed object.*

2 *It then builds a framework, linking the radials at the hub.*

3 *A wide spiral of dry silk is laid down on the radials, working from the centre outwards.*

4 *The spider attaches a spiral of sticky silk, starting from the outside.*

Strong and flexible
The dragline silk of a golden orb spider is the strongest natural fibre known; it can also stretch by a third before snapping. This strength and flexibility make it the perfect material for webs, securing all but the strongest victims.

STICKY SITUATION *Once tangled in the sinews of a web-casting spider's 'net', an insect is unlikely to escape.*

CAMPING CATERPILLARS ARE LEFT IN PEACE

Webs festoon the higher branches of fruit trees in much of the USA. This is where adult Eastern tent caterpillar moths lay their eggs and build a thick communal tent.

At first, the 100 or so newly hatched caterpillars shelter beneath the tent, feeding on buds, leaves, and empty egg-cases. When the food runs out, they move to the top of the tent to feed, where hungry birds can see them. Predators soon learn to leave them alone: as the caterpillars feed, they collect cyanide from the tree and if threatened regurgitate it as a foul-tasting liquid.

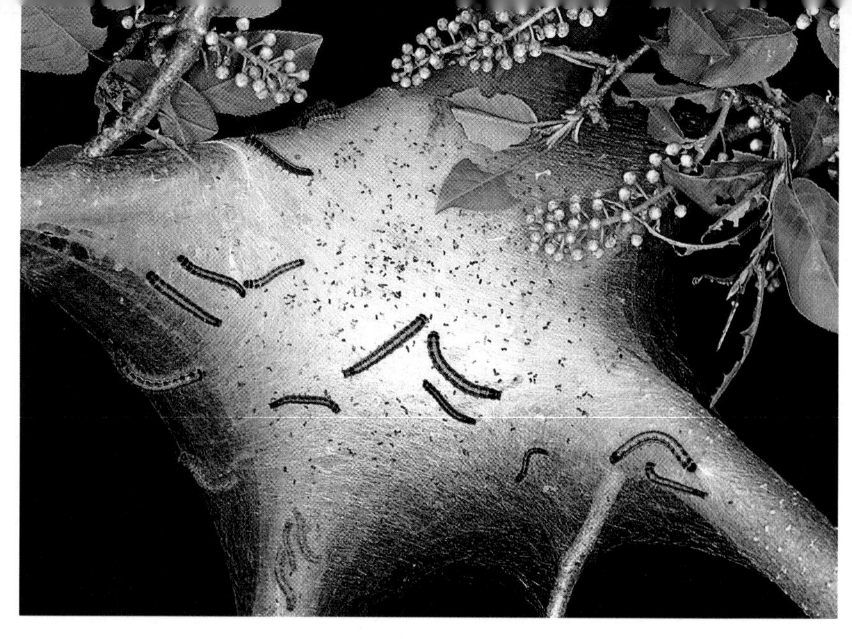

CATERPILLAR CAMP *Predators avoid foul-tasting Eastern tent caterpillars, which feast on fruit trees such as cherry, apple, and plum.*

BATS THAT CRAFT THEIR OWN LEAFY CANOPIES

A few bats choose not to seek out the ready-made shelters of caves, belfries, or attics. Instead, some species of American spear-nosed bat and the Sphinx's short-nosed fruit bat roost beneath a simple large leaf.

Honduran white bats nibble a row of holes on either side of the ribs and veins of the leaf. Eventually, the leaf collapses around them. Up to a dozen bats snuggle underneath, sheltered from the sun, wind, and rain, clinging with their feet to the bitten spots. At a height of 2 m (7 ft), the canopy is far enough away from the ground to be safe from predators.

LEAFY SHELTER *Honduran white bats, a species of spear-nosed bats, often return to the same campsite, but they rarely bed down under the same leaf-tent two nights in a row.*

TAILORBIRD SEWS A NEST FROM SILKEN THREAD

Spiders' silk is strong stuff, and the Indian tailorbird puts it to excellent use. Holding silk from a web in her beak, the female bird pulls together two leaves hanging close to each other and pecks holes in each, along the rim. Using one strand of silk per stitch, she threads the silk through the holes, drawing it tight to fasten the leaves together and tying a knot to stop it slipping out. Alternatively, she may curl a single leaf against itself.

Either way the end result, which can take up to four days to build, is a cup-shaped base for her nest made of living leaves that blend in well with the other leaves around.

TINY ACROBAT MAKES ITS HOME IN THE GRASS

The European harvest mouse weighs no more than a few grains of rice. As it scampers up stiff grasses and wheat to feed on seeds and grains at the top, the stalks barely bend. The mouse's tail can be used like an extra limb to help it to climb. This enables the female to build her nest high above the ground, well away from predators, but close to her food source. Using leaves and blades, which she nibbles so that they split into several lengths, she weaves a tennis-ball-sized nest and lines it with moss.

AVIAN ARTISANS PREFER THE KNOTTED LOOK

African weaverbirds do not weave, they tie knots. Building a nest that will lure a female takes practice and a male's first attempts often fall apart, but he soon perfects his skill (*see right*). Hanging beneath the knotted nest, the male attracts attention by flapping his wings. A female inspects the nest and if she likes it she pokes feathers into it to make an egg chamber. Once installed, the male builds a nest for another female.

THE WEAVERBIRD

Male weaverbirds, such as this lesser masked weaver from Kenya, use a variety of complex knots (*see below*) to build a sturdy, light nest. A male starts by stitching a swinging perch to a forked twig. He manipulates strips of vegetation to build this into a thick ring, using his beak to tie the knots while holding on with his feet. He then twists, knots, and loops more fibres around the basic structure until the nest is complete.

SPEEDY MOLE ENSURES FOOD FALLS AT ITS FEET

The European mole is a digging machine that can plough its way through 15 m (50 ft) of soil in an hour. Its rounded, spade-like stocky front feet are turned permanently outwards, and each paw bears five large claws. To excavate a new tunnel, the mole braces itself with its hind feet and pushes blindly through the earth, using its snout as a wedge and following with a powerful breast-stroke action. In the spring rutting season it covers up to 50 m (160 ft) underground.

The mole's nest is a football-sized, rounded chamber excavated a little deeper than the tunnels and lined with grass and dry leaves. The permanent foraging tunnels that radiate off it are the mole's larder. Worms, slugs, beetles, fly larvae, and other invertebrates break into these passages through the roof, walls, and floor. The mole patrols the labyrinth once every three hours. As it scurries along, it detects prey with its muzzle, which is highly sensitive to touch. It can swallow up to 55 g (2 oz) of worms a day, but if there are too many worms, the mole immobilizes the excess with a bite and stores them – sometimes in their thousands – to eat later.

When it is excavating a new tunnel, the mole digs upwards every so often to break through to the surface. The resulting molehills are ventilation shafts.

WIND TUNNELS *Moles dig upwards to create ventilation shafts: air sucked in travels through the nest-chamber and back up through another tunnel.*

HEXAGONAL HOME CRAFTED FROM PAPIER-MÂCHÉ

Strong, lightweight, and durable, papier-mâché is the building material of paper wasps. To create a new colony, the lone queen emerges from hibernation and selects a nest site, often under a branch or leaf, or beneath the eaves in a barn or garage. Then she makes paper, chewing mouthfuls of wood or other plant fibres mixed with saliva and spitting out soft, damp pulp.

The queen dabs a stalactite of paper to the roof of the nest site, which dries and hardens fast, forming a stem. She adds a few hexagonal cells to the bottom of this and lays a single egg in each one.

Soon, the new generation of wasps hatches and forms the queen's first work force. They get to work immediately, building more paper cells for the queen to lay her eggs, followed by the task of feeding the young in their paper cradles with pellets of chewed-up insect flesh.

PAPER-MAKING *Paper wasps build hexagonal paper cells, judging the angle and size of the walls by continually touching the walls of adjoining cells with their antennae.*

BUSY BUILDER *Beavers live in an island lodge in a lake, with submerged entrances.*

1 *If no natural lake is available, beavers create one.*

2 *They build a wooden dam.*

3 *The spectacular dam may span up to 30 m (100 ft).*

THE BEAVER'S MASTERPIECE OF ENGINEERING

No other mammal matches the beavers of North America and Europe for architectural genius. If there is no suitable lake on which to build their island den or lodge, beavers will make one by damming a stream.

This extraordinary piece of engineering begins when a pair of beavers sets up home in a wooded valley with a small stream. Having chosen their site, the beavers put their massive, razor-sharp incisor teeth to work. They gnaw through saplings and tree trunks more than 50 cm (20 in) thick in a matter of minutes. Saplings felled upstream are floated down along specially dug-out channels. The beavers ram a few sticks upright into the stream's bed and drag lengths of felled timber and leafy sticks across these supports. Using rocks, they weigh down the vegetation and bind it with mud from the riverbank.

Shaped with a steep upstream side and a more gently sloping downstream side, the dam is able to withstand the pressure of water in the newly formed lake. The structure may be up to 30 m (100 ft) long and 3 m (10 ft) high.

Beavers then construct the lodge, a huge, domed edifice of mud, reeds, sticks, and stones 1.8 m (6 ft) high and 12 m (40 ft) across. Inside, they hollow out the living chamber with a sleeping platform above the water level. To protect themselves from bears, lynx, and other predators, all the entrances are under water and the only way in is to swim.

If, after heavy rains, the lake threatens to overflow, the beavers drain off water by widening the spillways at each end of the dam. If the lake level drops so low that there is a risk that the lodge entrance will be exposed, they narrow the spillways.

SWIFT'S YOUNG GROW SAFE INSIDE A 'STOCKING'

The lesser swallow-tailed, or cayenne, swift builds a nest resembling a woollen stocking. Using saliva mixed with feathers and plant material, the bird, from South and Central America, creates a long sleeve with an entrance 1–5 cm ($^{3}/_{8}$–2 in) wide at the end. At the top of the stocking is an enlarged area with a ledge, where the swift lays her eggs. Sometimes she builds a nest in the shape of a tube, 24–48 cm ($9^{1}/_{2}$–19 in) long, attached to a vertical wall. She decorates the outside with feathers.

BEES USE GEOMETRY TO BUILD A SPACE-SAVING NEST

Honeybees need to build their comb economically. To produce just 55 g (2 oz) of wax from their glands, the bees have to eat 1 kg (2 lb 4 oz) of honey made from nectar and pollen and it takes about 7 kg ($15^{1}/_{2}$ lb) of honey to make enough wax to create the average hive.

Squares, triangles, and hexagons fit snugly and avoid wasted space. Bees favour hexagons, which have the shortest walls so less precious wax is needed. Secreting wax from glands in her abdomen, the worker bee lays down a thick ridge for the cell wall. To prevent honey and nectar dripping out, she tips the cells up 13 degrees from the base. The resulting comb is able to hold many kilograms of honey.

Castles of mud

The towering spires, domes, and pyramids that termites build are impressive enough from the outside, but on the inside the architectural sophistication is truly extraordinary. There are larders, gardens, air-conditioning systems, nurseries, living chambers, cellars, wells, chimneys, and royal chambers.

There are more than 2,000 species of termite, found mainly in tropical regions. They are social insects, living in colonies that range from a few dozen individuals to several million.

HIGH-RISE HOMES *Termite mounds dot the Australian outback. Each one of these insect skyscrapers may contain hundreds of thousands of the tiny herbivores.*

A royal kingdom

Termite colonies are ruled by the king and queen, which are the only fertile termites in the colony. They remain in the royal chamber, where the queen spends her life as an egg-laying machine. She is attended by worker termites, who look after the eggs and larvae and also maintain the mound. Soldier termites, with enlarged jaws, defend the entrances to the colony.

Inner-city designs

The blind worker termites construct their fantastic castles out of earth mixed with saliva, which sets like concrete. The walls can be 50 cm (20 in) thick, although the specifications vary with each species.

Living larders

Bellicose termites from Africa eat mainly dead wood, which is difficult to digest, so their droppings are still rich in nutrients. To avoid waste, the termites cultivate a fungus on their droppings, which breaks down the manure and after six weeks the termites can eat and digest the compost, fungal growths and all.

Air conditioning

An active colony produces a lot of heat, so termites have incorporated a cooling system into their design. Hot air rises through a large central cavity into upper porous chimneys, where hot carbon-dioxide-rich air diffuses out and fresh, oxygenated air diffuses in. The fresh, cool air then sinks to a cellar at the base of the nest. Sometimes termites dig a deep well down to the water table: the moisture helps the fungus to grow.

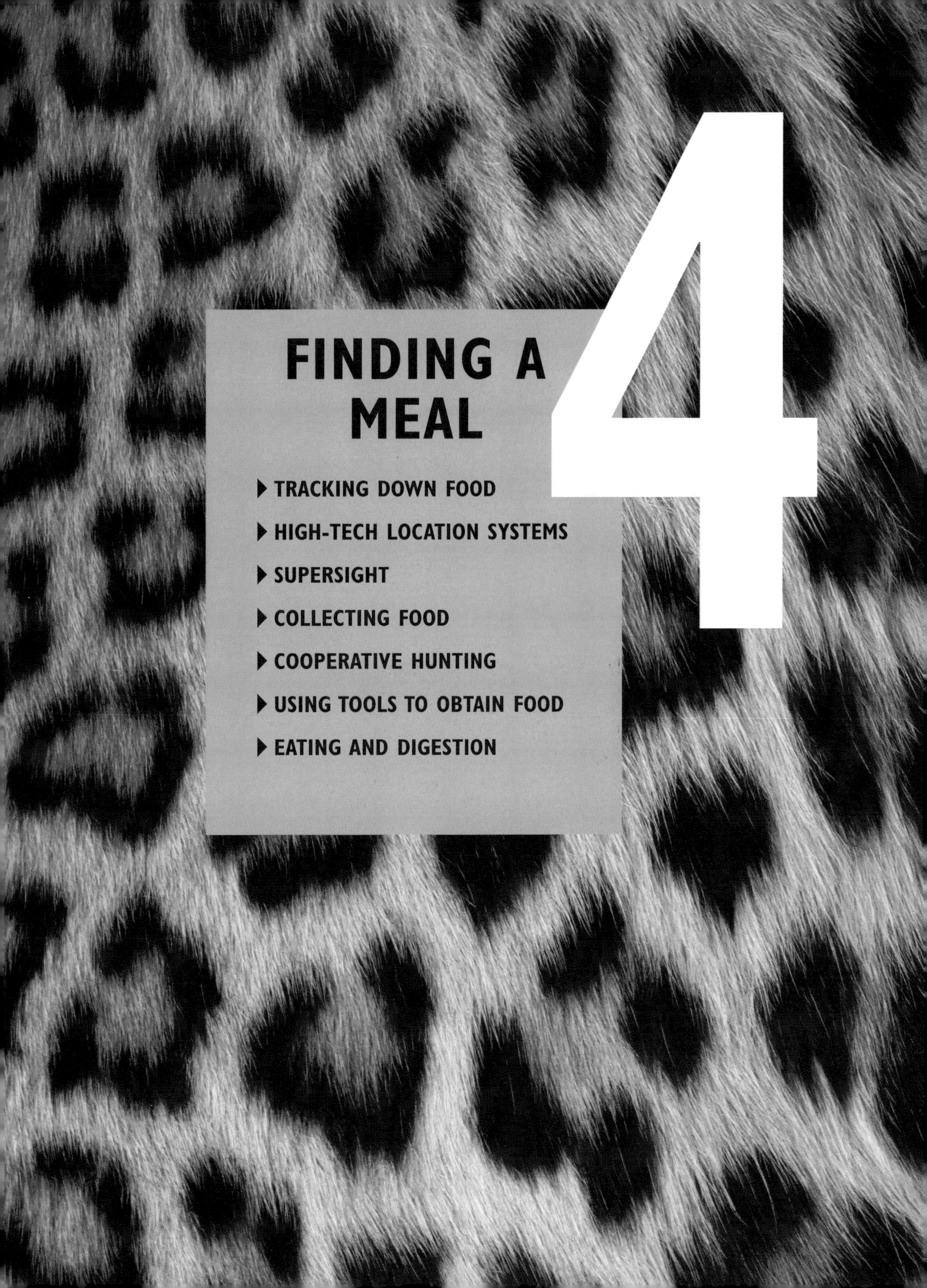

4

FINDING A MEAL

- TRACKING DOWN FOOD
- HIGH-TECH LOCATION SYSTEMS
- SUPERSIGHT
- COLLECTING FOOD
- COOPERATIVE HUNTING
- USING TOOLS TO OBTAIN FOOD
- EATING AND DIGESTION

TRACKING DOWN FOOD

Most animals feed regularly, but first they must locate a meal. To do this, they use a battery of senses – smell, sight, hearing, taste, and touch. Some animals rely more on one sense than another, and it can be very highly developed.

FORAGING KIWI FOLLOWS ITS NOSE

The kiwi, from New Zealand, is a curious bird. It cannot fly and its feathers resemble fur. It finds food like an insect-eating mammal, and can smell an earthworm 3 cm (1 1/4 in) below the ground.

At night, the kiwi, which is about the size of a domestic hen, roots around in leaf litter and soil with its long, sensitive bill. Nostrils at the tip are linked to tissues containing smell receptors at its base. This smell-sensitive membrane is folded, like the inside of a dog's nose, and nerves link to olfactory lobes in the brain that are larger than those in any other bird. With this sensory system, the kiwi can detect worms, slugs, and beetle grubs. Picking up the food with the tip of its bill, the bird throws it into the back of its throat in a series of jerks.

SHARP-EYED SCAVENGERS LOOK OUT FOR A MEAL

White-backed vultures in the African Serengeti rely on extremely good eyesight to find a meal. In the morning, they soar on thermals of rising warm air in order to survey their home area for food. They are scavengers and look for signs of animals that are dead or dying.

When a bird spots a carcass, it spirals slowly to the ground, but it is not alone for long. Soon an army of vultures descends on the food, all squabbling for a place at the table.

The birds approach the food cautiously at first, but once one has overcome its fear, the rest pile in for the feast. A small antelope can be stripped clean in half an hour.

STILL LIFE LEAVES FROGS HUNGRY

Frogs have large eyes, but cannot see their prey as a clear image. Instead, they detect the movement of small objects, such as insects or slugs, using special receptors in the retina at the back of the eye.

As soon as the prey shifts, the frog immediately recognizes it as potential food. It then flicks out its sticky tongue to trap the unfortunate victim.

THE GREY WOLF'S SUPERSENSITIVE NOSE

North American grey wolves have a sensitivity to smells 1,000 times greater than that of humans. They can detect the presence of a mother moose and calf more than 2.5 km (1 1/2 miles) away.

Compared to a human nose, a wolf's has about 50 times more cells that can detect scents and smells. These tissues are folded to present the maximum surface area, yet pack into a confined space; flattened out, they would cover an area the size of a large postcard. The nose itself is shaped so that air laden with smells passes efficiently over the smell-sensitive membranes. A healthy wolf also has a wet nose, which dissolves scent particles. This gives it more of the smell to analyze.

When a wolf passes downwind of potential prey that is well out of sight, its keen nose picks up the scent and it changes direction upwind to follow the smell back towards its victim. It then relies on its eyesight and speed to catch or bring down its prey.

KIWI PROBES BY SCENT *The kiwi emerges at night and forages for food, probing with its long, slightly down-curved, smell-sensitive bill.*

SURFACE SKIMMER *A black skimmer exploits the clear waters of its salt marsh habitat. Its bill snaps shut upon touching a fish in the water.*

THE BLACK SKIMMER'S KISS OF DEATH

The black skimmer's bill appears to have been built upside down. Its lower mandible is longer than the upper part, which enables it to catch food using the sense of touch.

The bird, found along the eastern coast of North and South America, flies low over relatively calm water such as a lake or a slow-moving river, usually at dawn or dusk when the wind is less strong. It opens its very thin bill so that the longer lower half cuts through the surface of the water with minimum resistance until it connects with a small fish. The bill then snaps shut automatically and the head drops down, trapping the fish sideways between the scissors of the bill. The fish is then taken to a nearby perch to be swallowed.

If the black skimmer hits an object such as a rock just below the surface, its head bends round and special muscles in its neck help to deaden the shock of the impact.

DRAGON TASTES THE AIR WITH A FORKED TONGUE

Indonesia's Komodo dragon can 'taste' the slightest trace of a smell – even a single molecule. The world's largest lizard feeds most often on carrion and can detect a rotting carcass from up to 400 m (440 yd) away. The Komodo achieves this not by smelling the air with its nose, but by tasting it with its tongue.

Like all monitor lizards, the 3 m (10 ft) long Komodo has a forked tongue, which flickers in and out continuously, picking up minute quantities of smells in the air.

When the tongue is withdrawn, it is pressed against a special organ, known as the Jacobson's organ, in the roof of the lizard's mouth, where a lining of cells sensitive to chemicals enables the dragon to decipher the smells.

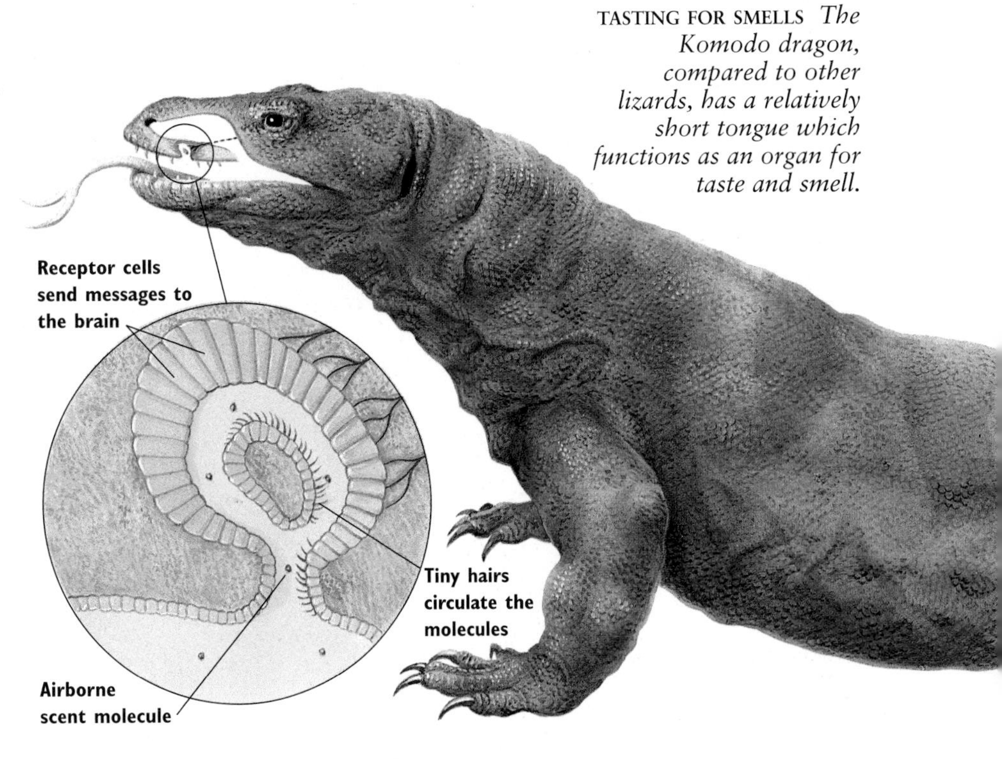

TASTING FOR SMELLS *The Komodo dragon, compared to other lizards, has a relatively short tongue which functions as an organ for taste and smell.*

GOOD LISTENER *Bat-eared foxes hunt singly, in pairs, or in family groups. They are usually nocturnal but sometimes forage during the day.*

BIG EARS LISTEN OUT FOR SOUND BITES

The bat-eared fox of southern and eastern Africa relies on sound to find a meal. Pointing its enormous ears downward, it listens for insects under the ground. The fox hunts at night during the hot dry season, but can be out and about during the day in the cooler winter.

Small family parties of four to six foxes spread out over the savannah to hunt. Each fox stops occasionally with its nose pointing at the ground and its ears tilted forwards, and listens, turning its head slightly from side to side to obtain a better fix. In this way, the animal can listen for termites scrambling about in their tunnels and even hear a dung beetle larva munching inside a ball of buried dung. Having located an interesting sound, the fox digs rapidly with its front paws and uncovers the prey.

Invertebrates, such as spiders, millipedes, worms, and termites, are common items in a bat-eared fox's diet, but mice, lizards, and small snakes also appear on the menu. The fox sometimes takes insects on the surface, but it finds them only if they move and make a slight sound. An insect frozen to the spot may be missed entirely.

SURFACE TENSION *Having detected an insect trapped in the water's surface film, the pond skater speeds across the water to catch its prey.*

CAVE DWELLERS HAVE NO USE FOR EYES

Troglodytes, or cave dwellers, rely on touch and the detection of vibrations to find food. Since they live their entire lives in deep caves, often in complete darkness, they have no use for vision. Many have no eyes at all or their visual organs are greatly reduced in size and function.

For example, the cream-coloured huntsman spider that lives in the caves of the Gunung Mulu National Park in Sarawak has two tiny, colourless eyes. To compensate, its front pair of legs have developed into 'feelers'. These are held out in front, like an insect's antennae, so that the spider can feel its way around.

Many of the underground spiders have extremely long, spindly legs for running across cave walls. One species in the caves of Gunung Mulu has a leg span of 15 cm (6 in).

READING RIPPLES AND DOING THE SUMS

European pond skaters are natural mathematicians. They find their prey by detecting ripples produced in the water by insects trapped in the surface film of a pond. The pond skater compares the time a ripple takes to reach adjacent legs or legs on opposite sides of its body and makes a lightning calculation of the direction from which it originated. Then it prepares to go hunting.

The pond skater, supported on air trapped by dense pads of hair on four of its six legs, speeds across the water, 'rowing' with its long middle legs and steering with its hind legs. It grabs the victim with its shorter forelegs, stabs the prey with its piercing mouth-parts and sucks it dry of body fluids.

SAVAGE SCAVENGERS *Most starfish are meat-eaters. These sea bat sea stars are scavenging on the decomposing body of a dead rock crab.*

BATS RELY ON WEATHER FORECASTS

The eastern pipistrelle bat has its own built-in barometer. With it, this North American cave dweller can sense whether or not the weather outside is favourable to go hunting for flying insects.

The bat has a sensory organ in the middle ear known as the Vitali organ, which can detect air pressure. When the pressure is high, bats remain at home because the insects they prefer to hunt are not on the wing. When the pressure drops, the bats know that their prey will be more abundant so they leave their caves and go hunting.

Averaging eight to nine catches a minute, a bat can scoop up more than a gram of insects in an hour – about one-fifth of its body weight. This makes the eastern pipistrelle one of the most efficient hunting bats.

THE STARFISH WITH A NOSE FOR FOOD

An aggressive predatory starfish lives in the pounding surf of the fjordland islands in southern Chile. It terrorizes the sea floor, eating just about any living thing it can catch.

The starfish, called *Meyenaster gelatinosus*, glides rapidly over the sea floor on its thousands of tube feet, holding on to prey with one of its five arms while following and grabbing something else with the other four.

The starfish locates its prey initially by 'smell', detecting chemicals in the water and homing in on their source until it makes contact. Then the sense of touch takes over. Creatures the starfish pursues are alert to its presence and can sense whether it is intent on feeding or not. Somehow, the potential victims seem to know when to run.

HIGH-TECH LOCATION SYSTEMS

Animals evolved sonar and radar systems millions of years ago to find their way or to detect objects in their path. They also developed infrared heat sensors and sensors that detect electrical activity in the muscles of their prey.

PLATYPUS HAS A NOSE FOR ELECTRICAL CURRENTS

The duck-billed platypus uses its snout to detect electricity. This enables it to find its food while it swims with its eyes, ears, and nostrils tightly closed.

The curious aquatic egg-laying mammal lives in the rivers of Australia. Lining its large leathery 'bill' are thousands of pore-like openings, each containing sensory cells that can detect minute electrical currents produced by the moving muscles of its prey, such as the flicking of a shrimp's tail, or the wriggling of a worm. Using its sensitive snout in this way, the platypus can scour the riverbed to satisfy its enormous appetite: it consumes almost its own weight in food each day.

With its bill the platypus can sense even weaker electrical fields associated with the movement of water over objects in the river, such as logs and rocks, helping it to find its way.

THE SPERM WHALE'S DEADLY SOUND WAVES

Sperm whales bounce sound off objects and analyse the echoes to locate prey. Their echolocation system is similar to that of dolphins, but even more powerful. They can stun their prey with sound. These giant toothed whales focus echolocation clicks through the gigantic, oil-filled spermaceti organ in the forehead. Among the clicks there are often loud 'gunshot' sounds. These ear-splitting salvos of sound energy create brief pressure waves which are thought to be used by the whales to stun their prey or kill it outright.

Sperm whales dive as deep as 2,500 m (8,200 ft) in search of squid, including the giant squid, the world's largest known invertebrate. The squid succumb to the high-intensity blasts, and while they are immobilized the whales slurp them up with ease. Squid measuring 12 m (39 ft) long have been found intact in sperm whale stomachs.

THE ELECTRIC EEL'S SHOCK TACTICS

The virtually blind electric eel uses electricity to find its way. It produces low-level discharges – about one pulse per minute – which create a short-lived electrical field around its entire body.

If the fish encounters an object or potential prey enters the field, it is immediately alerted and can then either avoid the obstruction or accelerate towards the prey. Its tail contains modified muscles – the electrical organs – that can stun or even kill a person with a discharge of several hundred volts, delivered not just once, but a hundred times a second.

The electric eel is not really an eel: it is a South American freshwater fish distantly related to the minnow. Unlike its tiny relatives, it grows to a maximum length of 2.75 m (9 ft), four-fifths of which is the tail.

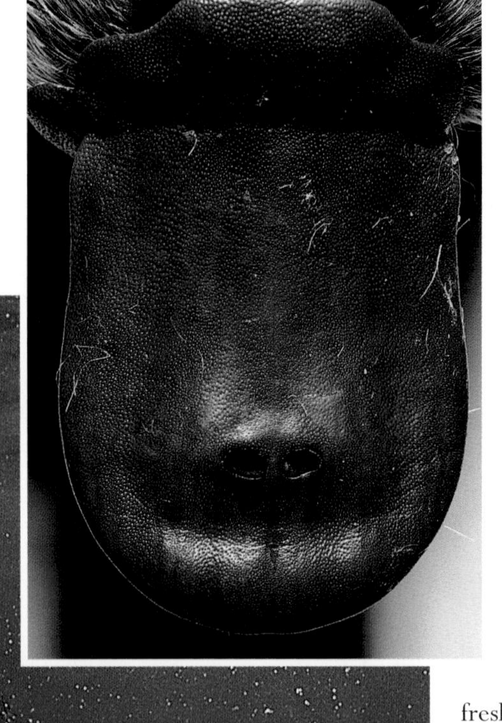

DUCK-BILLED DETECTOR *With its webbed feet and duck-like bill, the Australian platypus scours the riverbed for food. The tough, leathery snout is used to turn over stones in the search for shrimps.*

DOLPHIN ETIQUETTE *When bottle-nosed dolphins cross paths, they stop producing echolocation sounds momentarily to prevent interference.*

DOLPHINS' SOPHISTICATED SOUND SYSTEM

Dolphins 'see' with sound, to find their way and locate their prey. By sending out high-frequency sounds, or clicks, and analysing the returning echoes a dolphin can find prey, such as a shoal of fish. Its echolocation system also tells the dolphin about the size, shape, and structure of a target, and whether prey is calm or frightened.

No one is sure how the clicks are produced, but it is thought that they pass through the bulbous region of the dolphin's forehead (known as the 'melon') which acts like a lens, focusing the sound into a narrow beam ahead of the dolphin. The returning echo signal is picked up by the teeth, passing through the lower jaw to the ear.

HIGH-PITCHED SQUEAKS HELP TO FIND PREY AT NIGHT

Bats hunt at night, listening their way through the dark. A bat learns about its surroundings and its prey by emitting rapid high-pitched squeaks – or, in the case of some species of fruit bats, low-pitched tongue-clicks – and analysing the returning echoes. Its echolocation is accurate enough to track a mosquito.

Scanning systems vary between species. The little brown bat, for example, when hunting for flying insects, produces 5 to 20 very high frequency pulses of sound per second, rising to 50 per second when prey is within 90 cm (3 ft). As the bat locks on to its target, it learns more about the size, texture, speed, and direction of its prey by raising its pulse rate to 200 a second, the so-called 'feeding buzz'. As it reaches the prey, the pulse rate peaks and it catches the insect.

The greater horseshoe bat compares its outgoing and incoming signals and uses the Doppler effect to catch its prey. If prey is moving away, the frequency of the returning signal drops; if it is approaching, the frequency increases.

The greater horseshoe does not chase about like other bats but remains still, rotating its head and scanning the air with a beam of sound, like ground-to-air missile radar. When contact is made, the bat flies out, claims its prey, and returns to the launch site to eat it.

HOMING IN ON A MEAL *A hunting bat can tell in which direction its prey is travelling by the frequency of returning sound pulses.*

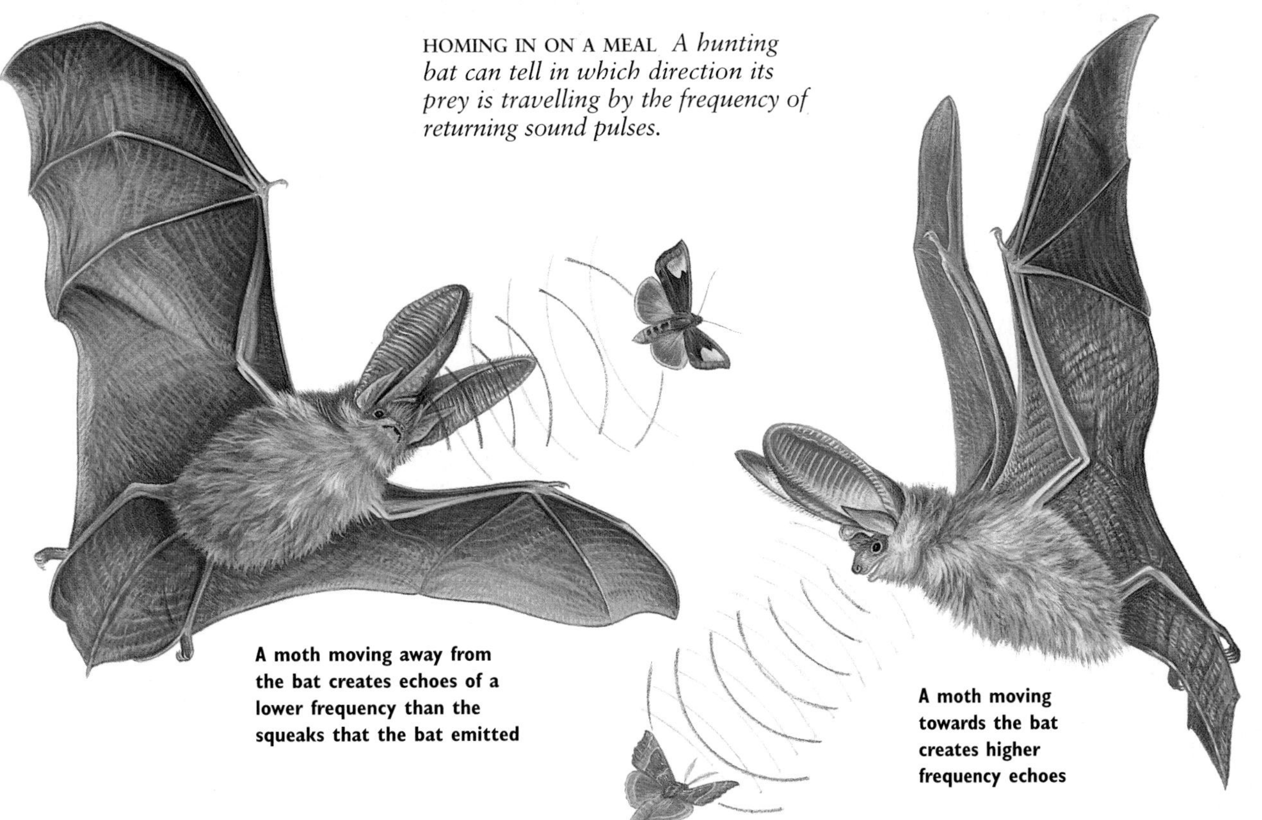

A moth moving away from the bat creates echoes of a lower frequency than the squeaks that the bat emitted

A moth moving towards the bat creates higher frequency echoes

OWLS USE FEATHERS AS A LISTENING DEVICE

Owls use their entire face to receive sounds. The heart-shaped disc of tightly packed feathers on a barn owl's face, for example, is designed to collect sounds and then focus them on the owl's ears, which are buried below the feathers on either side of its head. Owls hunt mainly at night and though they have good night vision, they rely on their ability to detect minute sounds to catch their prey (mice, voles, and shrews).

The owl can compare the sounds reaching each ear and work out the location of their source. In addition, one ear is slightly higher than the other so the bird can gather both horizontal and vertical information. With this arrangement, it can pinpoint the position of prey accurately in complete darkness. It then swoops down with silent flight, using its sensitivity to the slightest rustle made by its prey to align its talons to every zigzag movement the prey might make. At just the right moment, it grabs its victim and the hunt is over.

MIDNIGHT FEAST *A barn owl returns to its nest having located and caught a field mouse in pitch darkness. The owl detected the rustlings of the mouse as it ran through the grass.*

SLITHERING REPTILES HAVE HEAT VISION

Rattlesnakes and pit vipers can 'see' a warm-blooded creature such as a mouse in total darkness. The snakes have pits below their eyes, the loreal pits, containing membranes with nerve endings sensitive to infrared radiation, or heat. They can appreciate changes in temperature as minute as 0.005°C (0.009°F).

By comparing the signal reaching the pits on either side of its face, a snake can also pinpoint with some precision the location of its victim and strike accurately.

A mouse has a good chance to escape, however, for it must approach to within 15 cm (6 in) of a snake's snout before it can be detected.

HOMING IN ON A FEAST OF BLOOD

Ticks find a victim using a sense organ that detects smells, heat, pressure, and humidity. Known as the Haller's organ, it can be seen under the microscope as a small depression at the tip of the tick's first pair of walking legs.

Ticks are external parasites related to spiders and they have a penchant for blood, including human blood. A tick in need of a host climbs up a grass stalk and simply waits. When it detects the presence of a living body it suddenly becomes alert, and waves its first pair of legs in the air like antennae. If an animal or person should brush past, the tick immediately crawls aboard and, using special sensory pads on its mouth-parts, locates the best place to drill for a blood meal. It then pushes its hypostome, a drilling mouth-part lined with teeth, through the skin creating a space where blood can form a tiny pool. Its saliva contains chemicals that prevent the blood from clotting, so it can feed until it is full.

SHARK WATCH

Over the course of hundreds of millions of years sharks have evolved a remarkable array of prey-detection systems, probably the most diverse of any predator.

A shark can home in on its prey from more than 2 km (1 1/4 miles) away. This is because its ears are tuned in to low-frequency vibrations, such as those produced by an injured fish, which can travel great distances.

At a distance of 0.5 km (550 yd), a shark can smell blood or body fluids in the water and follow the trail back towards its source. It can detect one part blood in 100 million parts water.

At 100 m (330 ft) from its prey, the shark's lateral line – a row of fluid-filled sensory canals running along either side of its head and body – picks up changes of pressure in the water produced by the prey as it moves or struggles.

From about 25 m (80 ft), depending on the clarity of the water, a shark can see its prey, and some sharks see in colour, discriminating between blue, blue-green, and yellow. Shark eyes are ten times more sensitive to dim light than those of humans. This is due to a layer of reflective plates behind the retina that bounces light back onto the light-sensitive cells, enabling the shark to make use of every photon of available light. When a shark needs to rise rapidly from the murky depths to attack prey swimming in the brightly lit surface water, it avoids being blinded by covering up the reflective plates. Cells filled with the dark pigment melanin quickly move into channels on the reflective plates, preventing them from reflecting light. In dim light the cells migrate away.

When some species of shark attack, they protect their eyes either with a third eyelid, known as a nictitating membrane, or by rolling them back into protective sockets in the head. At this point, a shark is swimming blind and it brings into play yet another remarkable sensory system. Jelly-filled pits in the shark's nose contain cells that are sensitive to electricity. They are so sensitive that they can detect a change of a hundred-millionth of a volt per centimetre. This means that a shark can pick up the minute electrical currents produced by a fish's beating heart or the contraction of muscles in its tail.

The hammerhead shark, with its curious T-shaped head, is a prime exponent of this method of detection. It swings its head from side to side over the seabed, like a person with a metal detector, searching for tiny electrical currents. In this way it can locate the position of flatfishes, skates, and rays buried under the sand.

SEVEN SUPER SENSES *Sharks can detect food over long distances and home in on it with great accuracy in a wide range of conditions. They are even able to pinpoint prey that is completely still and hidden under sediment.*

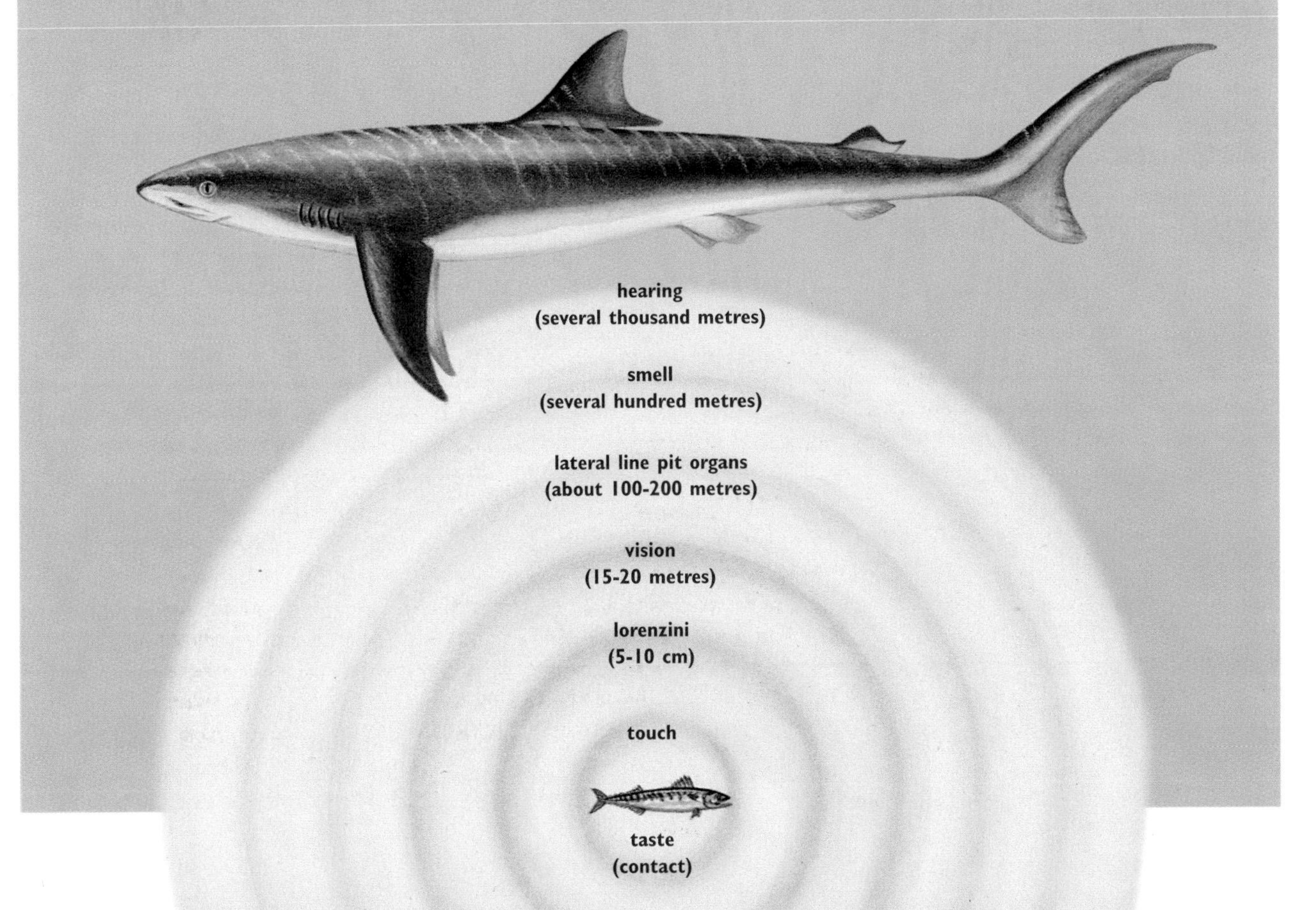

SUPERSIGHT

Over the course of some 600 million years, eyes have evolved into a diversity of shapes and sizes. Intriguingly, some of the smallest creatures on Earth turn out to have developed the most sophisticated visual equipment.

EIGHT EYES MAKE SPIDER A GREAT ALL-ROUNDER

Jumping spiders have six small and two large eyes, giving them almost 360 degree vision. Day hunters with good colour vision, they rely on sight to catch prey. They use their small eyes, with a wide field of view, to detect movement, and the large eyes, with a narrow field of view and sharp vision, for precise tracking.

To focus its large eyes on a target, the spider moves its retina while the lens remains fixed, and the eye colour changes: the spider is looking straight at you when the eye is darkest.

Able to identify prey at 30 cm (12 in), the spider stalks to within jumping distance, crouches, and leaps forwards up to 40 times its own body length.

UV-VISION HELPS HOVERING HUNTER TO FIND PREY

The Eurasian kestrel can see in the ultraviolet (UV) range of the light spectrum. Hovering high in the sky and scanning the ground below, it can spot the UV light reflected by the urine trails of its principal prey, mice and voles. The trails show up as dark streaks in the grass, allowing the kestrel to identify good hunting areas with plenty of rodent activity.

BIG-EYED AERIAL PREDATORS SEE LONG DISTANCE

Birds of prey have ultra-keen eyesight that helps them to spot small animals from great distances. A kestrel can spot a mouse in the grass from a height of almost 1.6 km (1 mile) and the powerful African martial eagle was once seen to take off from a hill 6.4 km (4 miles) away from a guineafowl, only to swoop down and capture the bird. A human could see the same-sized bird from about 1 km (1,000 yd) away.

Birds of prey have big eyes that face forwards, the best position for keeping their prey in view. It also means that the visual field of each eye overlaps, giving the bird binocular, three-dimensional vision, essential for judging the distance to its prey.

EIGHT-EYED HUNTER *The jumping spider sometimes uses 'sit and wait' tactics to detect and ambush its prey.*

The internal structure of the eye is highly specialized, too. The cornea is flat (unlike the curved cornea of the human eye), which increases the area of retina on which an image can fall. And with five times more light-sensitive cells on the retina than a human, a hawk's view of the world has a higher definition. In each eye, two foveas (points of sharpest vision on the retina) give a visual image eight times more acute than that of a human.

BEADY EYES KEEP WATCH FOR STARFISH

Scallops live on the seabed, where starfish are their deadliest enemies. They watch out for these creeping predators with the help of dozens of eyes. A scallop's eyes look like tiny bright blue beads, and they line the edges of both halves of its shell.

Scallop eyes are extremely simple and cannot make out much detail at all. But they can sense the shadowy form of a starfish as it inches towards its prey. If a scallop sees a starfish approaching, it responds by snapping shut its shell. This makes it shoot backwards and out of harm's way. Unfortunately, scallops have a 'blind spot' where the halves of its shell hinge together, which means that starfish can sometimes creep up unseen.

PRIMATE LISTENS THEN LOOKS THROUGH THE DARKNESS

The tarsier has huge eyes, each weighing more than its brain. Its eyes have developed in this way so that it can hunt on the darkest of nights. This tiny, insect-eating primate lives on the islands of South-east Asia, such as Borneo and the Philippines. It is the only exclusively carnivorous primate.

The tarsier locates food first with its ears, then it uses its eyes to guide its hands. It traps ants, beetles, and cockroaches by forming a 'cage' with its long fingers, dispatching its meal with needle-sharp teeth.

FIXED FOCUS *The huge eyes that dominate the face of a young female tarsier from the Philippines enable it to locate insects at night.*

LOOK BOTH WAYS *A pair of rotating eyes allows this flap-neck chameleon from Kenya to look for prey and watch for predators simultaneously.*

CHAMELEON CAN SEE IN ALL DIRECTIONS AT ONCE

The chameleon can look forwards and backwards simultaneously. So at any moment it can watch for prey in front and potential predators creeping up behind.

It has a pair of eyes set like cones, which can move independently of each other. While one eye is looking up, the other can be peering straight ahead or down, giving the chameleon the ability to look everywhere at once.

When prey is spotted, both eyes swivel to the front and focus on the same spot. The visual field of each eye overlaps, giving it binocular vision. This enables the chameleon to pinpoint its target and discharge its secret weapon, an extraordinarily long, sticky tongue.

THE FISH WITH FOUR EYES

In the lakes and lagoons of Central America, one fish has come up with its own very unusual way of spotting food. It swims with its eyes exactly level with the waterline, so that it can watch for food above and below the surface at the same time. It is known as the four-eyed fish, because each of its eyes is divided in two. The bottom half is designed to work in water, while the top half works in air.

Eyes designed for use in water need much stronger lenses than ones that work in air. But curiously, each of the fish's eyes has just one lens, shared by both upper and lower halves. So how does this one lens work for seeing in water and in air? The answer is that it has a specially flattened shape. Light arriving from water travels through the lens lengthways on, which gives the lens the power it needs. Light arriving from air travels through the thinnest part of the lens, so the power of the lens is much weaker. The result is a sharp image of two very different worlds.

CARIBBEAN SHRIMP CAN TAKE A CLOSER LOOK

The 5 mm (1/4 in) long opossum shrimp of the coral reefs of Belize carries a pair of binoculars. At first sight, its two compound eyes resemble those of other crustaceans, each consisting of hundreds of tiny facets that give the shrimp a 'honeycomb' view of the world. A closer look reveals that each eye also contains a larger lens. These large lenses have a field of view of 15–20 degrees, and give a resolution that is six times higher than the facets.

The large lenses point backwards while the shrimp is moving: it relies on the smaller lenses to find its way. When it spots a potential meal, the shrimp swivels its eyes so the large lenses point forwards, like a person using binoculars to take a closer look.

FARMING IN THE UNDERWORLD

We tend to think of ants as scavengers living off the carcasses of dead insects, yet many species farm their own food. From herding aphids to cultivating fungus, the aim is to produce a nutritious diet for the colony.

Honeydew, produced by sap-sucking insects such as aphids and mealybugs, is an important source of food for ants. Several species of European ants nurture aphids and 'milk' them by stroking them with their antennae to encourage honeydew production.

Although honeydew is high in sugars, it is generally low in proteins. To overcome this, one species of ant from Asia carries mealybugs away from older leaves to feed on young, protein-rich foliage. In this way the ant ensures that its honeydew has the highest protein content possible.

Perhaps the most bizarre ant farmers of all are the honeypot ants of Australia and the USA. Some of the workers are fed honeydew which they store in their extended abdomens. They hang from the roof of special galleries in the colony, like living storage jars, ready to provide for the colony in times of drought.

LIVING STORAGE JARS *Individuals in a honeypot ant colony are fed honeydew until they are engorged.*

STOCK RAISING *A red ant tends its herd of aphids. The ants 'milk' aphids to extract the honeydew that they produce after feeding on sap.*

Fungus farmers

The ultimate agriculturists are the leaf-cutter ants of South America. They cut up and collect leaves and other bits of vegetation which they carry back to the colony. On reaching the nest, the foragers place their loads onto the floor of one of the numerous underground chambers dedicated to fungus cultivation. Smaller ants snip the leaves into tiny pieces and even smaller workers chew and knead the snippets into moist pellets which they distribute around the chambers. This grey fluffy frass is the perfect growing medium for a type of fungus that resembles bread mould.

The fungus digests the cellulose and proteins in the vegetable matter, so making the nutrients available to the ants when they eat the fungus. The tiny worker ants 'weed' the garden of any alien fungi, and harvest chunks of the cultivated fungus for the larger ants to eat. The fungus, with honeydew collected from sap-sucking insects, feeds the colony's 5 million inhabitants.

LEAF SUPPLY *A column of leaf-cutter ants marches back to the nest with large sections of leaf to stock up the colony's fungus gardens.*

COLLECTING FOOD

Finding food is one thing, but collecting or capturing it is quite another. Many animals have adapted their bodies so that they can collect their food in the most efficient manner, using specialized mouth-parts, hands, or feet.

HOW A WHALE SWALLOWS FOOD BUT NO WATER

In a single gulp, the 27 m (90 ft) blue whale can swallow an entire shoal of shrimp-like krill. But it does this without washing it down with huge quantities of sea water.

Instead of teeth, hundreds of comb-like baleen plates grow down from the whale's upper jaw, which sieve out the solid food from the water. The plates are made of keratin and divide at the bottom into a fringe of bristles which overlap to form a coarse mat.

The whale approaches its prey, opens its mouth and thrusts its body forwards, engulfing the shoal in its huge maw. The result is a mouth filled with krill and sea water.

Pleats under its lower jaw enable the throat to expand to hold all the water. As the mouth closes, the pleats tighten and the tongue is raised, forcing water out through the baleen plates. The food is trapped on the inside of the comb and swallowed.

FISH'S JAWS SHOOT FORWARD TO SNATCH A MEAL

With its extendible jaws, the John Dory can gulp in prey in a fraction of a second. A native of European coasts, this 40 cm (16 in) fish approaches its victims almost unseen. Viewed head on, its body is tall and thin, like a discus moving vertically through the water.

When stalking prey, such as smaller fish and shrimp-like crustaceans, the John Dory swims slowly forwards, but as soon as it is within range it accelerates jerkily, and in an instant extends its jaws far forwards, grabbing the victim and swallowing it whole.

THE HARPY EAGLE'S SMASH AND GRAB

Swooping down through the treetops, the harpy eagle plucks a monkey from a branch. In a single squeeze of its formidable talons it crushes the animal to death. The harpy eagle lives in the tropical forests of South and Central America, where it is the top aerial predator.

The harpy has shorter wings and a longer tail than many other large eagles, enabling it to manoeuvre through the forest canopy at speeds in excess of 80 km/h (50 mph). Yet its 8 kg (18 lb) body weight makes it one of the world's largest eagles. It slams into its prey with an energy that rivals a bullet from a rifle.

Equipped with the most powerful feet of any known bird of prey, spanning 25 cm (10 in), and toes featuring massive dagger-like talons up to 4 cm (1 1/2 in) long, the harpy eagle can grab and despatch large mammals with consummate skill.

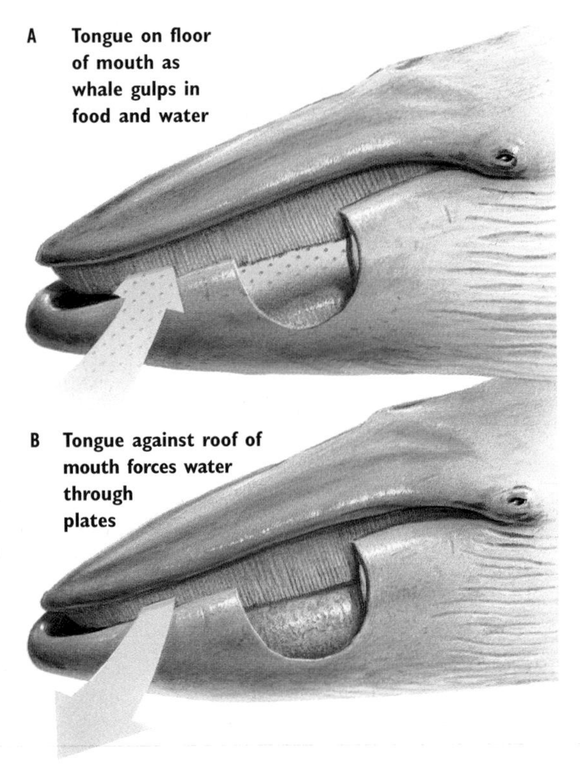

BALEEN SMILE *A barnacle-encrusted right whale exposes its baleen plates. Baleen whales feed by taking a gulp of water and krill (A) then forcing the water through the comb-like plates (B) so that only the krill is swallowed.*

LONG TONGUE *The convolvulus hawk moth hovers in front of a tobacco flower, and extends its proboscis to sip the nectar inside.*

MOTH WITH A DRINKING STRAW FOR A TONGUE

The hawk moth has the longest proboscis of all known moths and butterflies. It uses it to take a drink while still on the wing.

The European convolvulus hawk moth has a proboscis or 'tongue' that is 9 cm (3$^1/_2$in) long, twice as long as the moth's body. Its proboscis looks like a drinking straw and when not in use it lies coiled up under its head.

Hawk moths feed on plants with long, tubular flowers such as honeysuckle. Once it has found a suitable foodplant, the moth hovers in front of a flower, extending its proboscis to reach the sweet, energy-rich nectar at the base.

When hovering over a flower, hawk moths move their wings at such speed that, like a hummingbird, they are seen as barely a blur.

THE ELEPHANT'S VERSATILE TRUNK

An elephant can use its trunk to carry a tree, pick up a peanut, have a drink, or take a shower. The trunk itself is the modified nose and upper lip, lined with muscles that allow it to bend and grasp like an arm and hand. Unable to reach the ground or feed high in trees with its mouth, the elephant uses its trunk to break off branches, pluck fruits, pick off leaves, and pull up tufts of grass.

To drink, the elephant sucks water into its trunk then squirts it into its mouth. It uses the same technique to spray its body with water to keep cool or throw dust to keep down parasites.

The African elephant has two lips at the tip of its trunk; the Asian elephant has only an upper one. Sensory hairs lining the prehensile, finger-like lips allow small objects to be picked up.

ADAPTABLE TRUNK *An African elephant strips bark from a tree using its trunk like a hand to hold the food and deliver it to its mouth.*

AN ASSORTMENT OF BILLS

There are almost as many types of bills as there are species of birds, each one specially adapted to a particular way of feeding. The range of different bills seen in waterfowl, for example, is due to the different niches that the various birds exploit. **Flamingos**' bills have elaborate plates that sift algae and freshwater shrimps from the water, while the long thin bills of waders, such as the **snipe** and the whimbrel, probe wet mud and sand for worms below the surface.

The avocet's bill is long and upwardly curved, because it feeds by skimming off aquatic larvae from the surface of shallow water. **Spoonbills** also feed in the shallows and their bills are narrow and flattened for this purpose.

Long bills would be of no use to the **hawfinch**, which needs to be able to crack open seeds. Its bill is short and sturdy, as are the hooked bills of parrots and **macaws**, which feed on nuts and fruit. The **woodpecker**'s chisel-tipped bill has to be strong, since it is used to extract wood-boring insects from tree trunks as well as being drummed against trees in spring to attract a mate.

Birds of prey need to be able to tear through pieces of flesh, so their bills are strongly hooked and sharply pointed. **Herons** and bitterns are provided with dagger-shaped bills for seizing fast-moving prey, such as fish and frogs. Mergansers' bills have evolved in a different way: their saw-like bills are for grabbing hold of slippery fish that they chase under water.

The **pelican** has yet another way of catching fish, and this is reflected in its large, distinctive bill. It uses the membranous sac attached to its lower mandible like a fishing net, scooping up its catch.

Birds that feed on flying insects, such as **swifts** and swallows, have small bills with a wide gape, while specialized nectar-drinking birds, such as **hummingbirds**, have evolved long, thin, straw-like bills.

Though birds' bills are primarily adapted to feeding, they are not just for collecting food; birds also use their bills to preen and to make nests. A bird's bill can be a weapon, and it often plays a role in courtship displays.

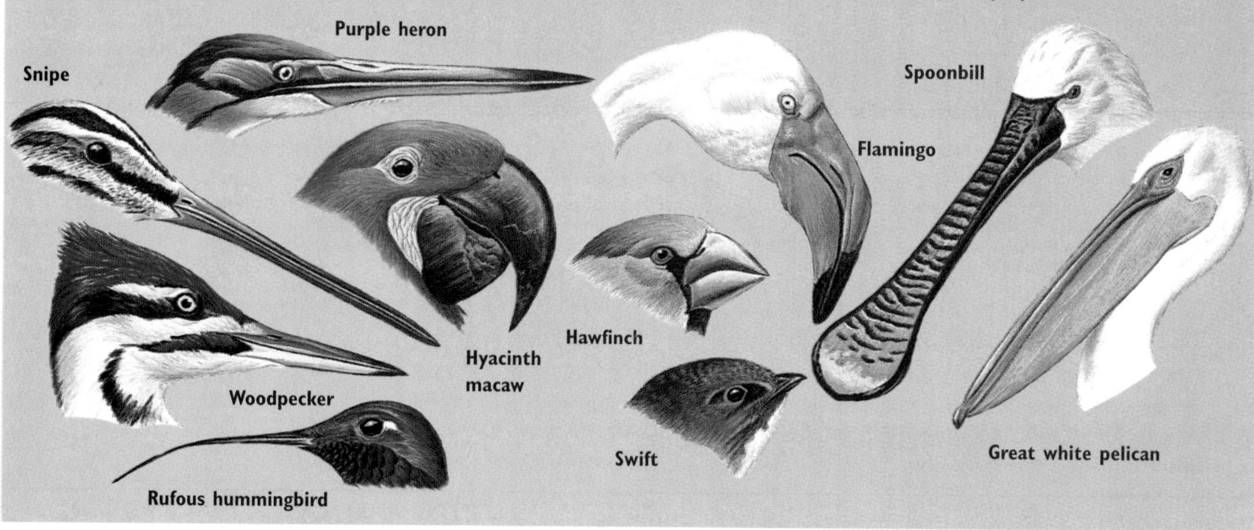

CUTTLEFISH'S MESMERIZING LONG-ARM TACTICS

Moving stealthily, the cuttlefish mesmerizes its prey with its ever-changing pattern of stripes, spots, and blotches. To add to the victim's confusion, its body changes shape and colour several times as it approaches, blending with the background. When the target, such as a fish, crab, or shrimp, is within range, the cuttlefish suddenly flings out two long tentacles from the circle of eight arms around its mouth and grabs the prey with the help of sucker pads on the end. The food is drawn back to the mouth where a horny beak chops it up and the pieces are swallowed.

DARTS DEAL DEATH IN AN INSTANT

Stinging cells on the tentacles of the jellyfish-like Portuguese man-of-war discharge at high speed. Inside each spherical cell is a hollow, whip-like thread that is stored inside-out and coiled when at rest. The thread resembles the finger of a rubber glove when pushed in. When prey, such as fish, blunder into the man-of-war's trailing tentacles, which form a curtain of death up to 9 m (30 ft) long, each cell is activated.

In just three-thousandths of a second, pressure in the cell rises, the thread is everted and its sharp, barbed tip is thrust into the victim at a speed of 2 m (7 ft) a second. It penetrates the skin, delivering a venom nearly as powerful as that of a cobra.

It takes 25,000 stinging cells to release 1 mg of venom. When the twitching prey is subdued, the tentacles contract and haul it up to specialized feeding cells underneath the man-of war's bladder-like float.

COOPERATIVE HUNTING

Social predators often hunt in groups. Together, there are more eyes to spot food, and by cooperating they can bring down prey much larger than themselves. A powerful group also ensures that other predators will not steal their food.

CLOSE-KNIT CLANS WITH STAMINA AND SPEED

Spotted hyenas will scavenge or steal anything edible. And with powerful jaws and sharp, strong teeth, they can tear flesh and sinews, and crush bones.

Accomplished hunters, hyenas rely on stamina and speed rather than stealth to bring down prey. A solitary hyena can chase a wildebeest for 5 km (3 miles) at 60 km/h (37 mph) and bring it down alone. And clans of 10 to 15 hyenas working together can tackle large prey, such as zebras. Serengeti clans of 50 or more have been known.

A hunt usually starts at sundown, when the clan begins to stretch and limber up. The animals head out across the savannah in search of prey, calling constantly to one another with screeches and cackles. Closing in on a herd of zebra, they run with their tails erect, scattering the animals. Soon they single out a sick, old, or young individual, which they overpower once it is isolated from the herd. In minutes it is disembowelled and dismembered, each hyena eating about 15 kg (33 lb) in a single sitting.

CHIMPANZEE HUNTERS WITH A LUST FOR BLOOD

Although chimpanzees are mostly vegetarian, they hunt live animal prey with spine-chilling efficiency. In the forests of West Africa, groups of half-a-dozen adult males hunt together at least once a week, often targeting troops of treetop-dwelling colobus monkeys.

The hunt is well planned, each chimpanzee with its role to play. The youngest usher the monkeys into the trap set by the others. Flankers on either side funnel them into the killing zone and ensure they cannot escape. And the oldest goes ahead, where he can ambush prey moving towards him.

The chimpanzees take up position on the ground in silence. The youngest of the gang race up the trees and begin the chase. The colobus flee, heading straight for the ambush ahead. At the right moment, the oldest chimpanzee shows himself and the monkeys, realizing they are trapped, panic. During this moment of indecision, the chimpanzees grab their prey. They announce their success with blood-curdling shrieks. The highest-ranking male then tears the colobus monkey apart, and distributes the pieces.

Chimpanzees also hunt young bush pigs, baby baboons, bushbuck fawns, mice, rats, and small birds – anything that is available or abundant.

PELICANS' SYNCHRONOUS SWIMMING FOR FISH

Florida's American white pelicans line up in the water to fish together 100 at a time. Working in an open horseshoe-shaped formation, they constantly dip their enormous bills below the surface and drive shoals of small fish towards the beach or into the shallows, where each can scoop up 1.8 kg (4 lb) of fish a day with ease.

The bill is used as a dip net. Groups of birds partly raise their wings and plunge their heads into the water. If successful, they raise their bills to spill out the water and swallow the fish.

MURDEROUS PACK *A clan of hyenas kills a wildebeest after separating it from its mother. They eat everything: only a bloodstain will be left.*

SPIDERS' ENORMOUS EXTENDED FAMILY

Communal living is explored to the full among certain groups of 'social spiders'. Found in tropical forests worldwide, they join forces to build webs more than 1 m (3 ft 3 in) long, they tackle prey together and even care for each other's young.

Most spiders avoid others of their kind because they may end up being eaten, but 'social spiders' are found in colonies of up to 10,000 individuals. The silk-spinning power of the colony means it can spin a larger web with greater catching potential. Hammock-shaped webs are suspended across rivers or in clearings by long threads attached to the vegetation.

By working together, the spiders can tackle larger prey. When a flying insect collides with the web, the vibrations of its struggles attract the attention of the closest spider. If the prey is too large or too difficult to subdue, reinforcements arrive to help. In this way, spiders just 5 mm (3/16 in) long can tackle prey more than 6 cm (2 3/8 in) long and the spoils provide more than enough food for the colony.

Unusually for spiders, the colony's young are cared for cooperatively, with adults feeding hatchlings that may not be their own.

KNOW YOUR NEIGHBOURS *Social spiders in the forests of Papua New Guinea identify members of the commune by using chemical signals.*

HUNTRESSES OF THE MASAI MARA

Among lions in Kenya's Masai Mara, the females hunt and the males keep watch over the pride. The females set out at sunset. They hunt as a close team and their ambush is carefully planned. Having identified a small group of wildebeest or zebra, one female makes her way unseen to the far side of the herd. The others slip down between the tufts of savannah grass, their bellies close to the floor, heads steady as rocks and eyes locked on to the target, usually a young, old, or infirm animal.

Gradually, they creep towards the chosen victim. Any sign of alertness and the pride freezes. Stealth is the lions' weapon, and surprise is vital. They must get to within 30 m (100 ft) of the prey before bolting from cover. A lioness at full tilt can reach a speed of 60 km/h (37 mph) for a short distance, but the prey can run faster and for longer.

In response to some hidden signal, the attackers burst from cover, driving the prey towards the ambush. The lone female erupts from the bush and charges the prey. She grabs it round the neck, while the others attack the rear, bringing it down. A suffocating bite to the throat sees the prey go

LIONS HUNT TOGETHER **1** *Subduing a zebra is hard for a lioness alone.*

2 *While one bites the zebra's throat, others try to bring the animal down.*

3 *Overwhelmed by the team of lionesses, the zebra falls.*

limp, in time for the pride's males to appear. With loud snarls, they scatter the lionesses and take over the kill. Only when they have eaten their fill will the females and youngsters feed.

CORMORANT RAFTS ON THE LOOKOUT FOR FISH

Blue-eyed cormorants gather to form enormous living rafts in the southern Antarctic Ocean. There they wait in their thousands for the shoals to pass below.

Every so often a lookout from the flock dips its head into the water and scans for signs of life. If prey is spotted, the lookout dives below and the rest of the birds follow. Fish and birds swim in all directions, and in the chaos a fish may manoeuvre to avoid one bird only to meet the bill of another.

DIVE-BOMBERS *Blue-footed boobies crash into the sea causing panic and confusion among a shoal of fish swimming just below the surface.*

BLUE-FOOTED BOOBIES DIVE-BOMB IN FORMATION

A flock of blue-footed boobies drops out of the sky and plunges into the sea to catch food. The flock circles above the ocean at a height of about 30 m (100 ft) and when a shoal is spotted the leader, usually a male, gives a whistle and all the birds dive together. Each bird locks on to a target under the water, and controls its descent using its wings, tail, and feet. At the last moment before entry, the wings are swept back to avoid injury.

Entry speed can be as much as 100 km/h (60 mph), the brunt of the impact taken by the bird's head, which is protected by a strengthened skull. The nostrils are closed to prevent water being forced into the mouth and throat, and the bird breathes through the corner of its mouth.

Boobies plunge only a few metres below the surface, grabbing a fish and then bobbing up and taking off to start the process all over again.

The birds, members of the gannet family, live along the Pacific coast of Central and South America. Like other gannets, they plunge-dive on shoals of fish or squid, but only blue-footed boobies seem to cooperate.

THE HAWKS THAT HUNT AS A PACK

Harris's hawks form scouting parties in Central and South American deserts. In groups of three to six birds, they search for prey, such as jack rabbits.

When a victim has been found, the hawks gather to make a kill. Some try to flush the prey from its hiding place while others surround it.

As the quarry flees, the hawks might take it in turns to dive-bomb it, until one eventually pins it down. The attackers divide the spoils.

BLOWING BUBBLES IN UNISON

Humpback whales collaborate when hunting to ensure the best possible catch. Up to seven whales will lunge together. They dive below a shoal of krill or small fish and shoot up to the surface, mouths agape, and so close together that they touch flippers, a behaviour known as 'echelon feeding'.

Humpbacks also work together to create a 'net' of bubbles with which they round up their prey. Forming a circle below the shoal, they begin to blow bubbles. These reflect light as they rise, frightening the prey, which concentrates into a tightening mass inside the cylinder of bubbles. The whales then scream to drive the shoal closer to the surface. Finally the whales rise through the column, huge mouths open, and burst into the air, their throats distended with water and fish or krill. They close their mouths, squeezing the water out through their baleen plates, and swallow their portion of the shoal. Whales often feed in the same group, practice making perfect.

DOLPHIN ATTACK *The water churns as several schools of dusky dolphins combine their herding power to drive fish close to the ocean surface.*

SCHOOL OUTING FOR HUNTING DOLPHIN

Hunting in schools enables dusky dolphins to feed for two to three hours at a time. Each morning in summer, schools of the dolphins move away from the coast of southern Argentina and head for deeper water in search of shoals of anchovies.

In groups of 15 or so, the dolphins swim abreast, about 10 m (33 ft) apart. Using underwater echolocation, they scan a 150 m (165 yd) wide swath of sea for prey, and when they come to the surface to breathe they watch for any sea-bird activity, a sure sign that fish are present.

On discovering a shoal of anchovies, several dolphins dive deeply and start to herd the fish towards the surface. Others swim round in ever-tightening circles until the anchovies are pushed into a tightly packed ball.

The dolphins then batter the fish with debilitating, high-intensity bursts of echolocation clicks, picking off those that become disorientated and snatching fish from the edge of the ball. The commotion caused by the leaping dolphins and diving sea birds that accompany them attracts other dusky dolphin groups. As many as 20 or 30 groups might congregate, with some 300 dolphins joining forces, their combined herding keeping the fish ball in place for several hours.

COOPERATIVE GIANT KILLERS

Killer whales, or orcas, employ deadly tactics when they join forces to attack. Young baleen whales are a common target. Off the Pacific coast of North America, orcas regularly take migrating grey and humpback whale calves, and occasionally charge down a blue whale.

Only adult orcas hunt; youngsters learn by watching. During the highly coordinated attack, some orcas flank the rapidly moving whale, as if herding it; others go in front, behind, and below, to prevent any escape. Another group swims over the whale and attempts to prevent it surfacing to breathe.

The orcas rasp their teeth along the whale's body, chewing off the dorsal fin and shredding the tail flukes to impair its movements. The large bull orcas pull off chunks of flesh and blubber. The chase can last for hours, until their prey is exhausted.

RIVER OTTERS' SOCIAL GRACES

Giant river otters are extremely patient with their fishing companions. If one should eat its food before the rest, it will wait for its fishing partner to finish its meal before moving on to hunt again.

Living in small family groups of eight or nine individuals in the lakes and rivers of the Amazon basin of South America, the giant river otter hunts in pairs or with the whole group, heading for deeper water when fishing together. The otter is active by day, diving to catch mainly bottom-dwelling fish in its powerful jaws, holding them in its forepaws and eating everything, including the bones.

USING TOOLS TO OBTAIN FOOD

At one time it was thought that humans were the only living creatures to use tools. Now we know that animals, including other mammals, birds, and insects, have discovered that tools can help them to acquire a meal.

THE SEA OTTER'S TRUSTY FLOATING ANVIL

The sea otter from the Pacific coast of North America has a clever way with shellfish. When the otter dives to the seabed to collect its meal of crabs, sea urchins, clams or mussels, it also picks up a stone.

At the water's surface, the otter floats on its back, anchoring itself with a strand of kelp, and places the stone on its chest, like an anvil. The shellfish is held in its front paws and hammered repeatedly against the stone until it breaks open and its contents are consumed.

The sea otter also uses a stone when diving for abalone, a shellfish that clings to the rocks on the seabed. The otter strikes the tool against the edge of the shell and it may require three or four dives before the shell is dislodged and the otter can claim its meal.

Otters sometimes have a 'favourite' stone, which they tend to use again and again. This is usually fairly flat, about 18 cm (7 in) in diameter, and is held in a flap of skin underneath the otter's armpit.

VULTURE LEARNS TO SMASH AND GRAB

Certain vultures in East Africa are committed stone-throwers. Faced with the tantalizing prospect of the substantial meal hidden inside an ostrich's egg, the Egyptian vulture must find a way of breaking through its tough shell.

Unable to pick up the large egg and hurl it to the ground, the vulture does the next best thing. It carefully selects a rounded or egg-shaped stone, picks it up in its beak and hurls it at the egg. The bird repeats this exercise several times until the egg is smashed. The skill is learned through trial and error. Young vultures may throw stones 70 times or more before abandoning the effort, and even the adults are never particularly accurate. So far, the best recorded score is 38 hits out of 64 attempts.

A vulture will occasionally choose the easier option of picking up a pelican or flamingo egg in its beak and dashing it to the ground to get at the contents.

STONE-THROWER *A juvenile Egyptian vulture learns how to throw stones at an ostrich egg to get at the protein-rich yolk inside.*

FORCED ENTRY *The thick shell of a sea urchin or clam is no barrier to a hungry sea otter, which uses a pebble to break its way in.*

SHRIKE SPIKES PREY TO EAT LATER

Also known as the 'butcher bird', the European red-backed shrike has a macabre habit. It impales any surplus food on thorns or barbed wire to be eaten later.

Creating a 'larder' in this way both prevents the food from moving and allows the shrike to tear away pieces while feeding. A typical larder might contain insects, lizards, and young birds. The shrike catches its prey like a hawk, first perching and watching for any sign of movement then swooping down and grabbing it in its slightly hooked bill.

MACABRE LARDER *A shrike impales a locust on a thorn. Adults feed their young by first dismembering prey that has been spiked in this way.*

Impaling food, a skill shared by the great grey shrike of Europe, Asia, North America, and the southern Sahara, is partly instinctual and partly learned by trial and error. A young bird will hold food in its bill and drag it along a branch in a random fashion. Should the food catch on a thorn or wedge in a fork, this helps the bird to learn about the right places to set up its larder.

SHORT BILL IS NO PROBLEM FOR INGENIOUS FINCH

The woodpecker finch of the Galapagos Islands uses tools to get at hard-to-reach food. Lacking the woodpecker's powerful bill and specialized tongue to dig out wood-boring grubs, the finch improvises to get at its meal.

At first, like a woodpecker, it uses its bill to prise off bark and pick off accessible prey. But if the prey is out of reach, the woodpecker finch visits a cactus plant and breaks off a spine. Holding the spine in its bill, the finch pushes it into the hole or crevice and winkles out any insects that might be hidden inside.

PRICKLY PROBE *The woodpecker finch uses a cactus spine to probe for wood-boring insects and other hard-to-reach morsels.*

THE GREEN HERON'S FLY-FISHING BAIT

The green-backed heron makes its own fishing bait, much like a fly-fisherman. It throws a piece of bait onto the water surface within stabbing range of its sharp bill.

The lures can be live baits, such as insects, or artificial ones fashioned from feathers, twigs, or even pieces of discarded biscuit. If the lure is too long, the bird snaps it in two. Any fish that ventures to investigate the scattered bait is caught in an instant.

WHOLEFOOD SNACK *A green-backed heron carefully manipulates a fish, head first, into its mouth before swallowing it down whole.*

HOW A MONGOOSE CRACKS EGGS

An anvil is the preferred tool of the Egyptian mongoose. The animal cracks eggs open by picking them up in its forepaws and throwing them between its back legs against a large rock until they break. It shares this technique with the banded and dwarf mongooses that are native to Southern Africa.

Eggs form only part of the mongoose's diet. Insects, grubs, earthworms, millipedes, lizards, snakes, and small rodents are more usually eaten, which the mongoose finds under logs and leaf litter.

WHEN SNAILS ARE ON THE MENU

The European song thrush smashes snails against a stone. A bird will return to a favourite stone for each feeding session, and broken snail shells often litter the ground. This and the sound of hammering are often the only signs that a song thrush is present.

A single large snail provides an ample meal, but the food preparation takes time and energy, so the song thrush eats snails only when more easily accessible food, such as earthworms, is in short supply.

SMASHING SNAILS *When food is scarce, the song thrush turns to snails, smashing their shells against a rock to get at the soft parts inside.*

EATING AND DIGESTION

Having found a meal, an animal must then eat and digest it to benefit from its nutrients. Some animals use special jaws, teeth, or other mouth-parts to reduce food to a manageable size; others deal with it only when it reaches their stomach.

GRUESOME HABIT OF THE VAMPIRE FINCH

On Wolf Island, in the Galapagos archipelago, the sharp-beaked ground finch has a taste for blood. Once content to divest sea birds of their parasitic flies and lice, the finch now pecks at the base of the birds' feathers and breaks the quills of nesting and roosting masked boobies, then drinks their blood. If that is not enough, the finch breaks into masked booby eggs and extracts the contents. Pressing its beak against a rock for leverage, it will kick the shell against a stone until it cracks.

SHARK TURNS UP THE HEAT TO PROCESS FOOD

The great white shark has a digestive secret in its stomach. After feeding, it raises its stomach temperature by about 6°C (11°F), which speeds up the digestion process. As a result, the shark is ready to feed again at the earliest opportunity.

The great white, which can be as long as 6 m (20 ft), is not the terrifying predator of popular fiction and modern cinema. The shark, found in tropical and subtropical inshore waters worldwide, eats spasmodically, depending on food availability, often going for weeks and even months between meals when prey is scarce. So it must be ready to feed at any and every opportunity.

Adult sharks prefer energy-rich foods, such as the blubber from living seals and dead whales. A 32 kg (70 lb) chunk of blubber provides enough nutrients and energy to keep a shark going for six weeks.

HOATZIN'S SLOW DIGESTION

The hoatzin of South America's tropical rain forests has a stomach like a cow. Most birds prefer energy-rich, easily digestible food, but the chicken-sized hoatzin feeds on green leaves which it snips off with its scissor-like bill. It must eat large quantities to meet its nutritional needs.

To digest the leaves, the hoatzin has an enlarged, thick-walled, muscular crop with a corrugated interior that resembles a cow's rumen. Horny tissues lining the walls grind up the leaves, while bacteria break down the cellulose, releasing nutrients and essential minerals. Food can take up to 48 hours to pass through the hoatzin's gut, slower than any other bird.

HAGFISH EATS FROM THE INSIDE OUT

Deep in the ocean's abyss, the hagfish attacks its prey from the inside. The primitive eel-like creature has no jaws, no stomach, and no eyes. Its mouth and nostrils are surrounded by up to six tentacles, and it is coated in slime.

The hagfish eats dead or dying fish, detecting them by smell. With its

BLOOD LUST *Sharp-beaked ground finches take advantage of a distracted masked booby in order to steal a small meal of blood.*

MORE THAN A MOUTHFUL *The egg-eating snake can swallow eggs that are larger than the width of its head.*

1 *The snake's jaw bones disengage so it can take the egg in its mouth.*

2 *The egg is swallowed whole, then sharp projections on the underside of the vertebrae crack the shell.*

rasp-like tooth plates it scrapes a hole in the side of the fish, and then loops its body into a knot which provides leverage to push itself through the flesh.

Feeding as it goes, the hagfish can disappear inside the fish's body, and continue eating its victim from the inside out. To deter other scavengers, it cocoons the carcass with thick slime that can clog the gills of fish.

EGG-EATING SNAKE SWALLOWS IT WHOLE

The East African egg-eating snake can swallow eggs up to twice its own width thanks to its hinged jaws. The snake can disengage its jaws to expand its mouth enormously. Once the egg is in the snake's mouth, the jaws, lined with small backward-pointing teeth, 'walk' the egg, moving first one side and then the other, into the gullet. The snake breathes by pushing its windpipe in and out of its mouth while swallowing.

With the egg in the gullet, strong neck muscles contract and spines on the neck vertebrae push down and pierce the eggshell, spilling its contents into the gut for digestion. The eggshell is regurgitated, and the jaws return to their normal position.

VAMPIRE BAT KEEPS ITS MEALS FLOWING

Vampire bats have anticoagulants in their saliva. These prevent a victim's blood from clotting, so the bat can enjoy a leisurely nocturnal meal.

Landing on or close to its warm-blooded host – often sleeping domestic livestock – the bat painlessly inflicts a small gash with its razor-sharp front teeth. It licks the blood from the wound, darting its grooved tongue in and out rapidly. It feeds for 10–15 minutes, usually taking about one-and-a-half times its own weight in blood. Blood is heavy and hard to digest so when it has finished feeding, the bat urinates to lighten the load and enable it to fly again.

Since blood is high in protein but low in calories, the vampire bat must feed every couple of days to survive. A female bat in dire need of a blood meal may be fed regurgitated blood by better fed females in the same colony.

SPIDER OPTS FOR A LIQUID LUNCH

Solid foods are not on a spider's diet sheet – it prefers liquid nourishment. A spider paralyzes or kills prey by injecting venom through a pair of fangs (chelicerae) on its head. Then it regurgitates stomach juices into the wound, introducing digestive enzymes into its victim's body. All the prey's internal organs break down into a liquid mush, which the spider sucks up through its small mouth opening using its 'sucking stomach'.

JUICE BAR *A hunting spider from Costa Rica pumps its digestive enzymes into a grasshopper.*

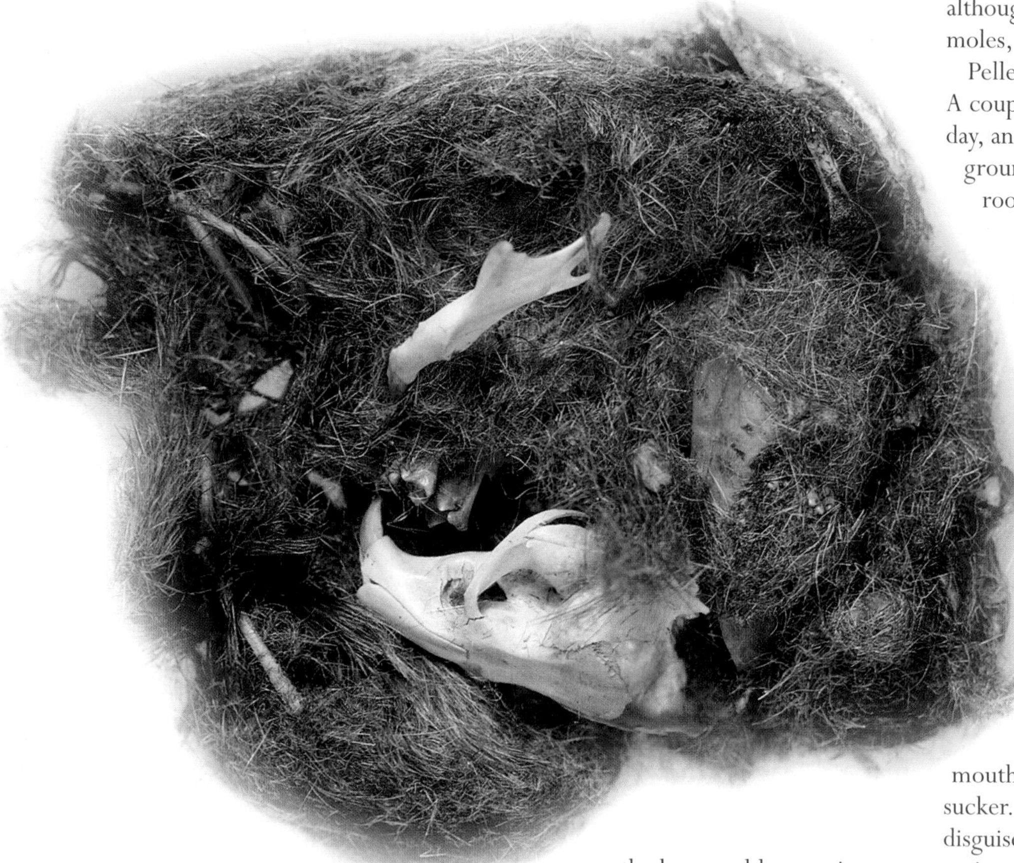

SPOILS OF THE NIGHT *A barn owl's pellet containing fur and bones reveals what it caught the night before – in this case, a small rodent.*

WRAPPING UP THE REMAINS

Owls bolt their prey whole, bones and all, but have to spit out the hard parts. Their digestive juices deal only with the nutritious soft portions. The indigestible bones, beaks, claws, teeth, fur, insect exoskeletons (tough, outer coverings), and even the tiny hairs in the body wall of earthworms are regurgitated and ejected as pellets.

To ease their passage, the pellets tend to have the hard parts at the centre with fur or feathers on the outside, and are covered by a thin layer of mucus. This makes them moist and soft at first, but they quickly harden. When those of the barn owl lose moisture, they acquire a distinctive glossy varnished appearance.

The pellet gives a good indication of what an owl has been eating, and different species produce pellets with assorted contents. Little owl pellets consist mainly of beetle remains, while those of the tawny owl often contain the bones and fur of mice and voles, although its diet can also include moles, shrews, small birds, and beetles.

Pellets also betray an owl's presence. A couple of pellets are formed each day, and many can be found on the ground under an owl's favourite roost or nest site.

FEAST OF BLOOD

The medicinal leech of Europe and parts of Asia seldom eats. It often has only one big meal a year. Blood, usually mammal blood and sometimes from humans, is its staple diet.

Attaching itself to its host with the help of suckers front and back, the leech makes a Y-shaped incision using three semicircular jaws armed with rasp-like teeth in a mouth at the centre of its front sucker. A local anaesthetic that disguises the leech's attack and an anticoagulant that prevents its victim's blood from clotting are secreted through each tiny tooth. The leech then sucks out the blood, which passes into its gut.

Enzymes, helped by bacteria in the gut, start the digestion process. Water, minerals, and salts are rapidly extracted, but the nourishing protein-rich part of the creature's blood meal may take up to seven months to digest. This is because the leech's metabolism is slow to break down the robust components of blood.

During the 18th and 19th centuries, the medicinal leech was used in bloodletting, a practice that was supposed to relieve a patient of 'vapours and humours'. Today, there has been a resurgence of medical interest in the leech's bloodsucking abilities. It is used to remove blood clots.

BLOODSUCKER *A medicinal leech goes to work – it can be used to get rid of a blood clot beneath the skin.*

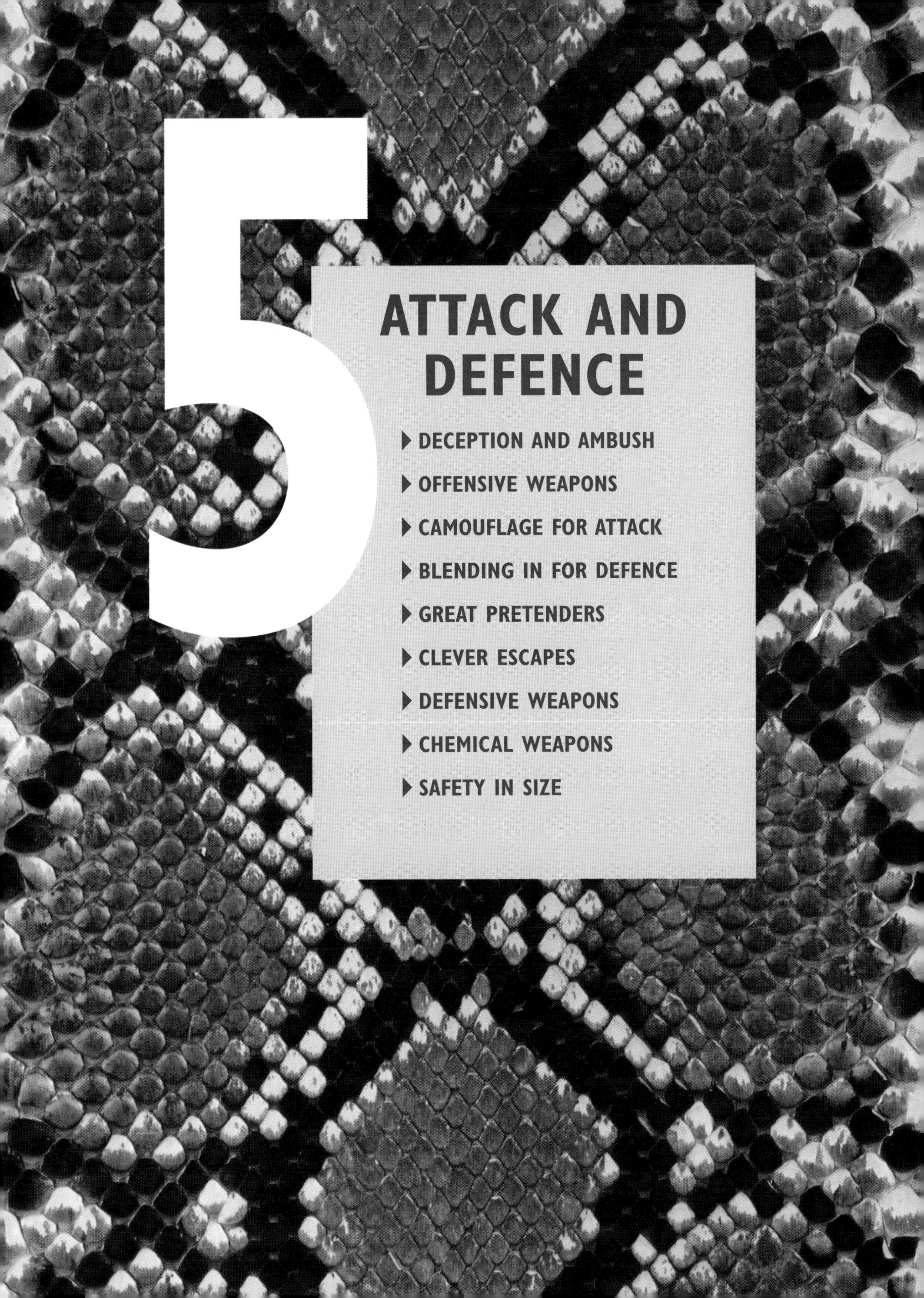

5 ATTACK AND DEFENCE

DECEPTION AND AMBUSH

Stealth, invisibility, disguise, pretence ... using strategies more sophisticated than any thriller writer could ever dream up, creatures that fill their bellies by ambush and bluff have some very nasty tricks in store for their unwary victims.

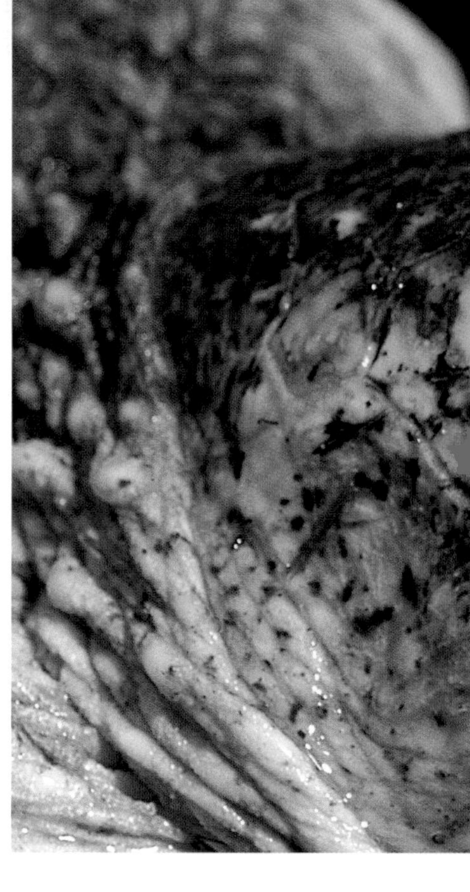

THE JUMPING SPIDER WITH A PERFECT DISGUISE

Some spiders look like an unpleasant-tasting ant to prevent birds from eating them. In the case of one species of jumping spider, the disguise is so perfect that it can move unnoticed even among the aggressive green weaver ants upon which it preys.

Spiders have two segments to their bodies, but this jumping spider's abdomen is nipped in to a tight waist to create the illusion of a three-segmented ant's body. Two spots of colour mimic the ant's large compound eye. The spider also copies the way ants move, imitating their antennae by waving its two front legs in front of its head. This leaves three pairs of legs, the same as an ant.

THE TURTLE WITH AN ALLURING TONGUE

A small, bright-pink ribbon of flesh adorns the alligator snapping turtle's tongue. Every so often, the turtle makes this ribbon of flesh twitch and when potential prey investigate the pink 'worm', all the turtle has to do for a quick meal is to snap shut its sharp horny jaws.

The turtle lies in wait, mouth wide open, at the bottom of a river or pond, its mud-coloured carapace blending in well with the vegetation.

This particular turtle from south-eastern USA is the largest freshwater turtle in the world. It can grow to 70 cm (28 in) in length. The only other snapping turtle that commands respect from all those who know it is the common, or American, snapping turtle. It eats anything, including carrion and other turtles. If the victim is too large to be swallowed whole, the turtle uses its jaws to hold it while tearing it apart with its forefeet.

MASTER OF DISGUISE *This jumping spider – from South-east Asia – disguises itself as a green weaver ant. Spiders use disguise to get a meal, or to confuse predators.*

FEARSOME BITE *The American snapping turtle has extremely powerful jaws and has been known to attack swimmers.*

LYING IN WAIT IN THE DEATH PIT

Lacewing larvae wait for prey at the bottom of small conical pits, concealed by grains of sand. Only the jaws of the ant lion – as lacewing larvae are known – stick out, ready for action. The sides of its lair are excavated at an exact angle, so that when a small creature such as an ant tumbles over the rim, the ant lion can pull the sand away from below the victim's feet, and the struggling insect finds it impossible to get a grip on the slippery grains.

Unable to climb out, it slides inexorably towards the ant lion's jaws. The ant lion seizes its victim and sucks it dry, discarding the empty skin. Sometimes, if its prey is taking too long to slide to the bottom of the pit, the ant lion becomes impatient and bombards it with sand. This is the only time in the ant lion's life that it eats; once it becomes a winged insect it ceases to eat altogether.

BLACK HERON'S FEATHER CANOPY FOOLS FISH

To hunt, the African black heron turns itself into an umbrella. In shallow saline and fresh waters in Africa, the heron crouches forward, lifts its wings above its back and stretches the wing tips downwards so that they brush the surface of the water. This creates a canopy under which the bird tucks its head and scans the water below.

The shade may help the heron to see into the water by preventing sunlight from sparkling on the surface. Fish and other aquatic prey, tricked into thinking that the shadow is cast by a rock or some other form of shelter, take refuge beneath it, just where the heron's long, dagger-like beak can stab at them.

CASTING A SHADOW *Using its wings to cast shade over the water, the African black heron creates a false sense of security for the fish below.*

THE SNAKE THAT MAKES GREAT USE OF ITS TAIL

To catch their prey, many snakes opt for the sit-and-wait strategy, for which they need to be well camouflaged. Several members of the pit viper family also have highly visible lures at the end of their tails. The tip of the Australian lowland copperhead snake's tail, for example, looks just like a bright, fat, juicy worm.

The otherwise well-camouflaged snake, hidden among leaves, raises the lure and twitches it to entice a frog or lizard closer. The snake pounces. When no suitable prey is in the immediate area, the snake will wiggle its lure on the off chance that it might attract something from farther away.

JAWS *Shooting up from deep water, razor-sharp teeth bared, the great white shark captures its prey with a single, massive bite.*

STEELY HUNTER THAT PICKS VICTIMS FROM A DISTANCE

The biggest predator in the world has senses so sophisticated it can stalk its prey from afar. The great white shark, which lives in warm waters worldwide, can hear low-frequency sounds, such as those of a struggling fish, from 1.6 km (1 mile) away. It can detect tiny concentrations of blood or other body fluids from 400 m (440 yd). In the murky depths it can see movement from 8 m (26 ft).

The shark is superbly camouflaged: its steely grey blends with the ocean floor. The shark regularly glances up, looking for the silhouettes of its prey – porpoises, seals, other sharks and rays, turtles, bony fish, and dolphins.

A great white shark can maintain its body temperature at 14°C (57°F) above that of the water. Its muscles, stomach, brain, and eyes function more efficiently at higher temperatures, which means that the shark is capable of sudden bursts of speed and power.

Once it has found a victim, the shark rockets up to seize its prey from below. To protect its eyes from the victim's struggles, the shark rolls them back into their sockets. Receptors in its snout detect the tiny electrical currents made by the prey's contracting muscles and guide the shark to its target. It thrusts its jaw forward, exposing rows of sharp teeth. After delivering a single disabling bite, the shark releases its victim, probably to avoid injury to itself. Then it grabs the prey, shaking its head vigorously to rip off chunks of flesh.

SLIT-SPIDER HANGS IN WAIT

The dunes of Australia's Simpson Desert offer no natural crevices, so the slit-spider builds its own. It makes a broad horizontal cut in the sand, with a low, gradually narrowing chamber behind. Since sand is highly unstable, the spider excavates at an angle of 20 degrees so the burrow lies within the firmer subsurface area.

The slit, 10 cm (4 in) wide and 16 mm (5/8 in) deep, acts as a trap for ants. The spider builds a tiny sand dune in front by sweeping loose sand from the burrow. An ant climbing the gently sloping surface slides down the steeper slitface and into the crevice. The waiting spider, hanging upside down in its burrow, pounces.

HOW LEOPARDS COMBINE STEALTH WITH SPEED

Due to adaptable hunting habits, the leopard is the most widespread member of the cat family. Found in Africa and Asia, the leopard can reach a speed of 70 km/h (44 mph), but since it cannot sustain this for more than 200 m (220 yd) it relies on stealth to get close to its intended victim. Leopards rest on the branches of acacia trees and scan the plains for prey. If an animal passes the tree, the leopard drops on top of it.

Hunting alone at night, it stalks its victim to within a close range. Then, with a short, fast rush, it attacks. If the kill is too big to eat at once, the cat will drag the remains up into the fork of a tree, safe from other predators. It can haul a carcass the same weight as itself up to a branch 9 m (30 ft) high.

OFFENSIVE WEAPONS

Peaceful, calm, gentle Nature? Never! In the battle for survival, evolution has created a deadly armoury of weapons. Swords, arrows, electric shocks, chainsaws, noxious chemicals, and poisons are all part of the arsenal.

THE MOST POWERFUL JAWS ON EIGHT LEGS

The world's strongest biting apparatus relative to body size belongs to the sun spider. It can crush small mammals, lizards, and birds to death with its jaws.

Sun spiders grow up to 7.5 cm (3 in) long. Unlike true spiders, or their other relatives scorpions, they do not produce venom but rely on force alone to kill their prey. Desert dwellers, they hunt at night and have massive appetites. As long as food is available, they will keep hunting and eating until they can hardly move. One particularly gluttonous individual, barely 2.5 cm (1 in) long, was recorded eating 100 flies in a single 24-hour sitting.

Sun spiders, also known as wind spiders or wind scorpions, hold another record: they are the fastest land invertebrates, capable of reaching speeds of 16 km/h (10 mph).

COOKIE-CUTTER SHARK TAKES A NIBBLE

Cookie-cutter sharks take biscuit-sized portions out of their unwary victims. These sharks are among the smallest at 50 cm (20 in) long. With luminous underparts, cookie-cutters can make themselves almost invisible by adjusting the amount of light they emit. When a whale, dolphin, or other large marine creature looks up, it sees what appears to be the silhouette of a fish. The larger animal moves closer to investigate. The cookie-cutter darts forward and sinks its interlocked teeth into its victim. Then, rather than going for the kill with one huge bite, it twists around to rip out a plug of flesh.

DEADLY BOX JELLYFISH KILLS IN MINUTES

A box jellyfish contains enough venom to kill four humans in less than 3 minutes each. The venom stops breathing, destroys red blood cells, and damages skin tissue. Found around the Australian coast, the jellyfish's intended victims are the sea creatures on which it feeds.

The box jellyfish is named after its square-shaped bell from which dangle tentacles up to 3 m (10 ft) long, covered with stinging capsules. When the tentacles brush past prey, they stick fast while the capsules shoot out a toxin. The jellyfish then hoists its paralyzed victim into its body cavity for digestion.

FEELING FOOTSTEPS *Sensory organs under its rear legs allow the sun spider, of Africa, to home in on its prey by detecting ground vibrations.*

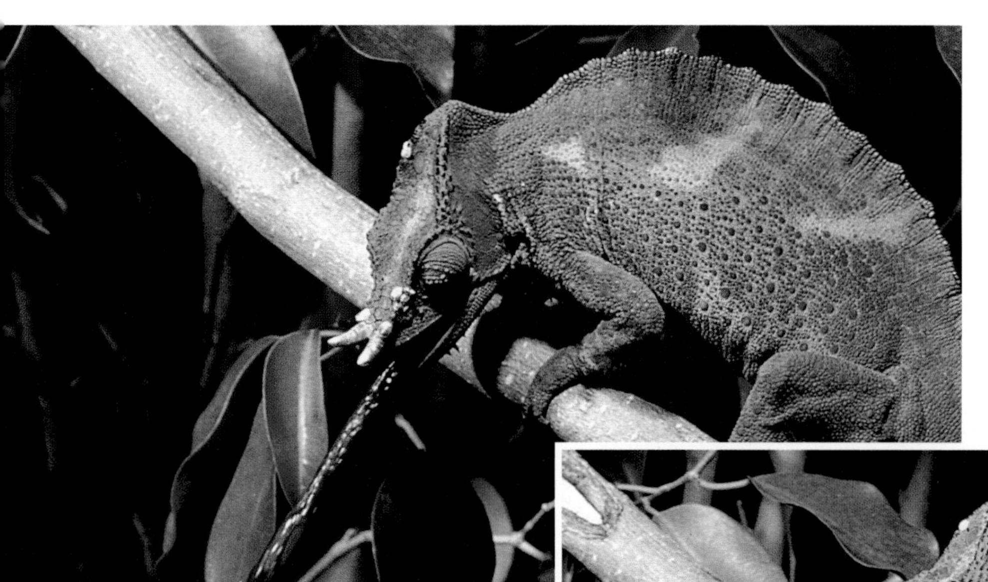

SHARPSHOOTER *Jackson's chameleon catches dinner from a distance with its telescopic tongue.*

1 *The chameleon catapults its tongue towards an unwary cricket.*

2 *The tongue retracts.*

3 *Sticky saliva, minute hooks, and the tongue's grasping tip hold the cricket.*

4 *The hapless insect is consumed.*

A TONGUE-LASHING FROM JACKSON'S CHAMELEON

Millimetre by millimetre, the Jackson's chameleon creeps along a branch towards its victim, using its toes, which face opposite ways to grip tightly. It closes in on an insect, spider, or scorpion unseen, its flattened, green body blending with the vegetation. Its eyes, mounted on turrets, swivel independently to provide three-dimensional vision, which allows the creature to judge distance when stalking a meal.

The chameleon uses its whip-like tongue to disarm its victims. When chasing a wasp, it strikes the insect on the lower abdomen, just below the sting, and swiftly pulls it, sting first, into its mouth. Immediately, it bites off the sting and spits out the wasp to ensure that the sting has, indeed, gone. If it catches a scorpion, it will remove the sting and claws before eating it.

The chameleon, from the tropics of East Africa, uses its tongue in defence, too. If it finds itself unable to run from a hungry viper, it will strike the snake on the head with its tongue to stun it before the viper can deliver the first bite.

UNDERWATER AMBUSHER'S MASK OF DEATH

Young dragonflies grow up under water, where they stealthily stalk their prey. Unlike the adults, they do not catch their food with their legs. Instead, they use a piece of equipment called a mask, which can shoot out in a split-second to make a kill.

The mask consists of a two lethally sharp hooks, attached to the end of a long, strip-like mouthpart with a central hinge. The hinge lets the mask fold up, so that it can be stowed away beneath the head. But if the young dragonfly encounters something edible, the mask suddenly shoots out, and the hooks sink into the unfortunate victim, preventing it from swimming away. Using this weapon, young dragonflies can even catch fish – including ones larger than themselves.

THE TIGER SHARK'S TERRIBLE CHAINSAW

Shoes, petrol cans, car number plates, nails ... the tiger shark will swallow just about anything. It is a living rubbish bin with fins. And what it cannot swallow whole, it rips up.

Found in both shallow and deep waters all round the world, the tiger is one of the most dangerous of all sharks because of its indiscriminate eating habits. It can grow up to 6 m (20 ft) long and has a set of vicious, serrated teeth that are curved and notched.

It uses them like a chainsaw to saw through shell and mammal bone, cut sea turtles in half, or dismember floating whale carcasses.

The tiger shark has a large appetite and favours jellyfish, lobsters, crabs, sea squirts, shellfish, birds and snakes, iguanas, turtles, other sharks, fish, seals, and dolphins. Land animals that end up in the sea have also found their way into tiger shark guts, including horses, cows, monkeys, rats, and even human parts. It suffers no ill effects from its curious diet and is even able to spit out its stomach and turn it inside out to get rid of anything it cannot digest.

3

CROCODILE'S SPINNING UNDERWATER DEATH-ROLL

Floating just below the surface, a crocodile looks nothing more than a piece of wood. The drab, greeny brown skin of the largest living reptile's back and tail, deeply creased like the ridges and furrows of a log, is enough to deceive its prey, particularly in muddy water. Logs naturally drift towards the shore, and so even animals as alert and nervous as deer, monkeys, and herons cannot guess that a crocodile is watching intently from just a metre away. In spite of being up to 4.5 m (15 ft) long, the crocodile makes not a single ripple to give itself away. Only its nostrils and eyes show as it watches and waits in the shallows for animals to come to drink at the water's edge.

When they appear, the crocodile bides its time, swimming very slowly forward. Then, with a sudden upward thrust of its webbed feet and a slash of its immense, muscular tail, it rockets out of the water so quickly that the prey has no time to escape. The crocodile immediately drags its victim under water to drown it. If the catch is too big to swallow in one gulp, the crocodile spins in an underwater 'death-roll', the catch still in its mouth. The twisting movement helps it to tear off large chunks of flesh. Sometimes, two crocodiles will help each other 'unscrew' the flesh, one of them holding the prey while the other spins, or both spinning in opposite directions, as though they were wringing water out of a wet sheet.

STUNNING TECHNIQUE OF THE KILLER WHALE

Weighing up to 5 tonnes and measuring 8 m (27 ft) long, killer whales are not that nimble. They cannot twist and turn in pursuit of darting prey such as salmon. Instead killer whales, also known as orcas, have devised an ingenious way of catching a meal. They work in packs of about 20, frightening fish by leaping out of the water and slapping their tail flukes on the surface. The fish bunch together and the whales lunge towards the shoal.

When orcas in Tysfjord, northern Norway, were studied as they hunted herrings, they accompanied each lunge with a massive slap of their powerful tails, battering the shoal with a speed of up to 50 km/h (30 mph). About 16 fish were stunned with each slap. The technique more than compensates for the whale's poor manoeuvrability.

4

BOW AND ARROW *The archer fish fires a jet of water with great accuracy to dislodge an insect for its next meal.*

ARCHER FISH TAKES AIM WITH A WATER GUN

The archer fish sets its sights above the water when in search of dinner. Found in rivers in India, Asia, and northern Australia, it has large eyes that face forwards. Through the water's surface it can spot insects perched on branches overhead.

The fish shoots at these insects by using a groove in the roof of its mouth, and its tongue, which is long and thin at the front but thick and muscular at the back. Once it is within spitting distance, the fish presses its tongue against the roof of its mouth, turning the groove into a narrow tube, with the front of the tongue forming a valve. It jerks its gill covers shut and flicks the tip of the tongue, sending up a jet of water. The jet knocks the insect off balance into the water. An adult fish can score a hit from 1.5 m (5 ft), but young fish must learn to compensate for the distortion caused as light enters the water before they can strike with such accuracy.

INSECT ARMY MARCHES ON ITS STOMACH

When army or driver ants set out on a raid, they do so with military discipline. Scouts at the head of the long thin column lay down a chemical trail which the other ants follow. About 30 m (100 ft) from the main camp, the ants – which are found in Africa and South America – fan out to scour the forest floor for prey. Any centipede, spider, scorpion, nestling, mouse, or lizard that fails to flee the area is doomed. The ants flood over it, sinking their jaws in. Whatever is not eaten is carried back to camp to feed the queen and her carers.

While the army pillages the forest, the queen is busy laying 25,000 eggs. After a few days, these hatch into hungry grubs. At the same time, pupae from the previous generation start to split and new adults emerge. Suddenly, there are many more mouths to feed. By this time, the army has stripped the area of food and it is time to move on.

The army marches in long columns, grubs carried aloft. It travels for two or three weeks, until the grubs are fully grown and turn into pupae. Then the ants set up a more permanent camp. The queen resumes her egg-laying, and her soldiers set off again on their merciless raids.

FEMALE MOSQUITO'S BLOODSUCKING SYRINGE

The female mosquito's mouth is superbly suited to piercing skin and sucking blood. Having tracked down her meal by following the trail of carbon dioxide that animals exhale, she finds a patch of bare skin and stabs. Her mouth features four needle-like piercing stylets, so fine that the victim is often unaware it is being bitten. She uses these, guided by her lower lip, or labium, to puncture a hole in the skin. Two other mouth-

Blood is sucked up via the labrum

Anticoagulents are pumped down the hypopharynx

Four needle-like stylets pierce the skin

BLOODSUCKER *A female mosquito's mouth-parts are perfectly designed for piercing human skin, enabling her to access the blood beneath.*

parts work together like a hypodermic syringe: while the mosquito pumps anticoagulants into the wound along a tiny tube-like structure called the hypopharynx to prevent the blood from clotting, she simultaneously sucks blood up through the upper lip.

As she feeds, the mosquito's body swells so much that the skin becomes translucent as it stretches. Her abdomen glows crimson from the blood she has sucked up.

Male mosquitoes feed only on plant juices, because their mouth-parts are too weak to pierce animal skin. Females, which need a blood-meal before they can reproduce, are responsible for spreading diseases, including malaria and yellow fever among humans.

THE FISH ARMED WITH A DEADLY SAW

Whether being used for killing or digging, the sawfish's saw-like snout is a multipurpose tool. The sawfish uses its lethal weapon to slash rapidly back and forth through a school of fish, killing many of them, then returns to eat its mutilated victims at leisure.

It does not depend entirely on its saw to get food: the short, flattened teeth in its mouth are excellent for crushing the shells of crustaceans and molluscs.

Sawfish belong to the same family as skates and rays and are common the world over. At up to 7 m (23 ft) long including the snout, the sawfish is a dangerous target for any predator: the saw bears anything from 12 to 30 pairs of razor-sharp teeth.

UNIQUE WEAPON OF SOLITARY SWORDFISH

The function of the swordfish's rapier is a mystery. But it is thought that it may use it to plunge through shoals of fish, slashing, slapping, and stunning them.

Formed from the fish's pointed snout, the sword is a powerful, flattened weapon extending into a sharp spike. The fish itself is a huge solitary creature that can measure 4.5 m (15 ft) in length, but its hydrodynamic shape allows it to travel at speeds of up to 100 km/h (62 mph).

Sometimes speed is thought to be its downfall: broken swords have been discovered in whale blubber, and even in the hulls of ships, indicating that the swordfish was unable to swerve in time to avoid the obstacle.

THE BOXER CRAB PULLS ON ITS GLOVES

Boxer crabs punch their enemies in the face with living, stinging boxing gloves. The crabs belong to a group of crustaceans called decapods, which includes lobsters, shrimps, and crayfish. They each have five pairs of walking legs, the front ones usually modified into claws.

A tiny anemone is coaxed by the crab to sit on each of these front pincers. If a potential predator gets too close, the crab punches, thrusting the anemone with all its stinging tentacles into the enemy's face. Boxer crabs also employ the anemones' stinging cells to help to get food and use another pair of legs, also modified, to feed.

STINGING PUNCH *A boxer, or pompom, crab in Hawaii holds her anemone gloves aloft, ready to strike a stinging punch.*

CAMOUFLAGE FOR ATTACK

Many animals catch a meal by going undercover. Using elaborate combinations of colour, shape, and behaviour, they create extraordinary disguises. Some blend in with the background, others masquerade as different animals.

THE CARPET SHARK'S STONY LOOK

The spotted wobbegong looks so like an underwater boulder that fish sometimes try to shelter beneath it. Fleshy fringes resembling fronds of algae dangle from the lips and head of this Australian shark, drifting with the current. Its skin is mottled like rock and encrusted with algae to complete the disguise.

The wobbegong stays totally still, resting flat on the seabed – hence its other name, the carpet shark. When creatures swim close to its tasselled mouth, the wobbegong suddenly opens its massive jaws and sucks in the unsuspecting fish.

ASSASSIN BUG LIES IN WAIT WITH A CORPSE

Nymphs of the assassin bug have a macabre use for the empty carcasses of their victims – they employ them to entice more prey. One of the 2,500 species of assassin bug is *Salyavata variegata* from Costa Rica. It approaches a termite hill clutching the sucked-dry shell of a termite. A worker termite, coming out of the hill, sees only the husk of one of its own kind. Instinct instructs it to dispose of the body, so the worker hurries towards the corpse. The assassin bug then stabs the worker termite, injecting a paralyzing saliva to break down the body tissues so that it can consume its prey more easily.

TURKEY VULTURE IMPERSONATION FOOLS PREY

To get close to its prey, the zone-tailed hawk pulls on a cloak of disguise. It pretends to be a harmless scavenging turkey vulture. The hawk, from tropical America, has sooty grey plumage with pale bars on its tail and under its wings. Light markings on the underside of the primary feathers match the silvery sheen of the turkey vulture's wings. In flight it imitates the rocking motion of the turkey vulture.

Vultures eat only dead flesh, and so the small mammals and amphibians the hawk feeds on, such as mice, frogs, and lizards, ignore the bird despite having well-developed vision that enables them to see predators from a distance.

SHARK TACTICS *In Australia, tiny bait fish hide in the shelter of a cave, oblivious to the spotted wobbegong resting on the sea floor below them.*

THE CRAB SPIDER'S COLOURFUL COSTUME

Crab spiders catch their prey unawares by changing colour to match their lair, usually a flower. By controlling the amount of liquid pigment pumped from the intestine to the skin they can, for example, turn yellow when sitting on an ox-eye daisy, or pink on heathers or orchids.

The crab spider – so called because of its crab-like sideways scuttling movements – lies in wait, its front two pairs of legs, armed with bristles, held out in readiness. When a bee or butterfly approaches the flower, the spider pounces and grips it in a death-hug. It injects a nerve poison so powerful that it prevents a struggle and avoids attracting the attention of birds. Then it sucks its victim dry.

COLOUR CODED *Perfectly matching the colour of English ragwort, a female* Misumena *crab spider sucks the life juices out of a fly.*

LEAFY TRAP *With a head shaped like a fallen leaf, the gaboon viper waits for a passing mammal, bird, or frog.*

KILLER IN THE GRASS

As it stalks through the long grass, the tiger crouches against the ground to eliminate shadows, its thin orange, buff, brown, and gold stripes helping to break up the outline of its body. Large parts of the Indian forests where the tiger lives are covered in tall dry grass, so its stripy coat blends in well, mimicking the dappled sunlight of the forest.

To kill, the tiger must be no more than 18 m (60 ft) from its prey. Any farther, and it stands no chance of out-running its dinner. It creeps towards its victim, usually a large, hoofed mammal such as a sambar, chital, or wild pig. Then, at the last possible moment, it breaks cover and reaches its prey in just a few bounds.

HIDDEN VIPER LIES IN AMBUSH

The longest fangs in the world, sometimes exceeding 5 cm (2 in), are tucked into the mouth of the West African gaboon viper. Nestled among the leaf litter on the forest floor, the gaboon viper almost disappears, its shape broken up by bold markings of black, brown, buff, and grey, which merge with the surroundings. As an extra touch, triangular black patches on its body divert attention from the head and eyes and mask its deadly intentions.

Like all vipers, it relies on ambush to kill. Its lethal venom can immobilize prey such as medium-sized mammals and birds. The gaboon viper strikes rapidly and needs its long, sharp fangs to pierce through fur and feathers.

KILLER LACEWING DONS A WAXY COAT

Larvae of the green lacewing *Ceraeochrysa cincta* from Florida disguise themselves with wax to live incognito among their prey. They feast on mealybugs which carry long filaments of wax. To blend in, the larva uses its curved jaws to steal wax from the bugs to cover its abdomen. Moving from one mealybug to the next, it perfects its camouflage.

Master of disguise

It can be difficult to spot a mantis because it is skilled in the art of masquerade, mimicking the exact shape of a leaf or hue of a petal. The mantis camouflages itself both to hide from predators, such as birds and lizards, and to catch insects that unwittingly land on it thinking it is a flower or a leaf.

There are 1,800 types of mantis, mainly living in tropical climates. Their disguises are quite diverse – differing in colour, shape, and texture – but all are impressively detailed. For example, the orchid mantis has flanges along its legs and body that take on the exact colour and texture of an orchid flower. The mantis sits among the blooms, waiting for its insect prey to approach. To complete its disguise, the mantis may even rock gently from side to side, imitating the way a petal might flutter in a slight breeze.

Once an insect approaches, the 'flower' bursts into life and the mantis snatches the creature with its hook-lined forelegs.

The species of mantis shown, *Choeradodis rhombicollis*, is native to Costa Rica, where its flattened green body and stalk-like legs help it to hide among foliage.

BLENDING IN FOR DEFENCE

Just as some creatures use camouflage to kill, so others use it to avoid drawing attention to themselves. They try to look like their surroundings, using stripes, lights, or quick-colour changes, and adapt their behaviour accordingly.

STRIPES HELP GAZELLES TO MERGE INTO ONE

The buff coat of the Thomson's gazelle merges well with the dry, open savannahs of Africa. The gazelle also has conspicuous thick bands of black on each side, which break up the outline of individual animals when they are in a herd. To avoid drawing attention to a newborn calf, a female gazelle drops her fawn on bare ground and grazes 300 m (330 yd) away, suckling it for only a short time. The fawn lies still and low, hidden from lions and cheetahs.

FISH THAT GLOW IN THE DARK

Some fish, such as viperfish and the California midshipman, carry their own light bulbs, which they can brighten and dim to order. By day, so much sunlight filters into the water that they are silhouetted against the surface, making them a target for predators below. If the fish turn up their body lights, they are able to blend in with the ambient light and their outline melts away.

The bioluminescence is caused by a chemical reaction that occurs within the body tissues of the fish.

GROUSE PREFER WHITE FOR WINTER

What to wear: brown feathers or white? For the willow grouse (willow ptarmigan), it all depends on the length of the day.

As winter approaches north of the Arctic Circle, the long summer days begin to shorten and the willow grouse's hormones are activated. It begins to shed its all-brown summer plumage and gradually the bird turns white. By the time snow falls in the tundra, the grouse has transformed from brown to pure white, blending in with its new background.

When the days lengthen at the end of winter, the willow grouse moults. Its coat of new brown feathers and old white winter ones is once again in keeping with its surroundings, a patchwork of white and brown.

FLATFISH KEEPS A LOW PROFILE

Masters at colour change, flatfish blend in superbly with the grainy seabed. The flounder can even reproduce a checkerboard, a design not found anywhere else in nature.

When chased, a flatfish scurries away, whipping up a cloud of sand to confuse the pursuer. Then it dives quickly to the seabed, dropping down as flat as a pancake. Being so very nearly two-dimensional, little of its body sticks up from the seabed. As the sand settles, some lands on and around the fish's fins, blurring the sharp edges of its outline. The flatfish wriggles a little to flick more sand over itself as it changes colour to match the shade and contrast of the background.

THE FROGMOUTH'S DISPLAY OF WOODEN ACTING

Even though it rests in full view of its enemies, the tawny frogmouth of South-east Asia and Australia is rarely disturbed. By day the bird clings lengthways to a tree, head pointing straight up, eyes closed, and stays absolutely still. Its blotchy brown plumage blends in well with the trunk, so that the frogmouth resembles a broken stump. When night comes, it leaves the tree to hunt for insects, frogs, snails, and small rodents.

Another bird uses the same disguise. The bark-coloured potoo, found in the West Indies and Central and South America, also pretends to be a broken branch.

SAFETY IN STRIPES *The black flashes on a Thomson's gazelle make it hard to pick out a single animal from this herd in Kenya's Masai Mara.*

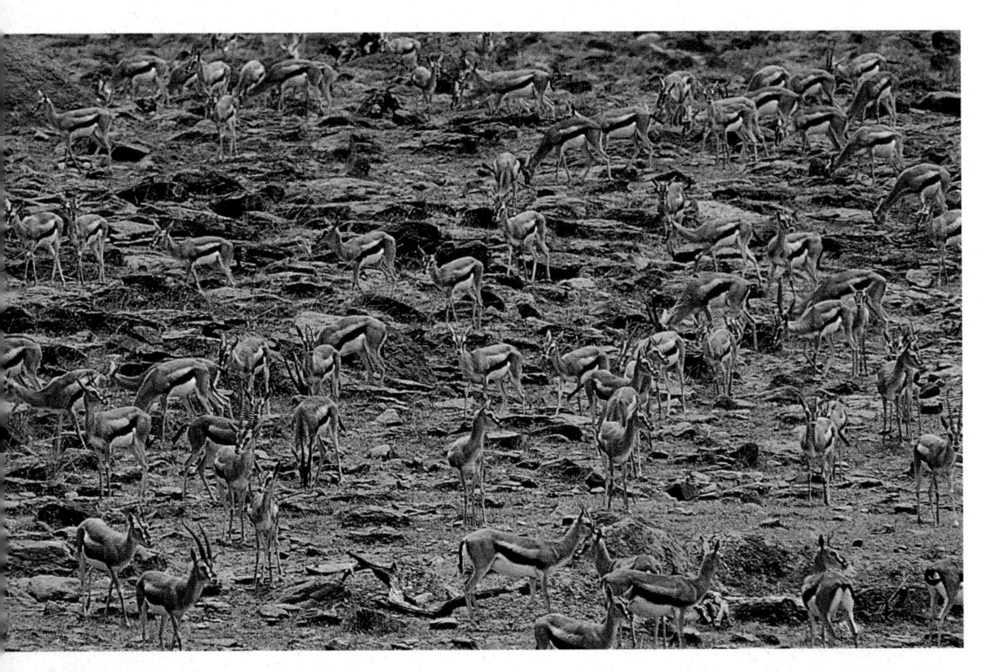

BRANCH LINE *To a predator, the tawny frogmouth nesting peacefully on the branch of a tree in Australia looks just like any other lichen-covered stump of wood.*

A BLOOMING GARDEN FOR THE SLEEPY SLOTH

When it rains, the sloth turns green due to algae growing in its coarse coat. This living camouflage enables the sloth to disappear into the lush vegetation, hidden from aerial predators. In dry weather the minute green plants lose their moisture and turn yellowy brown, enabling the sloth to merge with the dry tree trunks.

A resident of the South and Central American rain forests, the sloth spends more than 21 hours a day napping, hanging upside down from a branch. With a top speed of less than 1 km/h (1/2 mph), sloths have opted for concealment and cryptic movement as the best protection from predators.

CANNY GECKO TAKES UP TREE TRUNK POSE

Few lizards in the world can hide themselves as effectively as the nocturnal leaf-tailed gecko of Madagascar. During the day this gecko rests head-down, flattened against a tree trunk in the rain forest, changing the colour and the apparent texture of its skin to merge with the lichen on the bark. Even its large, creamy eyes are buff-streaked so they blend with the vegetation. The gecko tucks its back legs under its tail and spreads its feet against the bark, gripping on with sucker-like toes.

The best colour match alone cannot solve the biggest problem faced by any animal trying to disappear: outline and shadow. To blur both, the gecko's body is trimmed with a frilly border of loose skin appendages. Along with its leaf-shaped tail, these break up the shape of its body and the outline of its shadow. With these disguises, the gecko can bask in the sun in full view of, but unnoticed by, predators such as snakes and birds.

CAMOUFLAGE COAT *Algae and moths benefit from the three-toed sloth's slow pace through the rain forest. Undisturbed by abrupt movements, they thrive in its shaggy, matted fur.*

TAKING A NAP *Superbly disguised as part of a tree trunk, a nocturnal leaf-tailed gecko from Madagascar sleeps securely during the daytime.*

BABY WILD BOAR'S STRIPY COAT

Wild boar are born with stripes of buff and brown as extra protection against predators. The stripes are thought to be a form of camouflage to help them to blend in with the dappled lighting in the woodland glades where they live.

Baby wild boar live with their mother and several other adult females in a group called a sounder. Mature males forage alone and only join the females in the mating season, which is during winter in Europe.

STRIPY EFFECT *Wild boar piglets are difficult to see in the undergrowth because their stripy coats help to blur their outline.*

INSECTS IN PETAL FORMATION

Animals sometimes work together to create group camouflage, to fool predators. In Madagascar, tropical insects known as phiatids have brilliant coloured wings which, when massed together, fool bird predators into thinking they are just petals and of no interest. Individual *Ityraea gregorii* insects of the same family are born either red or green. Huddled together on a branch, they mimic the arrangement of colour in blossom and look like partly opened buds surrounded by green bracts.

GREAT PRETENDERS

Some animals successfully ensure their safety by being confidence tricksters. There are those that fool their enemies by looking like something inedible, and those that adopt the persona of a different, more dangerous creature.

HARMLESS CATERPILLARS TAKE ON A DEADLY DISGUISE

The hawkmoth caterpillar, from Central America, can quickly disguise itself as a tree snake. If its initial twig-like pose fails to ward off an approaching predator, the caterpillar effects a snake-like look: it shortens its head and turns on to its back, exposing a pair of convincing false 'eyes'. It flicks a 'tongue' in and out of the snake-shaped 'head' which is, in fact, its rear end.

Terror of anything snake-like is widespread in the animal kingdom and many harmless creatures have evolved snake-like characteristics. The great mormon caterpillar of Asia and northern Australia has eyespots on its rear end and a bright red forked 'tongue'. When threatened, the caterpillar of the Madagascar butterfly inflates its thorax, displays huge false eyes and sways in a snake-like manner.

DECAYING LEAF SURPRISES INSECT LARVAE

Some of the 'dead leaves' that float in South American rivers are, in fact, alive and well. Leaf-fishes have bodies that are flattened from side to side. Their colouring and shape matches a leaf almost exactly, and some even have a 'leaf stalk' growing out from the lower jaw. They usually hide under rocks or in crevices where they look like a dead leaf that has become wedged. When a suitable prey passes, the 'leaf' grabs it.

The most spectacular leaf-fish, *Minocirrhus polyacanthus*, lives in the Amazon and Rio Negro basins of South America. With markings that look like the ribs of a half-decayed leaf, it waits patiently on the riverbed for nymphs and larvae to approach.

PRAWN MIMICS ALL THE COLOURS OF THE RAINBOW

There is not a single colour in nature that the tiny hippolyte prawn cannot copy exactly. From bright green to purple, yellow to red or brown, it can transform itself, chameleon-like, into any hue.

Its technicolour versatility does not stop there. The hippolyte can handle patterns and textures, too. This crustacean, commonly found in tide pools, can not only mimic the exact shade of green of a dead leaf floating in the water, but also the leaf's vein structure and grain. Predators can stare at the hippolyte and never see it. At night it turns a transparent midnight blue, melting into the blackness.

SNAKE HEAD *From the triangular 'head' to the glint in the 'eye', the snake-mimicking caterpillar is one of the world's top impersonators.*

THE FROG THAT FELL TO EARTH

Deep in the Malaysian rain forest down on the forest floor, all is not what it appears to be. Invisible from above, there is a creature that mimics closely the inedible piles of fallen leaves that surround it.

The adult horned frog, often called a toad but actually a tree frog, spends the day crouched on the forest floor waiting for insects to emerge. It employs a combination of shapes and colours to mimic the leaf litter all around it, in order to remain as inconspicuous as possible to predators.

The frog has a mottled brown coloration. Leaf-like projections on its head, above its eyes, and on its elbows and legs help it to merge with its surroundings. Its mouth looks uncannily like a twig, contributing to the overall effect. As long as it remains motionless, it will be undetectable to birds circling above, and any other creatures, such as lizards and small mammals, that feed on frogs.

Similar species of tree frogs and toads are found in the forests of Madagascar and Brazil. Some have mouths that mimic branches and peeling bark.

WEEDY FAKE *The leafy sea dragon from Australia looks so much like real seaweed that small fish sometimes take refuge in its 'foliage'.*

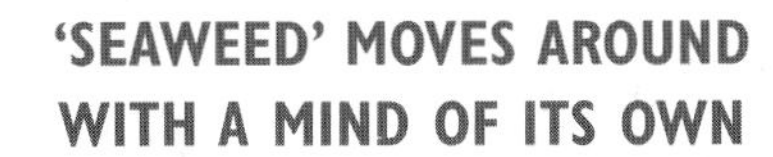

'SEAWEED' MOVES AROUND WITH A MIND OF ITS OWN

Not even seaweed looks as much like seaweed as the sea dragons of southern and eastern Australia. These relatives of sea horses are covered from head to tail in fleshy skin flaps which, like fronds of vegetation, float and sway with the swell.

Sea dragons live in coastal kelp beds where their camouflage makes them almost impossible to see. If they are spotted by a predator they have a second line of defence; a suit of bony armour and long dorsal spines, which makes them virtually inedible.

Sea dragons grow to 45 cm (18 in) long. Like sea horses, they feed on tiny shrimps and other invertebrates, which they suck in through their long, tubular snouts.

As if these fish were not unusual enough, it is the male sea dragon that broods the eggs. Fertilized eggs are fastened to a patch at the base of his tail and remain there until they hatch.

DOWN FROM THE TREES *The horned frog is a tree frog that lives on the ground. It spends most of its life stock still, lying in ambush for prey.*

DIRTY LOOKS *Disguised as a fresh bird dropping, the caterpillar of the king swallowtail butterfly rests undisturbed in Panama's rain forest.*

THE BIRD DROPPING THAT SPINS ITS OWN SPLASH

Droppings attract no interest from potential bird predators, and a few canny creatures have made the most of this fact. The dirty-white, lumpy spider from Malaysia, *Phrynarachne decipiens*, for example, fashions itself to look just like a blob of bird excrement. It even sits in the middle of a small mat of white silk, mimicking precisely the splash effect caused when a dropping hits the leaf. The king swallowtail butterfly caterpillar can also sit boldly in the open because it looks so much like an unpalatable dropping itself.

The longhorn beetle from Brazil takes the effect one stage further, exposing several large lumps on its legs that look just like the seeds that sometimes pass undigested through birds' intestines. In this way the creatures are effectively hidden from predators despite being in full view.

HOG-NOSED SNAKE IS A MASTER OF DISGUISE

If disturbed, the hog-nosed snake has two lines of defence: its cobra impression, or playing 'dead'. Its first bluff is to rear up menacingly, head and neck flattened, and pretend to 'strike' towards its aggressor. If the bluff works, the hog-nosed snake drops the act, deflates, and makes its getaway. If the bluff fails, the snake tries a different tactic. It pretends to drop dead, lying on the ground, floppy and motionless, belly-up, its mouth sagging wide open, and its tongue lolling out, letting out a stench of decaying flesh. When the danger has passed, the snake cautiously lifts its head, checks the surroundings and slides away.

The snake tends to overact the death-throe part by wriggling every so often to stay 'dead'. If its aggressor nudges it upright, it quickly flips itself back over again to resume its dramatic, motionless death pose.

STILL AS A CORPSE *Many predators will not eat carrion, which is why the hog-nosed snake finds it so effective to play 'dead'.*

BEWARE THE BLACK-AND-YELLOW OUTFIT

Wasps can inflict a nasty sting, and their black-and-yellow outfit is copied by bees, moths, flies, and other insects to scare predators. Among the most accurate mimics are hover flies. On a summer's day, up to 20 different, harmless wasp-like hover flies may be found in a European garden. As well as having the same colours and being the same size as the wasp, the hover fly even sits on flowers in a wasp-like pose. Its black thorax gives the impression that the abdomen ends abruptly in a narrow 'waist', just like a wasp's.

RED BATS HANG LIKE LEAVES

Red bats look, at first glance, like a row of dead autumn leaves in the forest. To enhance their leaf pretence each hangs by one stalk-like leg, while the white tips of its thick red fur give the bat a frosted appearance. This colouring blends in well with the trees and shrubs of southern Canada through America to

Argentina, where the bats spend the day. Sometimes red bats roost alone, but often they gather in small groups.

If disturbed by a passing bird such as a blue jay, they sway from side to side as though being blown by a breeze. They grow to about 9 cm (3 1/2 in) long, with a tail of 7.5 cm (3 in).

THE FISH THAT MASQUERADES AS A HOME HELP

Disguised as the little cleaner wrasse of the Indo-Pacific, the sabretooth blenny has discovered how to find a good meal. The wrasse offers a much-needed service to bigger fish such as parrotfish. It nibbles parasites, dead skin, and other debris from their skin, swimming safely right into the larger fish's open mouth and out through the gills to pick off lice. In return for this service, cleaner wrasse never get eaten.

Occasionally, though, a big fish gets a nasty surprise. What looks like and swims like a helpful cleaner wrasse can turn out to be a hungry sabretooth blenny. When the blenny gets close, it will suddenly chew a large lump out of the fish's fin or rip off a scale.

CUBS FIND SAFETY IN BADGER COATS

Cheetah cubs do not need to hide from their predators thanks to furry coats that seem to mimic those of fierce honey badgers. All big cat cubs are small, defenceless, and vulnerable to being picked off by predators such as eagles. For safety, leopard cubs, like their parents, have blotches of dark brown which serve as camouflage, but cheetah cubs are quite conspicuous, with dark bellies and pale backs. It is thought that young cheetahs mimic the ratel or honey badger, a small carnivore no larger than a jackal that fearlessly attacks any predators. In this disguise the cubs are less likely to be targeted by hunters.

COPYCAT *The aerial view of an infant cheetah's back may resemble that of an aggressive honey badger, deterring would-be predators.*

ROYAL IMPOSTOR *A single experience with a foul-tasting monarch (left) is so unpleasant that a bird will make a point of avoiding all butterflies featuring similar markings, including the harmless viceroy (above).*

BIRDS BEWARE THE FOUL-TASTING MONARCH

Any bird that tries to eat a noxious North American monarch butterfly spits it out immediately, sometimes vomiting. You may find a monarch with a beak-sized hole in its wing – evidence that a bird has attempted to eat it, but has been put off by the taste. The bright colours of the monarch's wings also act as a warning signal and birds quickly learn to avoid all monarchs.

But it is not only the monarch butterfly that benefits from this chemical defence and wing coloration. The totally unrelated viceroy butterfly has an almost identical wing pattern and so it is also avoided by the birds. Although it is totally harmless, its clever mimicry prevents it from becoming a snack.

Monarchs become noxious by feeding on milkweed when they are caterpillars. The milkweed contains chemicals called cardenolides, which persist in the adult butterfly. These chemicals are toxic to vertebrates, such as birds and lizards, although the concentration each butterfly contains, and hence the potency, varies.

Only about 30 per cent of monarchs are in fact unpalatable. But as long as birds get the occasional reminder of the monarch's foul taste, they will leave all members of the species alone, including its mimic, the viceroy.

MIMIC OCTOPUS HAS MULTIPLE PERSONALITIES

Able to change from a stingray to a sea snake to a blenny fish, the mimic octopus is the world's most extraordinary impersonator. It has at least 15 characters in its repertoire. Like all octopuses, the mimic has great control over its colour, flushing and fading in rapid succession. It can also change its shape and behaviour to prevent detection or scare predators.

One of its most impressive displays is as a sea snake. If threatened, the octopus plunges six of its arms down a hole and waves the other two, bent into an undulating S-shape. To imitate a flounder, the octopus flattens its body, tucks its arms behind, and fades to a light brown. It pumps water from its funnel to jet forward following the seabed, imitating the flounder's swim.

Since its discovery in Indonesia in the 1990s, the mimic octopus has also been seen masquerading as a lionfish, sand anemone, mantis shrimp, and hermit crab, among others.

CLEVER ESCAPES

If a disguise fails and a predator attacks, the best ruse is to buy time. Animals have an amazing array of tricks designed to startle, frighten, and surprise aggressors. In the confusion the animal has the chance to make a quick getaway.

LEGLESS LIZARD GOES FOR BROKE

Self-sacrifice can be a means of survival for lizards. In moments of extreme danger most lizards – including the greater earless lizard – will, as a last resort, voluntarily shed their tails. While the aggressor turns its attention to the convulsively twitching tail, the lizard scuttles away unharmed. The predator gets a snack, the lizard keeps its life. Eventually, a new tail grows. It is as good as the first, but it cannot be shed.

One lizard goes to extremes. Pallas's glass snake, the largest in the world at 1.5 m (5 ft), is a legless lizard found in the Balkan peninsula, central Asia, and parts of Russia. Instead of shedding its tail in one piece, the glass snake 'shatters' it into several pieces. Each vertebra in its tail has a special fracture plane through which a wall of cartilage passes, creating a weak 'crisis' point. The muscles, blood vessels, and nerves are also constricted to allow an easy, painless break.

Pallas's glass snake can cast off its tail at any point, or it can snap it at will into several long wriggling, writhing pieces. Since its tail is twice as long as its body, the glass snake appears to shatter into pieces, to the confusion of the predator.

FROGS PUT ON A BRAVE FACE

The South American frog, *Physalaemus nattereri*, has an ingenious method of buying time. It has glands on its rump prominently marked with massive eyespots. In times of danger the frog turns its back, raises its legs and points the staring fake eyes of an apparently much larger, more dangerous creature towards the aggressor. If this fails to deter, the frog secretes an unpleasant substance from the glands, which leaves a particularly bad taste in its attacker's mouth.

Some frogs and toads, whose most common predators are snakes, have another line of defence. They make themselves appear too big to swallow. Snakes are unable to chew and have to swallow their food whole. So frogs and toads inflate their bodies with air, stiffen their legs to stand up high off the ground and lean forwards. The European common toad often uses this technique against foxes.

INKY DARKNESS *As well as providing the giant octopus with a smoke screen, the cloud of ink forms a distracting shape in the water.*

INSTANT SMOKE SCREEN MASKS ESCAPE

The octopus squirts ink straight into its enemy's face just like other cephalopods such as cuttlefish and squid. When the dark fluid spurts out into the water, it creates an instant thick, black cloud. It is impossible to see through this, and by the time the ink has dispersed, the cephalopod has vanished.

The warty prowfish, in the coastal waters of southern Australia, also responds to danger by squirting. Instead of ink, it produces a smoke-like cloud of what is thought to be toxic fluid from a vent near its gills.

BREAK AWAY *Shorter but alive, the greater earless lizard discards a piece of twitching tail to distract its enemy while it escapes.*

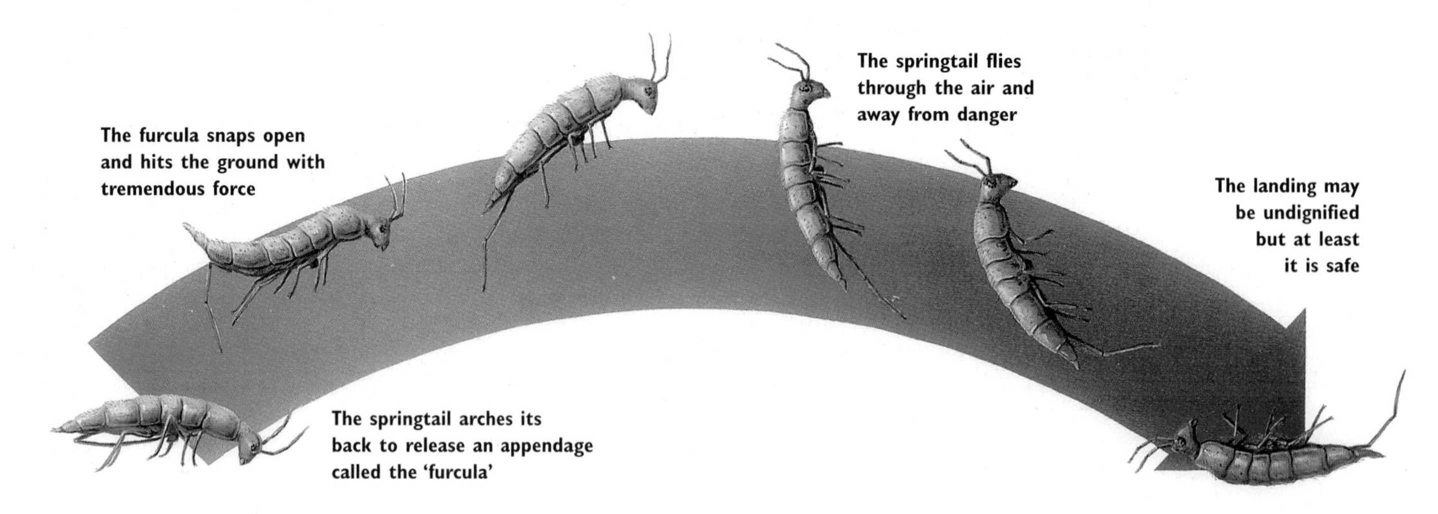

SPRINGTAIL ESCAPES ON A POGO STICK

Every day, in gardens all over the world, a miniature gymnast performs a routine of twists, high leaps, somersaults, and back-flips. The athletic springtail is so tiny – 5 mm (1/4 in) long – and moves so quickly that its show goes unnoticed.

Springtails have six legs, but they are not true insects. They belong to the oldest group of six-legged animals (Collembola) and are among the most abundant creatures in the world, living in plant debris, under bark, and elsewhere. Unlike true insects, they do not have wings, and so they cannot fly away from an attacker. Instead, they use a special appendage, like a built-in pogo stick, to cartwheel up into the air and away from danger. Springtails can leap 20 cm (8 in) high, which is about 40 times the length of their own body. To the predator, the springtail literally vanishes into thin air.

HIGH JUMP *A startled springtail releases a forked appendage (the furcula), which catapults it through the air and out of immediate danger.*

THE HORNY TOAD'S BLOODY REVENGE

Grotesque defences do not come much gorier than those of the horny toad. Being attacked is enough to make it spit blood. Horny toads are, in fact, lizards, about 4–13 cm (1 1/2–5 in) long, but with short, squat, rough-skinned bodies, they look more like toads, hence their name. Usually these North American lizards remain flattened against the ground to keep out of danger. But they have other, more startling defences. Spitting blood is the cleverest ruse of all. By restricting the flow of blood from its head, the horny toad bursts the small blood vessels in its eye membranes and shoots blood with great accuracy up to 1 m (3 ft 3 in). If the aggressor does not back off it is in for another shock. Most of these lizards have large spines sticking out of the back of their head which pierce the predator's stomach when swallowed.

WATER BIRD DROPS OUT OF DANGER

The moorhen is one of the world's most widespread water birds, although it seems to have few defences, apart from running or flying away. But it does have one remarkable defensive trick up its sleeve. If it is threatened, it heads for shallow water, and sinks like a submarine.

The moorhen in not a natural diver, so it does not swim off once it has sunk out of sight. Instead, it stays in one spot, breaking the surface with the tip of its beak, which allows it to breathe. To stop itself bobbing back up again, it forces the air out of its plumage, and also out of its internal air sacs – hollow spaces that are connected to its lungs. After a few minutes, it slowly surfaces, in the hope that the danger has passed.

SPITTING BLOOD *The jet of blood squirted from the eyes of a threatened horny toad is enough to blind an attacker temporarily.*

WATER BABY *To the astonishment of its pursuer, the basilisk lizard of the tropics runs across water, thanks to its specially adapted feet.*

SHRIMP BLINDS ITS ENEMY WITH LIGHT

When attacked, the luminous *Acanthephyra* shrimp of the North Atlantic dazzles its enemy with a show of flashing lights. Like a celebrity blinded by a wall of photographers' flashbulbs, the predator is startled for just long enough for the shrimp to make its getaway. The shrimp's illuminations are produced by chemical secretions.

THE RAZOR SHELL'S DEFT SLICING ACTION

Of all the creatures that burrow in sand, the razor shell has the fastest disappearing act of all. Found on beaches around the world, the razor shell is about 25 cm (10 in) long and lies vertically in the sand. Its secret weapon is its foot, which is concealed between the two halves of its shell. At the slightest sign of danger, detected through vibrations, the razor shell kicks out, and the pointed tip of the foot slices down through the sand, pulling the shell rapidly behind it.

GUT REACTION ENTANGLES SEA CUCUMBER'S ATTACKER

Few animals employ defence measures as desperate as the sea cucumber's. If threatened by a predator such as a loggerhead turtle, the sea cucumber simply contracts its body, vomiting some of its entrails out through its mouth and expelling the rest through its anus. The attacker, shocked and distracted, becomes entangled in the sticky mess of fine tubules, sometimes so much so that it drowns. The sea cucumber, meanwhile, crawls away. Its guts grow back within a few weeks.

LIZARD WALKS ON WATER

The basilisk lizard of tropical America pulls a stunt that leaves its pursuer high and dry: it runs across water. Travelling at speeds of up to 12 km/h (7½ mph), the basilisk can sprint across a lake without sinking. This spectacular feat has earned it the name 'river-crosser' or 'Jesus Christ lizard'. It runs almost upright, buoyed up by toe fringes which increase the surface area of its feet. Once safely out of reach, the lizard slams on the brakes and sinks into the water, where it stays submerged for a couple of minutes.

MOTHS SOUND ALARM AS BATS APPROACH

Many animals warn potential predators of their foul taste by their display of garish colours, but few use sound as an alarm. An exception are the foul-tasting Arctiid moths. The moths respond to the sound of an approaching bat with rapid clicks that 'tell' the bat not to bother eating them. They have evolved ear-like structures that vibrate in response to bat calls.

Tiger moths also react by making clicks similar to those of bats. These are emitted when a bat is about 50 cm (20 in) away and appear to jam the bat's information processing system.

DEFENSIVE WEAPONS

Instead of trying to make a clever escape, some animals protect themselves by wearing armour and carrying weapons. Deadly daggers, fearsome spears, needle-sharp spines, and venomous stings are all used in self-defence.

THE THORNY DEVILS' PRICKLY BLUFF

By virtue of its mass of large spines, the bizarre-looking thorny devil appears bigger and scarier than it really is. Otherwise known as a moloch, the thorny devil is just 15 cm (6 in) long. Most predators looking for a meal in the scrubs of Australia's deserts leave it alone as the spines make it too uncomfortable to eat. Covering its plump body and tail, the spines are an effective bluff. If threatened, this harmless, slow-moving Australian lizard tucks its head between its front legs, leaving the prickles to do their work.

Formed from enlarged, modified scales, the spines are not only useful in self-defence. Each one is scored with thin grooves spreading out from the central peak. During the night, dew condenses on the scales. Drawn by capillary action along the grooves, the precious moisture soon ends up in the thorny devil's mouth, enabling it to survive in its desert habitat.

THORNY PROBLEM *The thorny devil moves slowly and runs awkwardly, relying on its array of prickly spines to protect it from predators.*

POTTO HIDES WEAPONS UNDER A FUR COAT

Beneath its soft fur, the vulnerable-looking potto conceals a fearsome weapon, a phalanx of spines. This quiet, secretive night-dweller, a primitive primate, relies on moving slowly through the branches of the tree canopy where it lives in West Africa. If confronted, it clamps its hands and feet to anchor itself to the branch and faces forwards, head bent down below the shoulders to present the back of its neck to its aggressor.

Under a shield of thickened skin lie four to six bony knobs, the tips of specially elongated spines extending from the neck vertebrae. The potto lunges at the predator, striking it with its secret weapon with the intention of knocking it off the branch.

SALAMANDER'S SKELETON HAS A CUTTING EDGE

Only the line of warty knobs along the sharp-ribbed salamander's flank gives away its extraordinary defence mechanism. Each protrusion marks the tip of one of the salamander's rib bones. If it is provoked, for example, by a snake, the salamander pushes its robust ribs against the taut skin. It turns them, in effect, into prickles.

For extra defence, the sharp-ribbed salamander can poke its ribs right out through the skin like a row of teeth, so that part of its skeleton is outside its body. The ribs are sharp enough to draw blood, which deters predators.

Measuring 30 cm (12 in) long, this nocturnal creature lives in ponds, ditches, or slow-moving rivers in Spain, Portugal, and Morocco.

SOUTH AMERICAN GIANT WEARS A SUIT OF ARMOUR

The giant armadillo, which is as big as a sheep, is the most heavily armoured animal alive. Its tough, impenetrable body shield forms as the young armadillo's soft, pink skin hardens.

This carapace comprises strong, bony plates, called scutes, covered in a layer of horn. Broad, rigid shields cover the shoulders and hips, and bands across the middle of the back are connected to the underlying skin. The armadillo's tail, the top of its head, and the outer surfaces of its limbs are also armoured.

BEETLE WILL NOT BE MOVED

One tiny beetle can resist for up 2 minutes a force capable of moving 60 times its weight. If threatened, the blue palmeto beetle simply clamps down hard on the palmetto palm leaf on which it lives.

The 3 mm ($^1/_8$ in) beetle's secret anchor is in its feet. Each sole is covered with minuscule hairs, about 10,000 a foot in total. Each hair is split in two, and the tip of each fork holds a droplet of oil, which helps it to stick.

Usually, only a few of these hairs are in contact with the leaf, but at the slightest sign of threat, the beetle squats and glues itself down with all 120,000 hair-ends.

HAIRY MOMENTS AS RIVAL RHINOS JOUST

For something made entirely of strands of hair, a rhino's horn is a formidable weapon. People have been killed by getting in the way of a charging adult.

Unlike reindeer or antelope horns, rhinoceros horns do not have any bone in them. They are made exclusively of keratin (the chief constituent of fingernails, hoofs, and hair), but the hair-like fibres are so tough and compacted that the rhinoceros has a truly dangerous weapon at its disposal.

The horn (two horns in the case of the Sumatran and African rhinoceroses) emerges from a rough patch of the skull. It can grow up to 1 m (3 ft 3 in) long and is usually used in territorial confrontations between adult males. Often these meetings are resolved without any physical contact, but if a fight does break out, both African species (white and black rhinoceroses) fight with upward headbutts, using the horn to stab the aggressor.

The leathery skin on the animal's flanks offers little protection and if the fight becomes vicious, sparring rhinoceroses can inflict gaping wounds on one another.

PAINFUL PRICK *The barbed quills of a crested porcupine, such as this one from Kenya, can cause serious infection and even death.*

THE RODENT WITH A RATTLING COAT OF SPINES

The African crested porcupine is well protected from attack by big cats on the lookout for a snack. Its back is thickly covered with thin, lightweight quills about 50 cm (20 in) long. Each quill is cylindrical, formed of long, tough fibrous hairs, ending in a tip that is as sharp as a needle.

If the porcupine needs to scare off a predator, such as a leopard, it erects the quills, using the same muscles under the skin that an angry cat uses to bristle its fur. When that does not work, the porcupine produces a series of grunts and foot stamps, accompanied by a rattling sound as it rubs the quills together.

Then it turns its back on the predator and reverses into it at speed. The quills are loosely attached to the porcupine's skin and can easily become detached. Once deeply embedded in the predator's flesh, barbs on the quills make them difficult to remove and the animal may suffer a fatal infection.

ZIMBABWEAN CHARGERS *A female black rhinoceros will use her horns to defend her calf from a lion or hyena. Males use theirs to joust with rivals.*

INFLATABLE FISH'S SCARE TACTICS

At the slightest sign of danger, the porcupine fish transforms itself into a football-sized balloon covered in sharp spines. The modified scales on its body usually lie flat, but when threatened, the fish swells and the scales stand on end. Porcupine fish are the largest of their family at up to 90 cm (3 ft) long, and are related to puffer fish, which use the same ruse. They are found in temperate and tropical waters.

CATERPILLAR DONS A DECEPTIVE COAT

A number of caterpillars live boldly in the open, protected by a body armour of stinging hairs. Some of these urticating hairs, as they are known, have a poison gland at the base and shoot venom into any creature that tries to touch.

The spines on Australia's dazzling emperor gum-moth caterpillar, for instance, contain toxins. The Venezuelan emperor gum-moth caterpillar can inject a powerful anti-coagulant which may result in serious haemorrhage for animals or humans alike. Other caterpillars are not poisonous, but are covered with barbs that can cause severe skin rashes.

Hairy caterpillars cloak themselves in flamboyant advertising. Those that contain acrid-tasting substances announce their revolting flavour with brilliant skin colours of red, yellow,

BARBED BALL *When threatened, the porcupine fish inflates itself up to three times its normal size by pumping water into special sacs.*

CODED MESSAGE *Gaudy colours and a shield of spiky yellow hairs warn predators, such as birds, of the emperor gum-moth caterpillar's unpalatability.*

black, and purple. Plump, slow, and just the right size to make a juicy beak-full, caterpillars are extremely vulnerable to predation, particularly by birds. These defences work only if predators learn not to touch.

Some caterpillars try to foil predators by hiding among plant roots and stems or even right inside galls, seeds, and other plant tissue. Some build silk sleeping-bags covered in small grains of sand and plant debris, or camouflage themselves by trying to look like twigs, sticks or bird droppings (*see page 112*).

LETHAL WHIP OF THE STINGRAY'S TAIL

Being thrashed by a stingray's tail is not only extremely painful, it can kill. This effective defence weapon comprises a line of saw-like teeth which protrude backwards along the upper surface of the stingray's spine and tail. At the base of each tooth lies a venom gland. Human swimmers are at great risk from stingrays, for this slow, generally gentle creature is capable of a burst of speed when chasing fish or escaping, and if it feels threatened its tail can thrash out in self-defence. Bathers have been stabbed by its spines and the wounds can be severe as these needles have angled barbs, which make them very difficult to remove. Victims have been known to die.

Stingrays are found in warm tropical and subtropical waters all over the world with streamlined bodies which are flattened top to bottom, they spend most of their time lurking well camouflaged near the seabed, often in very shallow water. Even large specimens such as the Indo-Pacific stingray, which can reach more than 4.5 m (15 ft) in length, will lie in water less than 90 cm (3 ft) deep.

STING IN THE TAIL *Armed with a tail bearing venomous barbed spines, the ribbontail stingray cruises along a coral reef in the Red Sea.*

CHEMICAL WEAPONS

Whether used for attack or defence, the chemicals that exist in nature are some of the most deadly in the world. Natural poisons range from painful stings to venom so toxic that even the tiniest amount can kill a human in minutes.

BOMBARDIER BEETLE'S EXPLOSIVE REACTION

Swivelling its anus from side to side, the bombardier beetle takes aim. With an audible 'pop' it releases a volley of deadly chemicals from an anal gland straight into its enemy's face.

Some of the boiling hot, caustic liquid turns into a fine noxious gas. The attack leaves the predator dazed, confused, and perhaps temporarily blinded, giving the beetle a chance to make its escape.

If an insect, bird, or frog is foolish enough to ignore the beetle's gaudy warning black-and-red colours, it receives a nasty mouthful. It spits out the foul cocktail, but not before sometimes receiving minor burns.

The beetle can fire its poison juice up to 50 times from the highly mobile gun barrel at the tip of its abdomen.

SKUNK'S BIG STINK REPULSES ENEMIES

Any creature foolish enough to torment a skunk will be swiftly knocked back by a powerful, sulphurous stench. With its back towards its enemy and its tail raised, the skunk sprays its noxious fumes from two musk glands at the base of the tail, causing the attacker to retch and sometimes to vomit. If the vapour hits the eyes, temporary blindness can follow. The skunk is accurate to 2 m (7 ft) and the pungent stink can be detected up to 1.6 km (1 mile) away.

Most skunk species will give fair warning of their ability to defend themselves by snarling, stamping their front feet, and swaggering around. This, along with their black-and-white colouring, is enough to put off coyotes and most other attackers.

POISONOUS ENCOUNTER WITH A SHREW

There is enough poison in the saliva of the short-tailed shrew to kill a small mammal in a few seconds. The powerful neurotoxin, mixed with enzymes, comes from glands in the animal's mouth. It seeps through puncture wounds made by the shrew when it bites, and the shrew uses this nerve poison to catch fish, newts, and other small prey.

The North American shrew is only about 10 cm (4 in) long, with a tail of 3 cm (1 1/4 in), but its bite can leave a human in pain for several days.

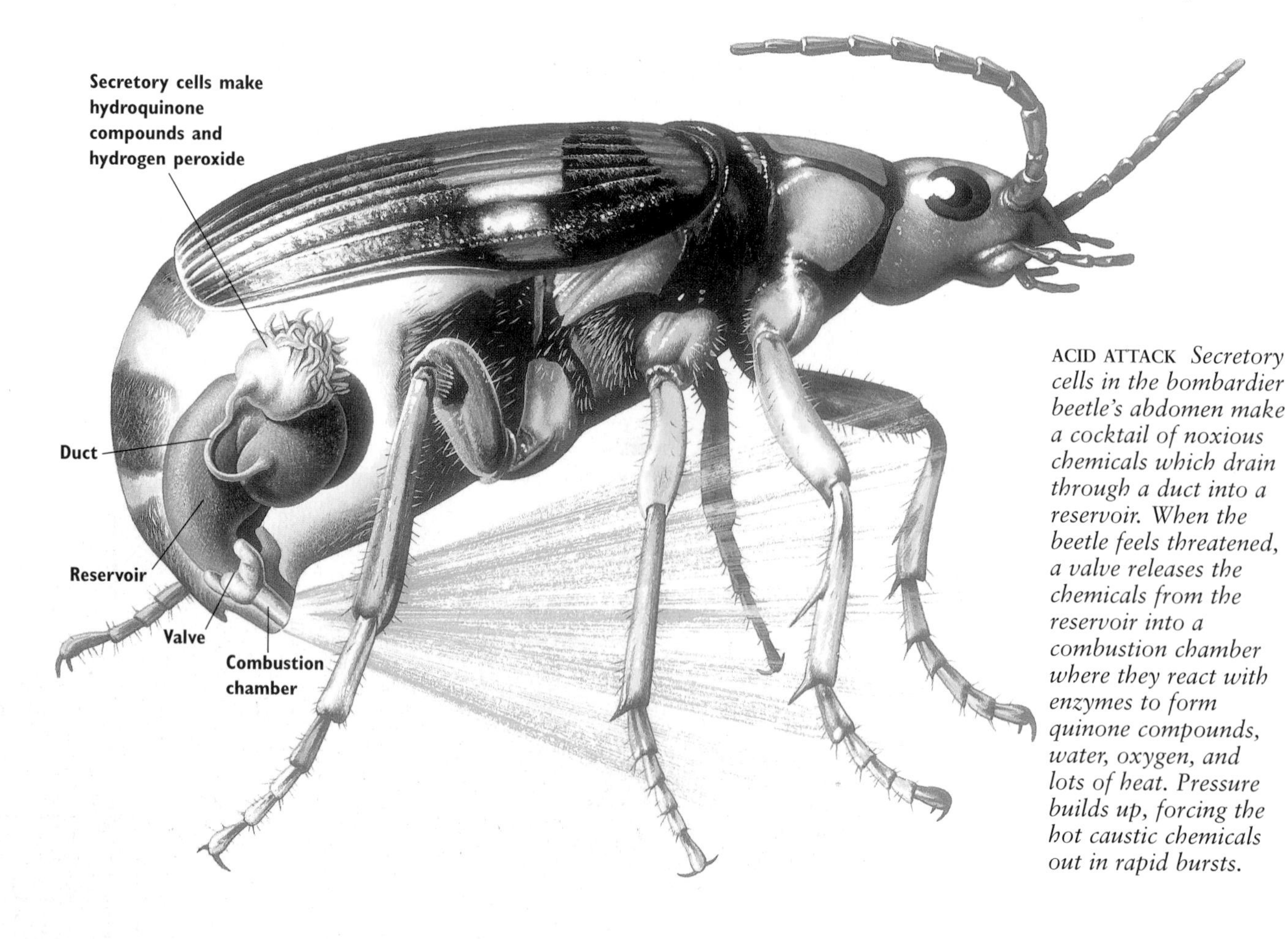

ACID ATTACK *Secretory cells in the bombardier beetle's abdomen make a cocktail of noxious chemicals which drain through a duct into a reservoir. When the beetle feels threatened, a valve releases the chemicals from the reservoir into a combustion chamber where they react with enzymes to form quinone compounds, water, oxygen, and lots of heat. Pressure builds up, forcing the hot caustic chemicals out in rapid bursts.*

LINE OF FIRE *Young cobras can perform this display shortly after hatching and their venom is almost as toxic as an adult's.*

SPITTING COBRA SHOOTS TO BLIND

Africa's most skilled and feared marksman is reluctant to fire, but when it does, it rarely misses. The spitting cobra shoots its jet of venom with deadly accuracy. It spits only in self-defence, preferring to inject venom into prey by biting in the same way as other snakes.

If the spitting cobra feels threatened, it rears up off the ground, spreading its hood in warning. Then, using muscular pressure in its powerful jaws, it pumps venom out of large glands on either side of the head. Fangs enclose the venom canals. The venom shoots out from a tiny hole near the tip of each fang in two squirts in rapid succession, turning into a spray of fine droplets.

Some spitting cobras can project the jet of burning venom over a distance of 2.5 m (8 ft), in an arc of up to 70 cm (2 ft). The venom is so potent that just 1 g (1/32 oz) of dried venom could kill 165 humans. If the venom hits an animal in the eyes the pain is excruciating, and the victim may be blinded.

BLACK WIDOW DELIVERS SUDDEN DEATH

Threatened by a sudden, accidental encounter, the black widow spider uses its deadly weapon of self-defence. Its bite is 15 times more toxic than that of a rattlesnake. The female's bite causes extreme pain, muscular contractions, nausea and, if untreated, death; the male is too small for his bite to harm humans, but its poisonous neurotoxin is effective against small mammals.

Several spiders are called 'black widows' in different parts of the world, but the true black widow is found only in North America. It usually has a leg-span of 2.5 cm (1 in), but can sometimes reach up to 6 cm (2 1/2 in). This passive spider avoids confrontation if it possibly can, lying low in meadows and pastures and slipping into cracks and crevices, out of the light, around human habitation.

Some black widows have been accidentally introduced into other countries by crawling into cargo ships.

FRESH FOOD *The western black widow spider eats only living prey such as the cricket being devoured, right.*

NATURE'S MASTERPIECES

Frog's poisoned chalice

Vivid pink, black-and-yellow, acid green, rich maroon with metallic blue spots: arrow-poison frogs of Central and South America come in a dazzling range of fantastic colours.

The frogs, often carrying tadpoles, travel boldly in the day on the rain-forest floor, protected by their beauty. For the brilliant colours proclaim a warning: look, but do not touch. The gaudy skin of some of these little frogs contains the most powerful known venom in the animal kingdom. Just 0.00001 g (0.000004 oz) is enough to kill a human. Some Native American communities have for centuries used poison, called homobatrachotoxin, extracted from the frogs' skins to dab on the tips of their arrows and blowpipe darts. It is so highly concentrated that there is enough poison in the skin of a single golden arrow-poison frog for 40 arrows. Monkeys, deer, and other small mammals are paralyzed in seconds. The frogs' poison also has other properties, including a painkilling chemical that is more powerful than morphine. The frogs do not succumb to bacterial diseases, since their skins are also thought to contain strong antibiotics.

GRASSHOPPER GETS ITSELF INTO A LATHER

Hissing loudly, the luridly coloured lubber grasshopper foams at the mouth and thorax in self-defence. To dissuade a predator, such as a lizard or bird, from being too curious the grasshopper switches on this astonishing chemical defence mechanism by mixing air with a solution of repellent chemicals.

The foam comprises thousands of tiny bubbles. As these burst, they release a pungent gas which envelopes the grasshopper in a protective chemical cloud of mist.

Should that fail to deter the aggressor, the lubber grasshopper makes one last attempt to defend itself by vomiting a droplet of foul-smelling, chemical-rich fluid from its mouth.

BITTEN TO DEATH *The bulldog ant, a native of Australia and Tasmania, first bites its victim, then stings it with formic acid. Just 30 stings from a bulldog ant can kill a human.*

SLOW LORIS INFANT GETS A LICKING

Leaving a vulnerable baby alone at night seems a foolhardy thing to do, but the slow loris, a nocturnal primate from South-east Asia, takes adequate precautions. At dusk, when it is time to go off in search of food, the mother parks the infant on a branch so that she can hunt more effectively. Before she goes, she licks the baby from head to toe, covering it in toxic saliva. The youngster instinctively stays put, clinging motionless to the branch until its mother returns, but its coating of poisonous saliva is a secondary defence against arboreal predators.

The slow loris lives up to its name, creeping quietly through the branches in the forests of South-east Asia from Bangladesh to Vietnam, Malaysia, Sumatra, Java, and Borneo. It feeds on a variety of insects. Its saliva is toxic enough to induce mild shock in larger mammals.

CHEMICAL COUNTERATTACK *If its brilliant colours fail to warn off a determined attacker, the lubber grasshopper puts its second line of defence into action and oozes a foul-smelling froth from its body.*

DEADLY ANTS ON ACID

The bulldog ant of Australia and Tasmania is the most dangerous ant in the world. It bites and stings at the same time. The powerful jaws slice through its victim's flesh, making an incision into which the ant squirts its stinging formic acid. The poison shoots out from a modified egg-laying structure at the tip of the ant's abdomen, and the sting is as powerful as that of a wasp's.

Bulldog ants are relatives of fire ants. They have excellent vision: the workers of some species literally bound through the air towards human intruders. Like all ants, they use chemicals for communication, orientation, and defence.

CATERPILLARS MURDER THE MESSENGER

Caterpillars of the European pine sawfly use a chemistry trick to silence their enemy. When a wood ant comes across a likely food source, the usual pattern of behaviour is to rush back to the colony to get help from the other ants, leaving a chemical trail to enable them to find their way to the food source. But when a scout finds a cluster of sawfly caterpillars, it is in trouble. The caterpillars dab the ant's head and antennae with a tiny blob of gum that they produce in their guts. The gum is made by mixing pine-needle resin with a chemical that the ants themselves use as a danger signal.

The gum confuses the ant scout so that it cannot find its way back to its nest. Even if it does, the trail it lays is so heavy with warning messages that no other ants will follow it back to the caterpillars.

TERMITES SQUIRT GLUE FROM A HEAD-GUN

Termites have an arsenal of chemical weapons more varied than that of any other animal. To defend themselves against raids by ants, their main enemy, one species, the African *Macrotermes*, injects a wax-like mixture into its attacker. When the termite bites, it also smears an anticoagulant over the ant's wound, so that the ant bleeds to death.

Other termites use a squirt-gun on their heads to shoot toxic glue at the enemy. *Nasutitermes* termite soldiers have an extended snout connected to a special head gland that secretes the glue. When threatened, the termite squirts the resinous substance to entangle the enemy. To avoid poisoning themselves, termites detoxify their chemical weapons using enzymes.

SAFETY IN SIZE

The world's heavyweights fear no predators. One of the simplest methods of avoiding becoming another creature's meal is to be big – really big. Some animals achieve this by working together; as a group they are too 'big' to attack.

MUSK OXEN FALL INTO FORMATION

At up to 410 kg (900 lb) in weight each, and with massive horns, adult musk oxen make formidable opponents. Wolves are the only predators feared by the oxen as they graze on the sparse tundras of Greenland, northern Canada, and Alaska. They face little danger, unlike their calves which are at great risk.

Oxen travel in herds of one or more adult males and several females. They cover long distances in search of food, their thick, layered coats protecting them against the bitter winds and snow blizzards.

As soon as oxen spot a pack of wolves hunting, they stop feeding immediately and huddle together in a defensive ring, standing shoulder to shoulder in a tight circle around the vulnerable calves, facing outwards towards the wolves, horns lowered.

The impenetrable circle is known as a 'phalanx' after the battle formation favoured by the ancient Macedonian infantry. The only chance the wolves have of breaking through to the calves at the centre is to harass the adults until one of them breaks rank. A wolf that misjudges its moment is likely to be gored to death.

BATTLE LINES *Working together, a herd of musk oxen bearing scimitar-like horns presents a formidable defence. The boss on an adult male's forehead can be 10 cm (4 in) thick.*

FEARLESS HEAVYWEIGHT CRUISES UNMOLESTED

With a maximum recorded length of 33 m (110 ft), and weighing 190 tonnes, the blue whale is the largest animal ever to have existed on Earth. Almost no other creature dares attack a healthy full-grown blue whale because of its

sheer size, and so it swims unafraid in oceans throughout the world. Its only real foe is man, who has hunted it to the brink of extinction for its meat and blubber.

A blue whale may live for up to 65 years, reaching an average weight of 100–150 tonnes. A quarter of its body length is taken up by its vast head, and its heart alone can weigh 980 kg (2,000 lb) – as much as a small car.

The blue whale's throat can stretch to hold 5 tonnes or more of food and water when fully expanded. Only by taking so much water into its mouth can it filter out the 4 tonnes of fish and krill it needs to eat a day.

BIG BABY *At 7 m (23 ft) long, even newborn blue whale calves are big enough to have few enemies. Once weaned at eight months they gain up to 90 kg (200 lb) a day.*

DAGGERS DRAWN *The double-wattled cassowary will, if threatened, leap feet-first at its enemy to stab with the long sharp claws that grow on two of its toes.*

THE CASSOWARY'S VICIOUS DAGGER FEET

Any creature bold enough to confront a cassowary risks being stabbed – possibly to death. The flightless double-wattled cassowary is armed with a dagger on each of its feet in the shape of a long, straight spike growing from its modified inner toes.

Normally shy, this massive bird is extremely dangerous when threatened, especially during the breeding season. It will kick at any intruder with its long, strong legs in self-defence, and people have been known to die after a vicious attack.

The cassowary grows up to 1.5 m (5 ft) tall and can weigh more than 55 kg (120 lb). It belongs to the same group as emus. It lives in the jungles of Australia and New Guinea, defending a territory of 1–5 km² (1/3–2 sq miles), and feeds on fruit.

CALVES WATCHED OVER BY FORMIDABLE GUARDIANS

Adult white rhinoceroses are so big and powerful that they are not at risk of attack from anything. But their calves are more vulnerable and for the first three years or so they are small enough to attract the interest of lions and hyenas.

A female will warn a lion away from her offspring with a series of puffing snorts, distress squeals, and mock charges. If need be, she will charge for real. There is at least one report of a female killing a lion that tried to make off with her calf.

When alarmed, a group of white rhinoceroses may press their hindquarters together and face in different directions, presenting their deadly horns to any creature that dares to come too close.

THICK SKIN PROTECTS WHALE SHARK

The whale shark has skin as tough as armour. Not that it needs much protection, because at up to 12 m (40 ft) long, it is the biggest fish in the sea, and almost nothing attacks it. There are reports of whale sharks reaching 18 m (59 ft).

Its skin, like that of other sharks, is covered with tiny, tooth-like denticles, making it hard and tough. Ridges running along its back add to the skin's strength.

This massive, placid creature, with a flattened head and a huge, broad mouth, lives in tropical waters.

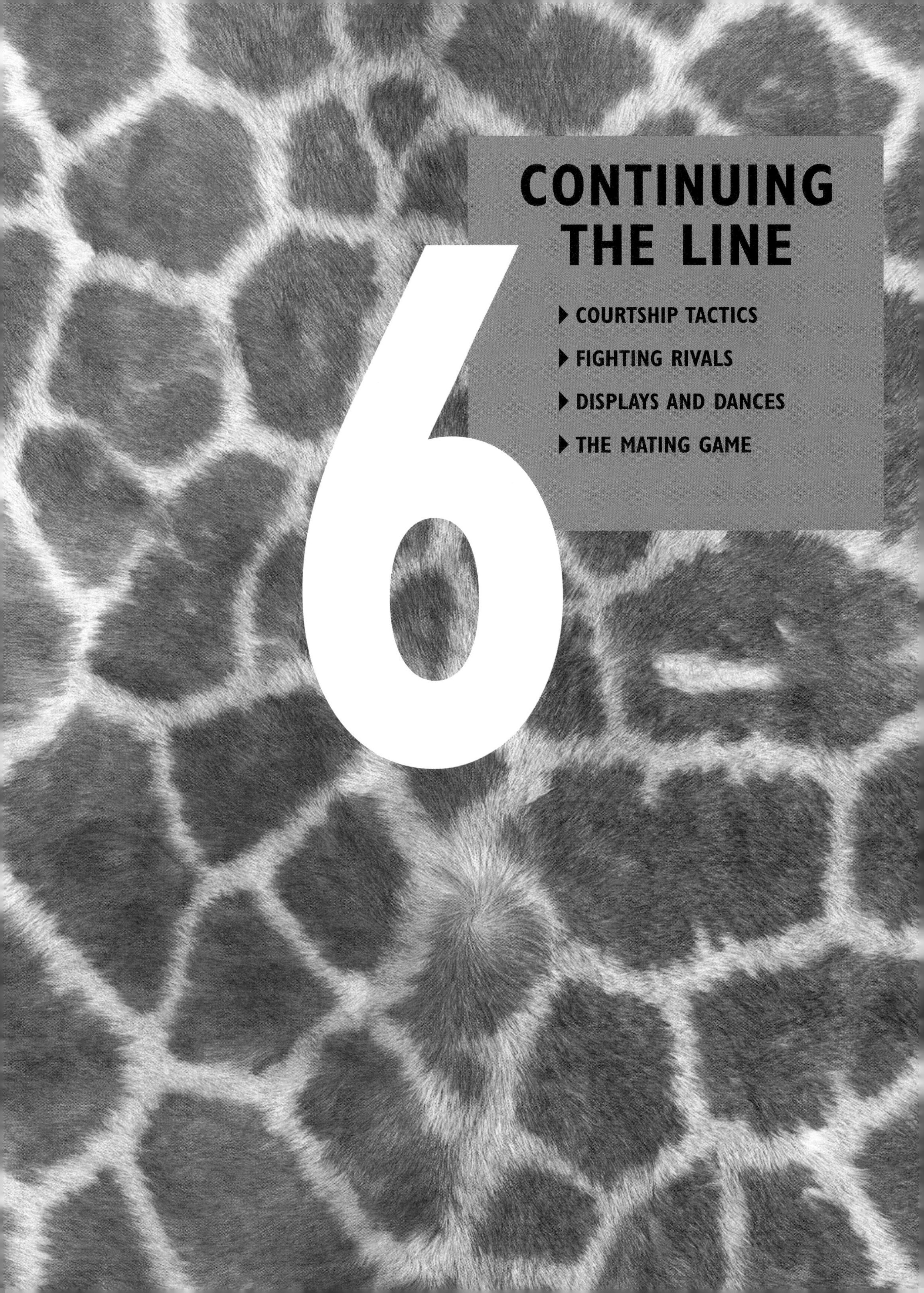

6 CONTINUING THE LINE

- COURTSHIP TACTICS
- FIGHTING RIVALS
- DISPLAYS AND DANCES
- THE MATING GAME

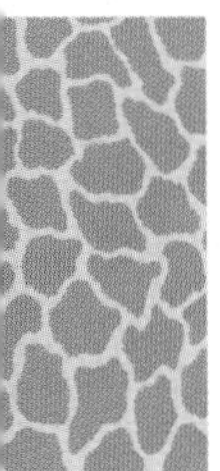

COURTSHIP TACTICS

Most animals can't breed alone; they need the help of a member of the opposite sex. Competition is fierce. To succeed in the mating game, animals have to use every possible colour, sound, and some very strange behaviour.

SCENT THAT IS IMPOSSIBLE TO RESIST

Male red-sided garter snakes begin a frenzied race to mate after just one whiff of a female's sex pheromones. In spring the snakes, which live in grasslands and moist habitats from Canada to Costa Rica, can be seen in tangled heaps as 100 males descend on one female.

Competition is so intense that the males do not even bother to feed after emerging from their six-month sleep. Instead, they wait for sleepy females to appear. The males detect the females' scent by flickering their tongues to 'taste' the air. After mating some males may even imitate the behaviour of receptive females to distract the competition. This annual orgy ensures that all the females are pregnant within half an hour of waking, and the young are born before the winter hibernation.

NO MORE TAKERS *After mating, a male red-sided garter snake leaves a wax-like 'plug' inside the female to prevent other males from mating.*

FIREFLY'S MAGICAL LIGHT-SHOW IN THE MANGROVES

Along the banks of Malaysia's mangrove swamps, trees sparkle as thousands of female *Pyrogaster* fireflies signal to their mates. There may be one creature on almost every leaf of trees stretching 100 m (110 yd) along the water, creating a spectacular effect as they flash about 90 times a minute in synchrony.

The pulsing light in the night attracts males from far and wide. Each of the 130 or so firefly species has its own code of flashes, chemically generated by special light organs in the abdomen. As a male flies by he flashes his 'calling card'. When a female recognizes his signal she flashes back a response and, if it is correct, he drops down to join her.

THE INDIAN MOON MOTH'S FINE SENSE OF SMELL

A male Indian moon moth can detect a female's scent from up to 5 km (3 miles) away. With the most acute sense of smell in nature, he is able to detect a single molecule of the female moth's sex pheromone in the air. To do this he uses huge, feathery antennae that are covered with tiny, sensitive 'chemoreceptors'.

The male flies towards the greatest concentration of scent, following it around obstacles such as trees in the forests of India, China, Malaysia, Indonesia, and Sri Lanka.

The male of the summer fruit tortrix moth has an extra skill: he releases his own pheromone, which blocks other males from picking up the scent of a female he is pursuing.

MATING FROGS MAKE AN AMPHIBIAN WALL OF SOUND

In many tropical and subtropical areas during the mating season frogs and toads may produce a deafening chorus of croaking. The call varies with the species and there is an enormous range. The spotted grass frog of Australia emits simple, brief clicks, while the trills of the great plains toad of North America last several minutes. The male Madagascar tree frog has a repertoire of 28 different calls, rather than the more usual five.

The calls serve to attract females to mate. Many frogs amplify their calls up to 100 times with huge resonating sacs that bulge from the corner of their jaws. The African common toad's call can carry nearly 800 m (875 yd).

CHORUS LINE *To call, the American toad inhales, seals its nostrils and forces air back and forth across its larynx using its bulging vocal sac.*

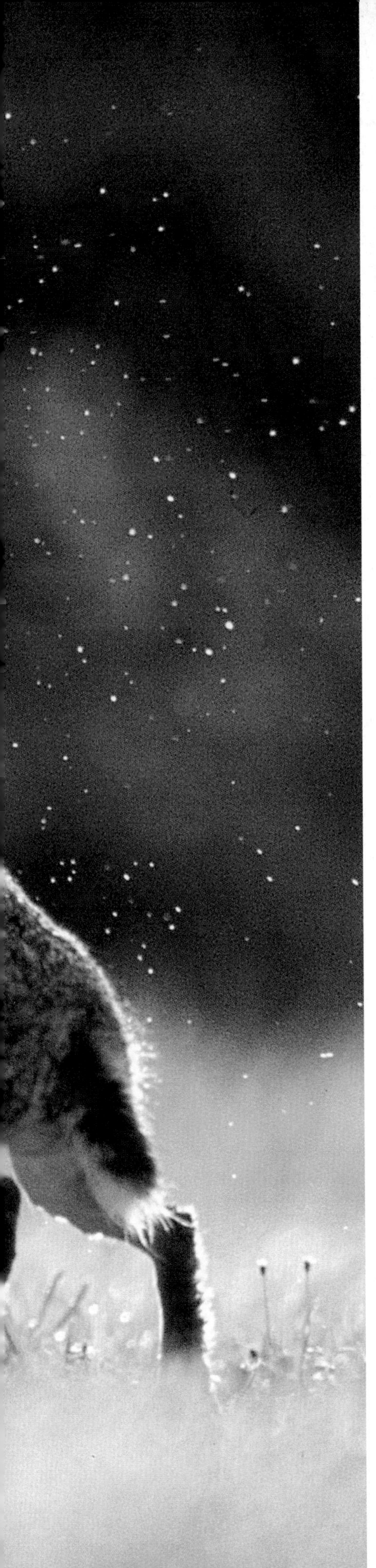

IN PURSUIT *A female Thomson's gazelle in Africa's Masai Mara walks while she mates, with the male following, balanced on two hind legs.*

MATING ANTICS OF THE THOMSON'S GAZELLE

To persuade a female to mate, a male African Thomson's gazelle has to perform a series of rituals. First, he approaches her and displays by stretching his neck out and herding her until she urinates. He sniffs and tastes her urine for chemicals that will tell him whether she is in breeding condition. If she is, he follows her from several metres away, standing tall and prancing with a sort of goose step, sometimes making a sputtering noise.

If the female tries to escape he herds her back. When she comes to a halt, he raises a foreleg, causing her to move forward. When the female is ready to mate, she holds out her tail.

FEMALES FIGHT OFF MAD MARCH HARES

Female hares refuse to be hurried when it comes to sex. An overzealous male may be seen off with a hearty thump. When a male tries to mate with an unwilling female (doe) she rears up on her powerful hind legs and boxes him with her front paws until he gets the message. Later, when the doe is ready to mate, the male can approach her without fear of getting bruised. But that does not stop males from trying their luck as early as possible, every spring.

CHAMPION BOXERS *Female European brown hares make good use of their long hind legs to rear up and box overamorous males.*

COURTING STINK-BUGS SEND GOOD VIBRATIONS

Southern green stink bugs use special vibrations to find each other amid dense vegetation. First, a male attracts a female to his shrub by sending out his scent, but that alone is not enough for her to find him.

Once she has landed in the vicinity, the female stays put and 'sings', using plant stems like telegraph wires. She vibrates her body, sending signals along the stems to communicate her whereabouts to the male.

The male responds to her signals by walking up the plant, touching the stem with his antennae, and calling back. He follows the vibrations until he reaches the female.

This species is found in Asia, Europe, and North and South America.

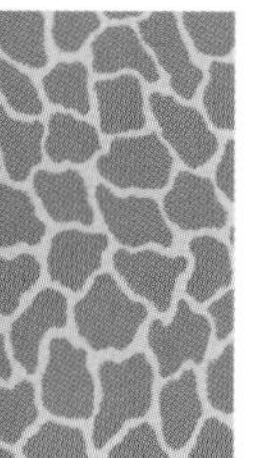

FIGHTING RIVALS

Some females do not choose their mates directly. Instead the males fight among themselves for the privilege. From harmless showing-off to full-blown combat that may end in death, males battle it out for the ultimate prize: fatherhood.

NATURAL ARMOURY OF THE FIGHTING BEETLES

Stag beetles are well armed for any duel: they use their greatly elongated mandibles to wrestle each other for females. The fighting males of Britain's largest beetle interlock their horns in a pincer grip, tussling with each other until the strongest is able to pull his rival upwards and drop him to the ground.

The enormous male Hercules beetle of Central and South America also carries his own weapon, in the form of a conspicuous horn emerging from the back of his 16 cm (6 1/4 in) body. Once he has a hold of his enemy he slams him down on his back, where he is viciously impaled on the horn.

JAW-BREAKER *Of the 900 species of stag beetle in the world,* Lucanus cervus *is the largest in Europe at 6 cm (2 1/2 in). Some tropical species are bigger, with jaws 10 cm (4 in) long.*

JOUSTING ZEBRAS FOLLOW TOURNAMENT RULES

When plains zebra stallions vie for a female their battles follow a set of precise, complex rules. On the savannahs of southern and eastern Africa, the encounter begins with one stallion glaring at his rival, his nostrils flaring, lips curled back to flash his awesome dental weaponry.

If that fails to scare the challenger, the two males circle round, each trying to bite the other's legs. One stallion will suddenly sit down on his haunches to protect his legs. This stage of the battle ends with both zebras squatting, still trying to bite each other. If neither wins this round, they move on to neck-jousting. Here, as at any stage, either stallion can surrender simply by galloping off.

Otherwise, the dispute escalates. Both animals scramble to their feet, rear up and kick hard with sharp hoofs, sometimes ripping bloody strips of flesh from the flanks or head of the opponent. This brawl continues until one of the zebras retreats.

DUEL TACTICS *Male plains zebras in Botswana compete for a female.*

1 *They try to bite each other's neck and flank.*

2 *Powerful back-kicks are aimed at the chest and face.*

CLASH OF THE BISON

An all-out battle between two male bison is one of the most dramatic sights in nature. After stomping around, these heavy beasts, weighing more than 800 kg (1,760 lb), hurl themselves at each other, their skulls crashing together. They grind their horns together, then circle each other, eye-to-eye. If one finds an opening and gores the other in its side, it can inflict a mortal wound. Disagreements occur during the

HEAD-BUTTING CHAMOIS DICES WITH DEATH

Most animals go to great lengths to avoid direct confrontation, but not the chamois goat. Even though it may be balanced precariously on a high mountain ridge, a male chamois will continue to attack a rival, using its strong, crooked horns to deliver potentially lethal blows.

This distant relative of the domestic goat, which lives in European mountains from the Cantabrians to the Caucasus, the High Tatra to the Central Apennines, will only back out when it looks as though it is going to be killed. When this happens, the loser lies flat on the ground, stretching its neck forward in submission.

breeding season and can be settled by threat and submission signals. A bull shows his dominance by bellowing, rolling, and arching his back. Two bulls stand in opposition, swinging their heads from side to side in a ritual 'nod' threat.

A bison can signal his submission by turning his head away or leaving the scene, otherwise the battle escalates.

CHAMPION NECK-WRESTLERS RULE IN THE GIRAFFE HERD

Young bull giraffes in sub-Saharan Africa engage in a ritualized, brutal neck-wrestling match to establish supremacy. The two rivals stand side by side and slowly intertwine their long necks, stretching to more than 4 m (13 ft) off the ground, pushing against each other.

If neither gives in, one giraffe bends his neck outwards and swings it back, like a golfer taking a putt. Then each male takes it in turn to whack his rival on the shoulders and neck with his blunt horns. In serious bouts of 'necking', bulls end up exchanging sledgehammer blows.

NECKING CONTEST *Male giraffes in Africa's Masai Mara fight each other when the hierarchy breaks down or a nomadic male arrives.*

UNDERWATER SKIRMISH *Aggression in the beadlet anemone is triggered by contact with another's tentacles.*

1 *Its swollen, bluish tentacles, full of stinging cells, extend to attack.*

2 *Under fire, the intruder retreats.*

ANEMONES' SLOW-MOTION UNDERWATER BATTLES

Although apparently delicate, sea anemones are effective fighters, armed with barbed stinging cells, which they fire at an intruder. Their underwater battles often pass unnoticed because they happen in slow motion. A 'strike' can take 10 minutes.

When the beadlet anemone, found throughout the eastern Atlantic and in the Mediterranean, senses the presence of another encroaching on its personal space, it reacts with aggression. Its poison-bearing tentacles, or acrorhagi, inflate into angry, bluish swellings. Then, it extends its soft body column upwards before 'hurling' itself towards its rival, ramming the line of swollen acrorhagi into the other's flesh and firing off a volley of potentially lethal stinging cells laden with poison. The fight ends when one of the anemones creeps away, tentacles withdrawn.

SPIRAL DANCE OF THE RIVAL ADDERS

Male adders perform an extraordinary display of gymnastics as a test of their strength. For most of the year these gregarious snakes, found in Europe and northern Asia, live in harmony. It is only in spring, during the mating season, that the males become aggressive. Even in combat, they do not try to hurt each other.

Their bodies entwined, two male adders lift themselves up off the ground, their heads slapping together, weaving and reaching up higher and higher until they are dragged back to the ground by gravity. If the adders are evenly matched, this wrestling can last several minutes (occasionally up to half an hour) and drain the energy out of both of them.

Finally, the out-wrestled male disentangles himself from the loops of the winner's body and races away. The winner heads back to the female.

COCK ROBINS SEE RED IN FIGHT FOR TERRITORY

For a European male robin, the colour red can be provocative enough to incite murder. To him, a flash of red signals another male on his territory. When a cock robin spots a rival male on his patch, he tries to scare him off by puffing up his chest feathers to expose his orange-red breast, face, and throat. If the trespasser does not flee, the owner of the territory attacks. In a full-scale battle, one of the fighters may be dead within a few minutes, its brain exposed by a flurry of vicious pecking. Up to 10 per cent of deaths among robins are thought to be caused in this way. So strong is the stimulus of the red breast that robins will respond to anything red, intruder or not.

RED ALERT *A male robin attacks a decoy. Cock birds may even behave territorially towards other bird species many times their own size.*

DISPLAYS AND DANCES

For females who choose between suitors, it's important to select the best potential father on offer. Males go to great lengths to attract attention. Colourfully attired, they dance, sing beautiful songs, and put on elaborate displays.

MALE FRIGATE BIRDS ARE FULL OF HOT AIR

As male great frigate birds huddle in bushes and trees on islands in the tropics they look like clusters of enormous crimson fruit. They try to attract the attention of females by puffing out their gaudy, bright-red throat pouches.

When a male has inflated his pouch to its full capacity – about the size of a man's head – he begins to vibrate it, clapping his bill noisily and waving his wings in an attempt to impress a passing female.

If he is lucky, one will land near by to take a closer look before making her choice of mate.

ALL PUFFED UP *It takes the male great frigate bird, seen here on Tower Island, Galapagos, several minutes to fill his lungs and blow up his throat pouch like a balloon, until it is drum-tight.*

BOOBY PLAYS THE FOOL FOR HIS MATE

The blue-footed booby's courtship dance is one of nature's most comical sights. The Spanish word *bobo*, 'clown' or 'stupid fellow', is probably the origin of the blue-footed booby's name.

This species, which lives on the western coast of America, has the most extraordinary feet. They are huge, webbed, and bright blue.

The male's courtship ritual involves much parading around, high-stepping, and lifting those enormous feet up and down to show them off to best effect to the female, who is clearly fascinated by them. This performance is accompanied by mutual preening, bill-fencing, and head-tossing. Later, the fantastic feet serve another purpose: the booby rests them over its eggs to keep them warm. The adult birds then support the young on the top of their feet for a month after hatching.

BOOBY SHOW *A male blue-footed booby on Hood Island, Galapagos, flashes his aquamarine feet in a slow high-step to impress a female.*

BIRDS OF PARADISE DRESS UP FOR SONG AND DANCE

Gaudy costumes and ecstatic dances are part of the extravagant show put on by birds of paradise to win female approval. The males are polygamous and attract females one by one, going to extremes to make themselves desirable. They have developed ornate plumage to impress the less vividly clothed females. The male blue bird of paradise, for example, has sapphire wings and 'goggles' of white skin around his eyes. Gauzy blue plumes sprout from his flanks and lower breast, and he spreads these into an azure fan while he hangs upside-down during his display. Two long tail quills wave above him as he trills like a drill, the sound throbbing in time to his pulsating plumes.

The King of Saxony bird of paradise has two long quills, each with a line of sapphire droplets; the superb bird of paradise has a expandable bib of emerald on its chest; and the 12-wired bird of paradise has an inflatable yellow waistcoat. From the magnificent bird of paradise's tail, two quills curve up in wide circles like the ribbon on a wrapped gift. There are golden feathers on his shoulders, and a green shield on his breast.

ATTENTION SEEKERS *Raggiana bird of paradise males show off their plumage as they flit from branch to branch, fluttering their wings.*

THE LYREBIRD'S TALENT FOR IMPERSONATION

Parrot screeches, other birds' songs, dog barks, a chainsaw... there is no end to the sounds that the Australian lyrebird will copy. One male was found to duplicate the calls of 16 other bird species, and the songs are often so loud that they can be heard 800 m (875 yd) away.

To perform, the terrestrial lyrebird builds a stage from vegetation. He stands in the centre, throws his tail plumes over his head like a parasol, and sings loudly. Females travel from court to court to watch. The males may spend half the day displaying.

GREAT CRESTED GREBE'S DAZZLING WATER DANCE

The mating dance of the great crested grebe is one of nature's most intricate *pas de deux*. The dance begins with two prospective partners facing each other and moving their heads rapidly from side to side.

Mated pairs progress to the 'discovery display' in which one bird spreads its wings while the other dives and surfaces with its beak pointing downwards. In the 'retreat display', one partner dashes across the water, away from its mate. The weed dance is the final, most intimate performance. Pressed together, the two grebes tread water and rise up, neck-to-neck, breast-to-breast, each flicking a beakful of weed from side to side.

Great crested grebes, inhabitants of Europe, are monogamous for life and this complicated dance serves to reinforce the bond between the pair.

SALMON GET THEIR TEETH INTO BREEDING

As salmon migrate towards their spawning beds, both sexes undergo dramatic physical changes in preparation to mate. The females' bodies fill with eggs, while the males develop new characteristics and mature into fighting machines.

Soon after leaving salt water, a cock Atlantic salmon's skin changes colour,

ARCHITECTS AND DECORATORS

Thatched huts, lichen-covered bridges, canopies, silvery stages, blue-tinted walls, maypoles, towers, and alleyways carpeted with sparkling jewels – these are just some of the bowerbird's fabulous constructions.

Bowerbirds, from the jungles of New Guinea and Australia, are the plain cousins of birds of paradise. Unlike their extravagantly plumed relatives, male bowerbirds do not advertise for a mate using their plumage. They build bowers instead.

The tooth-billed catbird (or stage-maker) begins by clearing an area of forest floor about 2.5 m (8 ft) wide. He snips off up to 36 leaves of the favoured wild ginger bush and lays them, underside uppermost, on his 'stage'. Then he sings loudly from a nearby tree, hopping down onto his stage to dance for any females that stop by.

Archibold's bowerbird uses a greater variety of carpeting, decorating the avenue he clears with piles of shiny beetles' wings, snails' shells, berries, and bird of paradise feathers. MacGregor's bowerbird builds maypoles out of sticks piled into a pyramid around the trunks of saplings and dances around them.

The golden bowerbird goes one better, building two towers and piling up a rampart of sticks 1.2 m (4 ft) high between them. Having decorated this saddle with tufts of pale lichen, he performs on it.

Perhaps the most gifted architect is the gardener bowerbird, which uses the trunk of a sapling to support the roof of a thatched, conical hut, 1 m (3 ft 3 in) high and 1.5 m (5 ft) across. In front, he lays a lawn of moss, and on this arranges piles of flowers and fruit.

The males' efforts are judged by the females who patrol the forests, examining the buildings. First prize for these promiscuous birds is to mate and the ultimate evolutionary goal of passing on their genes.

IN A BLUE MOOD *The satin bowerbird is particularly fond of blue. As well as shells, butterfly wings, flowers, and feathers, he collects blue pen-lids. He may even paint his walls with blue berries.*

taking on a range of reddish tones. His head elongates, his jaws enlarge, and a hook develops on the lower jaw.

Other species of salmon also undergo conspicuous changes. From a streamlined silvery fish, the breeding sockeye salmon's transformation is most dramatic: his back arches, his head turns green, his body scarlet, and he develops a fierce set of teeth.

When the salmon reach their spawning grounds, males battle over females, shooting through the water at speed to chase away rivals.

COURTING CRANES' WILD DANCE ROUTINES

Ground-living cranes have loud calls and elaborate, highly ritualized nuptial dances to seal their relationships. During courtship wattled cranes give a shrill call, in unison, each bird coiling its head over its back and then stretching its head and bill upwards. They dance wildly, jumping high into the air, bowing, tossing grass up, and calling.

Blue cranes also have an elaborate dance, running in circles, jumping with their wings flapping, and calling, while grey crowned cranes toss objects into the air. Cranes are inhabitants of North America, Europe, Asia, Africa, and Australia. They pair for life, and the calls and dances reinforce their bonds.

BLACK-CLAD WIDOW BIRD SURPRISES WITH A DANCE

The male Jackson's widow bird introduces himself to a likely female by suddenly jumping up from behind tall grasses. For a short while he indulges in an energetic, swooping, jumpy flight, dragging his long black tail behind him like a bridal train, before dropping down out of sight. The male's dance is designed to catch the eye of a female so he can lead her to his nest. An inhabitant of the plains of southern and eastern Africa, this thrush-sized bird's tail feathers grow to 20 cm (8 in) long.

THE DANCING STICKLEBACK'S RED HEART

A flash of red zigzagging through the water instantly arouses the curiosity of a female stickleback. When a male is ready to spawn, his belly, chest, and mouth lining turn crimson and he displays this to the female in a jerky dance.

Most fish shed sperm and eggs into the water and rely on the currents to bring them together, but the three-spined stickleback, a European freshwater fish 8 cm (3¹/4 in) long, puts a bit more effort into protecting its young. In the breeding season, the male builds a nest by digging a hollow in the mud and erecting a dome-shaped arch from weeds and mucus. After one female has been persuaded to lay her eggs in his nest, the male waits for another to come by and do the same. Then he guards the nest, constantly fanning oxygen-rich water over the eggs to aid development.

BIG HONKERS WIN THE MATING GAME

Female hammerhead bats are attracted by the deafening, metallic trumpeting of the males. The louder a male is, the more irresistible the females find him. This has led to the evolution of a massive larynx, which fills most of the male hammerhead bat's chest. Air passages in his nose, which amplify the call, help him to out-honk his neighbours as they hang upside-down over waterways in the tropical forests of West Africa.

The females fly back and forth, listening. Should one hover in front of a male honker, he ups his call rate. Finally she makes her choice of mate.

1 *When a female stickleback swims by, her belly swollen with eggs, the male leaps into action, zigzagging repeatedly towards her.*
2 *To signal her readiness to spawn, the female assumes a tilted-up position near the surface of the water.*
3 *The male then leads her down towards the nest, indicating the entrance with his snout.*
4 *The male turns on his side to display his red belly.*
5 *If the female will not spawn, the male may become frustrated and attack her, nipping her flanks.*

PARTNERS FOR LIFE *Albatrosses on South Georgia usually pair for life and 'divorce' only if they fail to breed. They can live for 80 years.*

6 *The female enters the nest, where the male prods her belly to stimulate her to shed her eggs (up to 100).*
7 *The female leaves the nest and the male goes in to release his sperm.*
8 *The male then tends the eggs. His movements help to oxygenate the water around them as they develop.*

WANDERING ALBATROSSES TAKE THEIR TIME

The wandering albatross has the longest courtship ritual of all birds, sometimes lasting eight weeks. Two albatrosses, standing face-to-face with wings outstretched, bow to each other while vibrating their bills. They clap their bills in a loud rattle, throw their heads back, exposing long white necks, then rub their bills together in the equivalent of an avian kiss. With wings still outstretched, they then circle around each other to complete the courtship.

ALLIGATORS MAKE WAVES

American alligators, native to the southern US, make waves with their bodies to attract females. Lying just below the surface of the water, a male alligator will vibrate his body to create concentric waves. In the breeding season, this may be followed by loud roaring. During courtship they also 'head-slap' the water, clapping their upper jaw on the lower jaw to make a loud crack.

MAYFLIES' ACROBATIC DANCE OF DEATH

In May, dense, shimmering clouds of mayflies dance above the surface of ponds and rivers. The insects, living in freshwater habitats worldwide, perform stunning aerobatics to attract a mate.

Mayflies spend most of their lives as aquatic larvae. After two or three years, they float to the surface of the water to breed and die in just one day. First they moult into a pre-adult, then minutes later into an adult, with 5 cm (2 in) long wings. When a female rises to join the swarm above the water, a male seizes her. They mate and the female returns to the water to lay her eggs. Then, having moulted, danced, and bred in a day, the mayflies die.

PEACOCK'S DAZZLING DISPLAY

Common peafowl originate from India and South-east Asia but the spectacular displays of the cock bird have long been a familiar sight in public parks and gardens in North America and Europe. However, there is more to the male's ornamentation than just mere decoration.

A peahen assesses her mate entirely on the size and quality of his tail. When the quills are fully fanned out, a male may show off up to 200 eye-catching blue-green spots. The deciding factor for females is how numerous and finely defined these spots are.

Each breeding season the male struts around his territory, waiting for females to admire his show. An attractive male has his pick of females and may mate with several in a season. This gives him an enormous reproductive advantage. But why do peahens choose males with big tails? One theory is that males with the largest tails are perceived as being the fittest for they are able to survive in spite of such a great physical hindrance.

THE MATING GAME

From hermaphrodites to sex changes, sperm-filled heads to child brides, the natural world comprises the most weird and wonderful sexual practices. In some species, males are literally devoured by their partners while mating.

SPECIAL DELIVERY ENSURES BETTER SURVIVAL RATE

Male cephalopods wrap their sperm up into neat little parcels and deliver them directly. These animals, including octopus, squid, and cuttlefish, carry the packages on a specially modified arm, the hectocotylus.

In the breeding season, a male octopus uses the arm to reach into his body and scoop out a mass of sperm parcels. During mating, the male pushes the parcels through the female's breathing tube, and into her genital duct. As the packets absorb water, a spring-like ejaculatory device in each uncoils and the sperm spill out over the female's eggs.

TOUCH AND FEEL *A male giant cuttlefish in the Indo-Pacific uses his specially modified arm first to stroke his mate and then to deliver sperm.*

DUSKY SLUG CHANGES GENDER WITH AGE

The dusky slug, which lives in Europe and North America, begins life as a male and turns into a female as it develops. Being a male when small makes it easier to deliver sperm, and being a female when fully grown helps the slug to fight other slugs for precious egg-laying sites.

Many molluscs are hermaphrodites – where an individual possesses both male and female organs. When two such individuals mate, both will produce a batch of fertilized eggs.

WHY CLOWNFISH CHANGE SEX TO REACH THE TOP

In a community of clownfish, which live on Australia's Great Barrier Reef, only the dominant pair are sexually active. If the dominant male dies, one of the sub-adults will take his place, but if the female dies, her mate – the dominant male – changes sex and becomes a female. Then a new sexually active male joins the new female at the top.

Among the blueheaded wrasse of the coral reefs in the Atlantic, should a dominant male die, then the largest of the dull-coloured females undergoes a magnificent transformation into a male.

HOW MALE SHARKS FLUSH OUT THE COMPETITION

Shark courtship is a complex manoeuvre requiring the male to 'dock' alongside the female. A male shark has two sexual organs to enable him to fertilize a female whether he docks on the left or the right. These 'claspers' are grooved, rod-like structures, which he inserts into the female. Using a muscle-driven hydraulic system in his abdominal wall, the sperm is transferred in a stream of water, which at the same time flushes sperm from previous matings out of the female.

MALE MANTIS IS ON FEMALE'S MENU

For praying mantis males, sex is a dangerous activity, since the females tend to eat them during copulation. However, the threat of death does not damage a male's sex drive. In fact, once the male's head has been chewed off, a nervous reflex causes him to copulate more vigorously. Only if a male approaches the female extremely carefully from behind, out of reach of her grasping forelegs, and then makes a hasty retreat, can he avoid being eaten.

MATE FOR A MEAL *Mantis females may eat their mates to supplement the massive protein intake needed for making up to 1,200 eggs.*

BABOON TAKES CHILD BRIDES FOR HIS HAREM

Male hamadryas baboons kidnap young females from their mothers to add to their personal breeding harems. These large monkeys live in bands of up to 70 individuals in the deserts of the Horn of Africa. Family units, headed by dominant males, are usually small, making it easier for aggressive males to snatch juvenile females. These males build their harems with one female at a time, creating new breeding units. Some males can dominate and control up to ten females.

The females in the harem are not closely related, so there is a lack of kin support. As a result, the dominant male gets away with treating them harshly; if a female strays, she is punished with a hard bite to the neck.

CAUTIOUS COURTSHIP OF THE MALE SPIDER

When a male web-spinning spider approaches a female, he carefully identifies himself to avoid being eaten. Otherwise, the larger, predatory female, found mainly in shrubs worldwide, might mistake the male for something to eat and dispense of him with one swift bite of her poisonous fangs. On reaching the web, the male plucks at a silk thread with his leg, sending the resident female a rhythmic, long-distance signal. The characteristic vibrations tell her that he is a mate and not a victim struggling in her web.

The mating that follows this cautious courtship may be brief. The male will have already spun a small silken napkin, deposited a drop of sperm on it and sucked it up with his palps, the feeler-like organs on his head. Mating involves poking one of these palps into the female's genital opening and squirting in the sperm. As soon as the male has done his job, he scuttles off before the female's passion turns to hunger.

Male nursery web spiders try wooing the females with a nuptial gift of an insect wrapped in a thick cocoon of fine silk. A chemical message in the silk may act like a name badge, helping her to identify the male as a mate. The gift packaging may also arouse the female and possibly distract her for long enough for the male to sneak in and inseminate her.

SPIDER'S GIFT *A male nursery web spider hands over his gift. He produces sperm instinctively and not always in the presence of a female.*

LORDING IT *Hamadryas baboons live in large bands made up of several family units. Each family unit is controlled by a single male.*

SWALLOWED SPERM IMPROVES CHANCES OF FERTILIZATION

One species of South American catfish has an unusual mating technique whereby the female drinks the male's sperm. When fish release their eggs and sperm into the water for external fertilization, they face a common problem: turbulent water washes many away. By attaching her mouth to the male's genital area, the female catfish ensures that fewer sperm are wasted.

While attached, the female will simultaneously release her eggs into a pouch she forms with her ventral fins. The female's gill covers remain tightly shut as though she is drinking. She swallows the sperm, with a little water, and passes it quickly through her intestine and out with the eggs.

THE DECORATED CRICKET'S TASTY GIFT

The desert-living male decorated cricket courts his mate by offering her a delicious meal. As the male mates, he transfers two things to the female: a capsule full of sperm and a large blob of gelatinous substance known as spermatophylax. The female eats the spermatophylax after mating. In other cricket species living in arid environments, this substance is known to be highly nutritious and it provides the female with the energy to produce more eggs. But the males of the decorated cricket are more miserly. The spermatophylax that they hand over is nutritionally valueless.

In spite of the sham, female decorated crickets seem content with the offering. It is believed that even though the gift does not contain many nutrients, it provides the female with a much-needed drink. Also, while she is occupied devouring the tasty snack she does not go off and mate with any of the male's rivals.

WINNER COMES LAST IN DUNG FLIES' RACE TO MATE

In the race to mate, male dung flies worldwide aim to be last in line. This is because the last male to mate with a female before she lays her eggs will displace the sperm from previous matings and fertilize about 80 per cent of the eggs.

Males hang about on fresh, wet dung where females lay their eggs. When a female arrives several males may try to mount her at the same time, each battling frenziedly to mate, clinging on to her with grim determination and fighting off rivals. It is a frantic struggle that the female could do without, because she needs to lay her eggs as quickly as possible, before the dung cools and is too hard for her to lay in. Also, the males can sometimes be so enthusiastic that they trample her into the dung.

Some males lie in wait at older dung piles, where there may be fewer females, but also fewer rivals, while others take up watch on newer dung, where competition is more intense.

PRIME SITE *Dung flies waste no time locating fresh droppings. They are on the scene quickly to capitalize on this temporary resource.*

VELVET WORMS APPROACH REPRODUCTION HEAD-ON

When a male of an African species of velvet worm is ready to mate he sticks a bundle of sperm onto any part of a female's body. The sperm are absorbed through the skin and swim to her ovary.

Velvet worms are also found in Australia, where some have bumps and syringe-like shapes on their heads. To mate, a male scoops sperm from his genital opening and carries it on his head until he meets a female. To deliver his sperm, he pokes his head into her reproductive canal (cloaca).

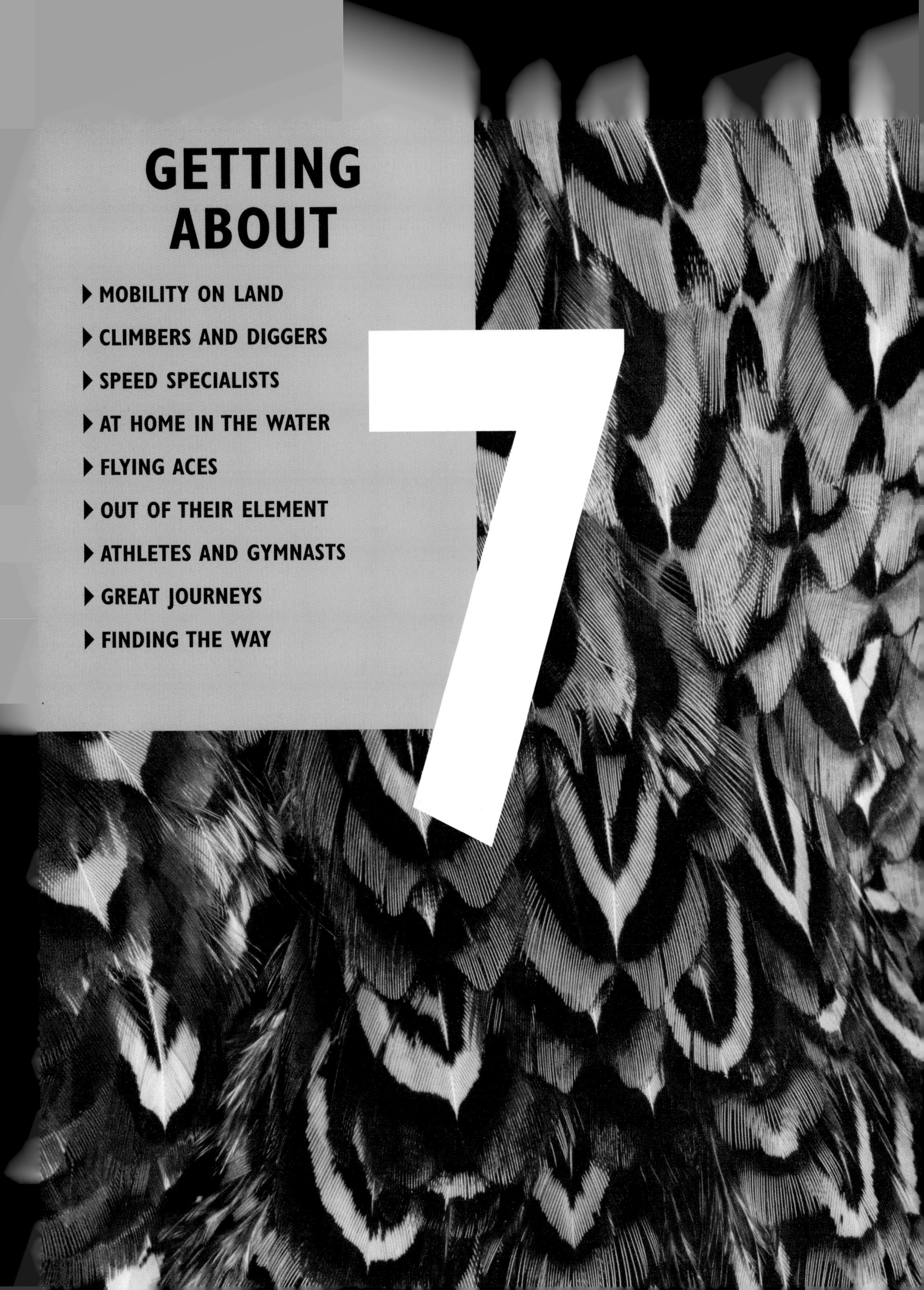

GETTING ABOUT

- MOBILITY ON LAND
- CLIMBERS AND DIGGERS
- SPEED SPECIALISTS
- AT HOME IN THE WATER
- FLYING ACES
- OUT OF THEIR ELEMENT
- ATHLETES AND GYMNASTS
- GREAT JOURNEYS
- FINDING THE WAY

7

MOBILITY ON LAND

Over millions of years, animals have evolved the means to move around, find food, escape from enemies, and seek a mate. On land, variations in body size and shape allow animals to crawl, run, or hop on legs, and even slide along on slime.

BIG GREY BOOMER IS CHAMPION JUMPER

When in a hurry, the Australian eastern grey kangaroo hops in powerful bounds, using its muscular, elongated hind limbs. Its much shorter forelimbs are held against its chest.

The creature's large, heavy tail acts as a counterbalance when this, the largest of kangaroos, is moving at speed – which in flat terrain can reach 50 km/h (30 mph). Horizontal leaps are generally about 7.5 m (25 ft) long, although the record distance is 14 m (46 ft). When bounding, an adult grey kangaroo, known as a 'boomer', rarely jumps higher than 1.5 m (5 ft), but when pursued by dingoes or human hunters it can leap over obstacles, such as farm fences, twice that height. One mature male, when being chased by farm dogs, cleared a pile of wood about 4 m (13 ft) high.

KANGAROO HOP *The grey kangaroo has long, muscular legs with tendons that act like springs. Hopping uses less energy than running on four legs.*

THE CRAB THAT CLIMBS TREES TO FEED

Several species of crab live on land rather than in the sea, and some even climb trees. The sharp tips to their legs enable them to grip and scale rough tree trunks. The most powerful climber is the robber, or coconut, crab of islands in the south-western Pacific and Indian oceans.

The largest specimens of this giant shell-less relative of the hermit crab have a 1 m (3 ft 3 in) leg-span and weigh up to 5 kg (11 lb). At night they emerge from burrows to feed on carrion, vegetation, and fallen coconuts. They climb sago palms to eat their soft fruit, and to escape from predators. The name robber crab is due to their habit of stealing shiny objects.

CLAMBERING CRAB *The robber crab escapes danger by climbing a coconut palm. When leaving the tree, it climbs down backwards.*

HIGH-JUMPING FLEA HAS A BUMPER STORE OF ENERGY

The humble flea is a world record breaker. From a standing start it can jump up to 130 times its own length – the equivalent of a person leaping over Paris's Eiffel Tower with 60 m (200 ft) to spare. And it can do this 500 or more times an hour almost nonstop for three or four days.

Such feats are possible because the flea's thorax contains an arch made of resilin, a protein which, when compressed, stores a huge amount of energy. At the point of takeoff, segments in the flea's thorax uncouple and the energy is released, thrusting the flea into the air. Extra boost is provided when the rear legs recoil against the ground. With this mechanism, a cat flea 3 mm (1/8 in) long can leap 34 cm (13 1/2 in) high.

LONG-DISTANCE WALKER TRAVELS IN GANGS

The Australian emu is a flightless bird capable of walking great distances in its nomadic life. With its powerful legs it moves at a steady 7 km/h (4 mph), accelerating away

HOW SNAKES MOVE

A snake can move in several ways, depending on speed and the surface over which it is travelling. The most common is **lateral undulation,** whereby the muscles of its back contract sequentially, causing S-shaped waves of bending to move along its body from head to tail. The head and neck set the direction and the bends follow their track like carriages behind a railway engine. Sea snakes have oar-shaped tails to help to propel them in water.

An alternative method is **sidewinding,** seen in Africa's horned viper and the sidewinder of North American deserts. Its body rolls sideways in a series of arcs along the ground, resembling a rolling spring.

For **concertina** movement, the snake alternately pulls up its body into bends then straightens it forwards – a method used by large puff adders in tunnels, and by tree boas when climbing.

Rectilinear locomotion is simply moving in a straight line – the preferred method of constricting snakes, such as the pythons of Africa and tropical Asia. At several points along the snake's underside, the belly scales are alternately lifted from the ground and pulled forwards and then pushed downwards and backwards. The scales dig into the ground, causing the snake to move forwards.

Although snakes look fast as they slither through the undergrowth or over the sand, they are actually remarkably slow. The rattlesnakes of North America normally progress at a modest 3 km/h (2 mph). The world's fastest snake, the black mamba from Africa, was once seen to chase a man at 11 km/h (7 mph).

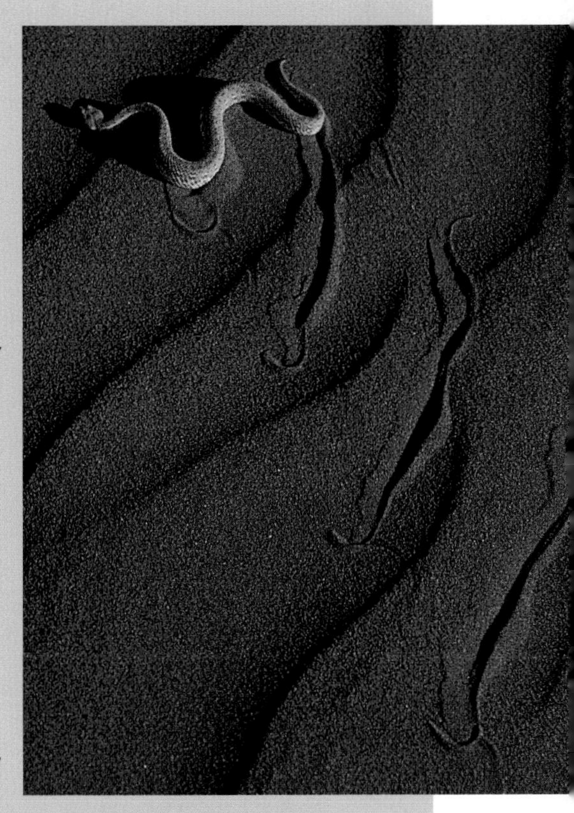

MAKING TRACKS *Sidewinding enables desert snakes to move relatively quickly while making only minimal contact with the hot sand.*

Lateral undulation (above) is the method most snakes use to travel. Bunching the body up then pushing the head out forwards is known as concertina movement (left). Rectilinear locomotion (below) is favoured by large snakes that kill by constriction.

from danger, such as dingoes, at 48 km/h (30 mph).

The emu travels in family groups or 100-strong gangs for hundreds of kilometres, following the rains that herald fresh vegetation and insects.

The 1.9 m (6 ft 3 in) tall bird has wings just 20 cm (8 in) long. They are raised in hot weather to expose bare skin, served by a network of blood vessels from which the emu loses heat.

HOW THE MILLIPEDE NEVER PUTS A FOOT WRONG

The millipede does not have a thousand legs, as its name suggests, but it does have many more than any other animal. Some giant millipedes have 400 legs and movement has to be carefully coordinated. There are two pairs of legs in each segment. At any given moment, most of a millipede's legs are in contact with the ground, but a series of waves of about 22 pairs of raised legs travels from front to back causing the animal to move forwards. The faster they move, the more legs are raised per wave.

Millipedes are found worldwide. They are generally slow-moving creatures, spending most of their time burrowing through leaf litter and soil.

WADDLING WAITERS AND TOBOGGANING EMPERORS

Penguins are well adapted for a life at sea, but out of the water some are capable of travelling great distances. While all species stand upright on their short legs and waddle, penguins such as the Adélie, chinstrap, and emperor can proceed more rapidly over ice and snow by tobogganing on their bellies, using their feet and wings for propulsion.

Adopting this method of transport the emperor penguin, moving at 4.5 km/h (3 mph), can cover great distances. It may travel up to 200 km (125 miles) across stable sea ice on its journey between the sea and its inhospitable Antarctic breeding sites. By tobogganing it uses much less energy than if it had walked – vitally important in the hostile subzero temperatures of the Antarctic.

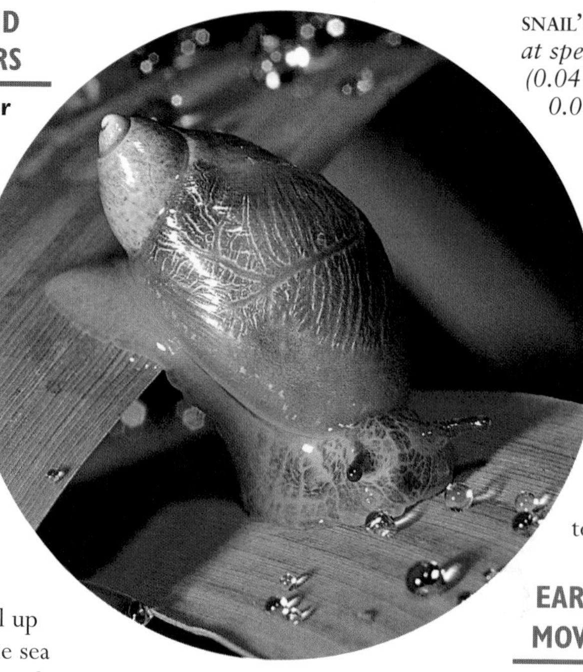

SNAIL'S PACE *Land-living snails travel at speeds ranging from 0.013 m/sec (0.04 ft/sec) to a very slow 0.0028 m/sec (0.0092 ft/sec).*

CREATURES THAT SLIDE ON PATHS OF SLIME

Slugs and snails glide very slowly over the ground using a single, large, creeping foot. The lower surface of the foot moves in a series of tiny waves produced by the muscles of the foot that pass from the back to the front. The crests of the waves are directed to the rear, so the slug or snail is propelled forwards. To help their progress they lay down a path of slime, the silvery trail seen on garden paths and plants, secreted from a gland behind the mouth. Slugs and snails are active in moist conditions at night, foraging on vegetation. In the day they usually return to the same protected spot.

EARTHWORM MAKES WAVES MOVING THROUGH THE SOIL

The earthworm bores its way through the soil. This fluid-filled, segmented creature, surrounded by muscles, moves by producing waves of contraction and relaxation which pass along the length of its body.

Movement through the soil is aided by a covering of sticky slime and eight tiny hairs that protrude from each segment and give extra grip on the sides of its burrow.

SNOWMOBILES *Emperor penguins toboggan across the ice from the sea to their nesting rookery at Kloa Point on the Antarctic mainland.*

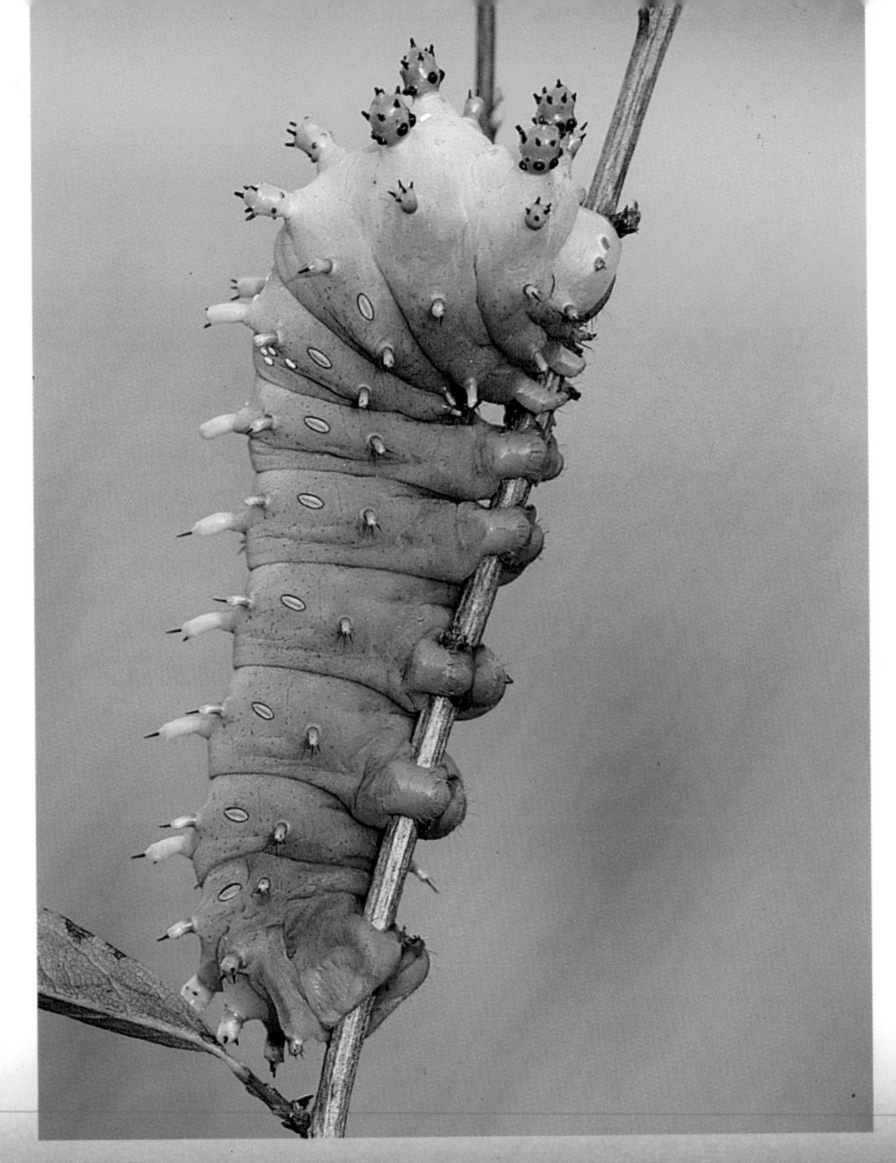

STALK WALKER *The North American cecropia moth caterpillar grasps a twig with its walking legs, using its sucker-like prolegs to support itself.*

HOW CATERPILLARS MOVE ON TRUE LEGS AND FALSE FEET

The caterpillars of moths and butterflies have true legs at the front and false feet at the back. There are three pairs of true legs on the thorax (the middle section of the body), each with a single claw. Although known also as 'walking' legs, these are mainly used for holding twigs and leaves. The six pairs of fleshy false feet, or prolegs, are located on the abdomen. Each foot is like a small suction cup with a ring of hooks around the lip. Using these appendages to attach to a stem or leaf, caterpillars move in a rippling fashion.

Muscles in the rear segments contract and blood is pushed into the forward segments. The true legs hold onto the forward position while the rear false feet disengage and the front muscles contract, pulling the rear segments forward. Some caterpillars spin a carpet of silk to help them to move over smooth surfaces.

NATURE'S MASTERPIECES

Powerful legs, giant leaps

The frog's short, tailless body and long, powerful hind legs make it a great leaper. The American bullfrog, found in the west of North America, can leap nine times its own body length, and the smaller sharp-nosed frog of South Africa, 40 times its own length – the equivalent of an Olympic long-jump champion leaping nearly 75 m (82 yd).

At rest, the frog sits on its haunches, its hind legs gathered up and its forelegs extended. When the frog needs to move, it prepares to jump. Its forelegs raise the front of its body and its ankles lift its rear legs. At takeoff, the hind limbs straighten and push against the ground, propelling the frog upwards and forwards at an angle of 45 degrees. Its eyes sink into its head for protection. On landing, the forelegs and chest take the impact and the rest of the body follows. The legs are gathered up and the eyes return to their normal position.

Not all amphibians can jump. The European common toad walks on all fours like many other vertebrates.

CLIMBERS AND DIGGERS

There are two ways of finding food while keeping clear of terrestrial predators: shinning up a tree or excavating a burrow. Many animals have adaptations such as prehensile tails or greatly enlarged feet to enable them to climb or dig.

LEAP IN THE DARK CATCHES INSECTS UNAWARES

Leaping between trees and branches, bush babies can cover 10 m (33 ft) in 5 seconds or less. With their well-developed hind limbs, large forward-facing eyes, and long tail for balance, they climb and leap with ease. Found mainly in the equatorial rain forests of West Africa, the bush baby uses its agility to move through the forest canopy at night in search of the insects on which it feeds.

When moving rapidly, it disturbs prey and is able to locate its position using its large, sensitive ears. Keeping its feet firmly fixed to the branch, it stretches out its entire body and snatches the insect with its hands. Such is the accuracy and speed of movement, that grasshoppers and moths may be caught on the wing.

AGILE LEAPERS *Pressing its delicate ears against its head and pulling its arms and legs towards its body, the lesser bush baby can leap 6 m (20 ft).*

MOLE RAT OBEYS THE RULES OF THE HIVE

The naked mole rat is a burrowing mammal that behaves very much like a bee. It lives in underground colonies that are dominated by a queen.

The queen is the only member of the colony to give birth to young. The other mole rats are her subordinates, undertaking tasks such as looking after the queen's young, finding food, and digging tunnels, a bit like worker bees.

Excavations are carried out in a regimented way. A mole rat at the front digs away the soil with his huge front teeth, while the others work like a conveyor belt to carry the soil away. The second in line gathers the loose soil then moves backwards through the tunnel, pushing the soil behind it. It passes underneath the other mole rats that are moving forwards to pick up their loads. At the end of the line, the load is kicked outside. The transporter then rejoins the line and scrambles back to the front of the queue.

JUNGLE ACROBAT *Spider monkeys can dangle by their tails from the slender tips of branches and so avoid heavier tree-climbing predators.*

MONKEY'S HELPING HAND

The spider monkey uses its long, prehensile tail like a fifth limb, to help it to clamber through trees. The tail is specially adapted for grasping and is used to grab branches in the tropical rain forests of South America where they live. The monkeys hang from the tail like a safety line to reach food or to lower themselves into muddy clay licks. The tip is as sensitive as the palm of a hand, so the animal is able to pick up food with its tail.

UPSIDE-DOWN SLOTH LIVES LIFE IN THE SLOW LANE

The slow-moving sloth of South America is completely adapted to an upside-down life. Even its fur grows down, with a parting along its stomach. Two-toed sloths move less than 40 m (44 yd) each day. They might remain in the same tree to feed for two consecutive days. Three-toed sloths are more active, and may change trees twice a day.

Once a week the sloth carefully crawls down to the base of the tree it is inhabiting to defecate. In this way valuable nutrients are recycled back to the tree.

SPEED SPECIALISTS

For a predator, acceleration is one way to catch prey, but prey can also use speed to get out of trouble. Fast-moving predators and prey try constantly to outdo each other in a race that means life or death.

THE FASTEST THING ON FOUR LEGS

Small-headed, long-legged and built for speed, the cheetah can accelerate from a standing start to 72 km/h (45 mph) in three seconds. Although it can quickly reach a top speed of 97 km/h (60 mph), the cheetah has no stamina. The chase must be short. An average pursuit is over a distance of 170 m (185 yd) and lasts no more than 20 seconds.

With such powerful muscles and no surplus weight, the cheetah – which lives in Africa – generates enormous amounts of heat. During a 200 m (220 yd) dash, its temperature can rise to 41°C (105°F), a level certain to cause brain damage if sustained for more than a minute or two.

To compensate, the cat adopts one of two hunting strategies. One way is to amble nonchalantly towards its prey, often a young Thomson's gazelle, freezing each time the target looks up. The other is to use stealth, stalking unseen through the undergrowth to within 50 m (55 yd), then taking off on its high-speed chase. Unlike other cats, the cheetah's claws are permanently exposed, giving it the equivalent of running spikes.

The prey takes flight and follows a zigzag course, but the cheetah anticipates its escape route and runs as straight as it can. When the moment is right it puts on a last burst of speed. It swipes the gazelle's back legs, straddles the struggling body, and suffocates it with a bite to the neck.

The cheetah will not feed immediately. It rests and pants in order to cool down. Then it bolts its food lest a leopard, lion, or hyena should chance by and steal its kill. The cheetah avoids competition with nocturnal predators by hunting during the day, at dawn or dusk.

BUILT FOR SPEED *The African cheetah has an exceptionally flexible backbone that enables it to increase its stride when running very fast.*

SPRINTING BIRD *The roadrunner is a member of the cuckoo family. It was popularized in a Warner Brothers cartoon because of its clownish gait.*

SWIFT HUNTER OF THE CHAPARRAL

The little roadrunner of North America is the cheetah of the bird world. This 50–60 cm (20–24 in) long hunter can run at speeds of more than 42 km/h (25 mph). It lives in the dry desert chaparral of the south-west, where it feeds on lizards, snakes, and insects. It is able to fly, but pursues its prey on the ground. It runs with its neck pushed forwards, its wings partly open to act as stabilisers, its legs going 12 steps a second, and its tail used as a rudder that can turn the bird through 90 degrees without it slowing down.

FAST-MOVING SUN SPIDERS

Desert-living sun spiders are the fastest animals on eight legs. They move in short bursts of speed, which can exceed 16 km/h (10 mph). Although usually active at night, an individual caught out in the day will run between shadows.

Sun spiders appear to have ten legs, but in fact they only have eight. The first pair are modified appendages, or pedipalps, that act as 'feelers' for killing the insects on which they feed. The legs of the largest species may span up to 15 cm (6 in).

WINNING WAYS OF THE OCEAN'S FASTEST FISH

The sailfish, a 2.4 m (8 ft) long speedster, has been clocked at 110 km/h (68 mph). When in pursuit of high-speed prey, such as tuna, mackerel, and squid, the fish flattens its enormous bright fan-like dorsal fin against its back and holds its pectoral fins against its sides. This streamlines its body into a living torpedo. Sailfish live in tropical and temperate waters worldwide. They sometimes hunt in groups, and by raising their dorsal fins, they frighten the fish on which they prey into a tight ball and can pick them off one by one.

CHAMPION RACERS

It is difficult to assess the speed of animals. They do not fly, run, or swim in a straight line and some are so small they are barely visible at high speed, so world records are claimed and discredited regularly.

While the peregrine's stoop at speeds around 240 km/h (150 mph) makes it the fastest animal on Earth and the spinetailed swift achieves a flapping and soaring speed of 170 km/h (106 mph), several other birds achieve remarkable speeds in level flight. A magnificent frigate bird has been credited with 154 km/h (95 mph).

The cheetah may be the swiftest land mammal at 97 km/h (60 mph), but the gazelles it catches reach 80 km/h (50 mph) and can run for longer without overheating. The North American pronghorn can out-distance any predator by running at a constant 72 km/h (45 mph) for several kilometres. In England, a frightened red deer stag running down a street in Stalybridge, Cheshire, was caught on police radar doing 68 km/h (42 mph).

In the sea, the sailfish reaches 110 km/h (68 mph) in short bursts, but the bluefin tuna is not far behind with a top speed of 104 km/h (65 mph). The fastest fish over sustained distances are the marlins, which can swim at 64–80 km/h (40–50 mph).

The Australian dragonfly *Austrophlebia costali* is the current titleholder of fastest flying insect with a top speed of 98 km/h (60 mph), but claims of a male horse-fly pursuing a female at 145 km/h (90 mph) may topple the champion.

ART OF FLIGHT *A peregrine pursues a pigeon. Both birds fly fast, but the peregrine swoops in from above and behind to hijack its prey in midair.*

DEADLY COMBINATION OF SPEED AND SURPRISE

The speed at which the peregrine drops out of the sky to catch prey makes it the fastest animal in the world. The bird's normal cruising speed of about 65 km/h (40 mph) more than doubles to a wind-whistling stoop of 160–240 km/h (100–150 mph) when diving after prey.

A peregrine sets up an ambush in the sky to intercept its prey. It waits in the clouds about 1.6 km (1 mile) above the ground until it spots a passing target. Then it flies out of the sun, folds back its wings and plummets. Seemingly out of nowhere it strikes down on its victim with a single blow, circles, and then drops down to retrieve its meal. The peregrine is found all over the world, inland and on the coast.

THE SPEEDY INSECT THAT OUTRUNS ITS PREY

Tiger beetles are among the fastest animals on six legs. Over a distance of 30 cm (12 in), a North American tiger beetle has been timed sprinting at 50 cm (20 in) a second – the equivalent of 54 body lengths a second – making the beetle 10 times faster than the world's top human sprinter.

A voracious predator, the tiger beetle uses its speed to catch ground-living insects such as ants, spiders, and other beetles. Its chase is jerky: the beetle runs so fast that it cannot gather enough light to form an image in its large eyes. It must stop, look around, check the position of its prey, and then go. Even so, it is so fleet of foot that it can overtake its prey and capture its meal with its dagger-like jaws.

BEETLING ALONG *Large eyes, powerful mandibles, and long legs enable the sand tiger beetle of North America to hunt prey at high speed.*

HEATING CONTROL KEEPS SHARK ALERT FOR THE CHASE

The ocean's fastest shark, the shortfin mako, can maintain a speed of 50 km/h (30 mph) over 0.8 km (880 yd). It can accelerate away in even faster bursts. To do this, it possesses a physiological trick that many other sharks do not enjoy.

Although all sharks are cold-blooded creatures, the mako maintains the temperature of its swimming muscles, eyes, and brain about 5°C (41°F) higher than the surrounding sea water. This ensures that it is alert for any opportunity to feed and ready for a high-speed chase.

Found worldwide in both tropical and temperate waters, it preys on other fast-swimming species, such as tuna, swordfish, and blue sharks.

BEACH ATTACK *Killer whales on the coast of Patagonia have learned how to use their speed to surprise and pluck sea lions from the beach.*

KILLER WHALE IN HOT PURSUIT

When pursuing prey, the killer whale travels at 64–80 km/h (40–50 mph) making it the fastest member of the dolphin family. It is propelled by the up and down movement of its powerful tail. When not in pursuit, it cruises at 10–13 km/h (6–8 mph) and can cover up to 160 km/h (100 miles) a day.

Its powerful acceleration makes it an effective predator of fast-moving dolphins and seals.

INSECT HUNTER STAYS COOL ON THE WING

The dragonfly *Austrophlebia costali* is one of the fastest known flying insects. It can swoop downhill at 98 km/h (60 mph), but more usually dashes about its territory at about 58 km/h (36 mph). An active hunter, it intercepts other insects on the wing.

Dragonflies may be fast flyers but they flap their wings relatively slowly, at about 30 beats per second (compared with a hoverfly at 200 bps or a honey bee at 300 bps). Their body temperature varies with the air temperature. If their muscles are cold, they are unable to fly, so they need to warm up first by basking in the sun or shivering their muscles.

Once airborne, dragonflies tend to overheat, so they make 15 second glides to help to cool the body. They can also divert warm blood from the thorax (the body part to which the wings are attached) to the abdomen, where it cools before returning.

THE SWIFT'S HIGH-SPEED COURTSHIP RITUAL

During its courtship displays, the white-throated spinetailed swift has the fastest flapping flight speed of any known bird. With its long, thin, crescent-shaped wings it can fly at 170 km/h (106 mph) and soar aerobatically on thermal currents.

At lower speeds the bird, which lives in Japan and southern Asia, scoops flying insects from the air for food. It spends much of its life on the wing, including mating and sleeping, and lands only to raise its brood.

AT HOME IN THE WATER

Animals are found at all depths in water, some on or near the surface where the struggle for food is fierce, and others deep down where there is less competition. To survive, they have developed many different ways of floating and moving.

CORMORANT'S SPECTACULAR DIVES FOR FOOD

Cormorants are strong diving birds that can swim as deep as 55 m (180 ft) and remain below the surface for more than a minute. They propel themselves under water with paddle-like feet set on legs supplied with large and powerful thigh muscles. Found in oceans, seas, and lakes worldwide, cormorants hunt actively under water for flatfish, eels, shrimps, and squid, regularly reaching depths of 10 m (33 ft). Specimens have been caught in trawl nets set at 37 m (120 ft). After diving they stand with their wings spread out in order to dry their plumage.

WHY THE SPERM WHALE CAN DIVE TO THE OCEAN'S DEPTHS

The sperm whale is the deepest diving mammal in the oceans of the world. It is the largest of the toothed whales and can dive from the surface to a depth of 1,000 m (3,300 ft) at a rate of 170 m (560 ft) a minute.

This gigantic creature may owe this ability to the structure and content of the spermaceti organ that fills most of its huge head. Inside is a web of tubes filled with clear spermaceti oil. By circulating sea water through the organ, the whale can cause the oil to solidify, helping it to sink. At depth, heat from the muscles is circulated to liquefy the oil and help the whale to rise again.

The sperm whale preys on sharks, deep sea squid, bony fish, and octopus, eating about 1,000 kg (1 tonne) a day.

HUGE SEALS DOZE UNDER THE SEA

When diving to great depths, northern elephant seals sleep during their journey up and down. They descend to 330–800 m (1,083–2,246 ft) where they remain for 20 minutes or more feeding on fish and squid. At these depths there is little competition from other fish-eaters. Some seals reach 1,500 m (4,920 ft) and stay below for up to 2 hours.

Their lungs collapse as they dive, their heart rate slows from 55–120 beats per minute at the surface to 4–15 bpm at depth, and they take cat naps on the way down and up.

UNDERWATER ATHLETE *A white-necked cormorant dives in Africa's Lake Tanganyika, where it pursues freshwater fish, including cichlids.*

ICE DIVER *A Weddell seal dives under ice at Signy Island in the Southern Ocean. It eats fish such as Antarctic cod, which it finds in the depths.*

DEEP-SEA DIVER OF THE ICY ANTARCTIC

The Weddell seal is specially built to swim at great depths under the Antarctic ice in search of a meal of fish. One individual was followed to a depth of 600 m (1,970 ft) in an excursion lasting 73 minutes.

After a few 15-minute exploratory dives to locate prey, the seal descends directly in a straight line for the first 70 m (230 ft) with a few powerful swimming strokes. The seal is able to stay under water for so long because it can store large amounts of oxygen in its bloodstream – up to five times more than the average adult human being.

DEEP-DIVING PENGUIN WITH A BUILT-IN WEIGHT BELT

The emperor penguin has small stones in its stomach that help it to dive, much like a scuba diver wearing a weight belt. With this extra weight, the bird can descend regularly to 18–21 m (60–70 ft) to catch fast-swimming squid and fish. It can stay under water for up to 20 minutes. At the surface, the emperor travels by 'porpoising' at 11 km/h (7 mph). It uses its paddle-like flippers for propulsion and steers with its tail.

USING JET PROPULSION TO GET OUT OF TROUBLE

For high-speed escapes, squid rely on jet propulsion controlled by a nervous system with the largest nerve cells found in any animal. A hose-like siphon just below the head squirts water at high pressure, propelling the squid at up to 33 km/h (20 mph). Baby squid instinctively switch on their jet propulsion when they are startled by anything unusual. There is a critical period during their early life when they must learn not to jet away from everything they meet.

DESIGNED TO FLOAT

The materials from which animals are made – body tissues, bone, cartilage, shell, muscles, and skin – are more dense than water, so they have a tendency to sink. To combat this problem aquatic animals have developed various features that make them buoyant, and allow them to use the minimum of energy to maintain their position in the water.

Sharks have aerofoil-like pectoral fins that give them lift as long as they are moving forwards. In whale and basking sharks – the two largest fish in the sea – the oil-filled liver occupies most of their body cavity and accounts for up to 25 per cent of their body weight, allowing them to float motionless at the surface. Coelacanths and deep-sea lantern fish keep afloat by having waxy materials in their bodies. Most bony fish have an air-filled swim bladder which adjusts to maintain buoyancy.

The Portuguese man-of-war has a purple bag-like float that sits on the surface of the sea, while its tentacles trail through the water for 9 m (30 ft).

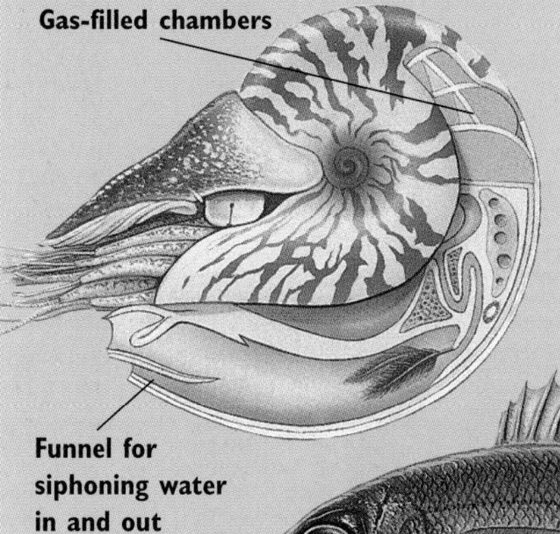

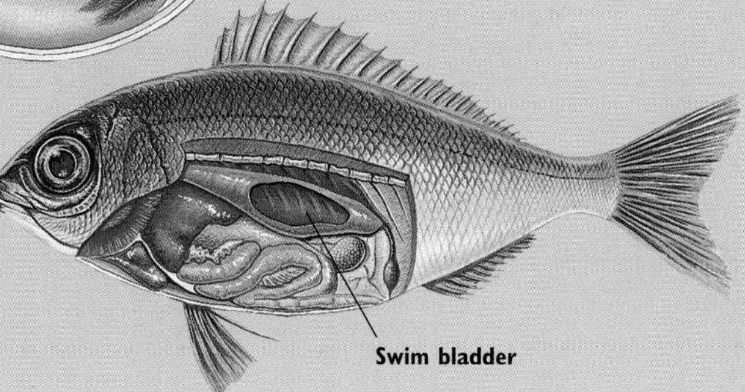

The pearly nautilus maintains buoyancy by adjusting the ratio of gas and fluid contained in its shell chambers. Bony fish have a swim bladder filled with oxygen-rich gas. A fish can vary the amount of gas held in its bladder, so controlling its buoyancy.

HOW ANIMALS MOVE UNDER WATER

Aquatic animals use all manner of methods, including paddles and propellers, body undulations, and jet propulsion, to force their way through water.

Jellyfish and the larvae of sea anemones move with a pulsating bell, octopus and squid squirt water from a funnel, scallops clap their shells together, and lobsters and prawns escape predators by bending and flexing their tails and shooting backwards – all forms of jet propulsion.

Many bony fish, sharks, and sea snakes move with S-shaped undulations that pass from head to tail pushing the animals forwards. Some microscopic animals move with a whip-like flagellum that also works by passing an S-shaped wave from base to tip. Skates, squid, and cuttlefish create S-shaped waves in the fins alongside their body. Other fish, like gurnards, 'crawl' across the sea floor on modified pectoral fins, and batfish 'hop'.

Of the amphibians, newts use their flattened tail to move like fish, but frogs and toads push with powerful back legs and webbed feet. Some aquatic birds, such as cormorants, have webbed feet that move alternately when at the surface and together under water.

Crocodiles have powerful tails which they move from side to side. Marine mammals, such as whales and dolphins, move with an up and down movement of the body, while the tail flukes are kept horizontal to give maximum thrust.

Penguins and sea turtles literally fly under water. The movement of their flippers is similar to the flapping of a flying bird's wing, except that there is power in both the upstroke and the downstroke.

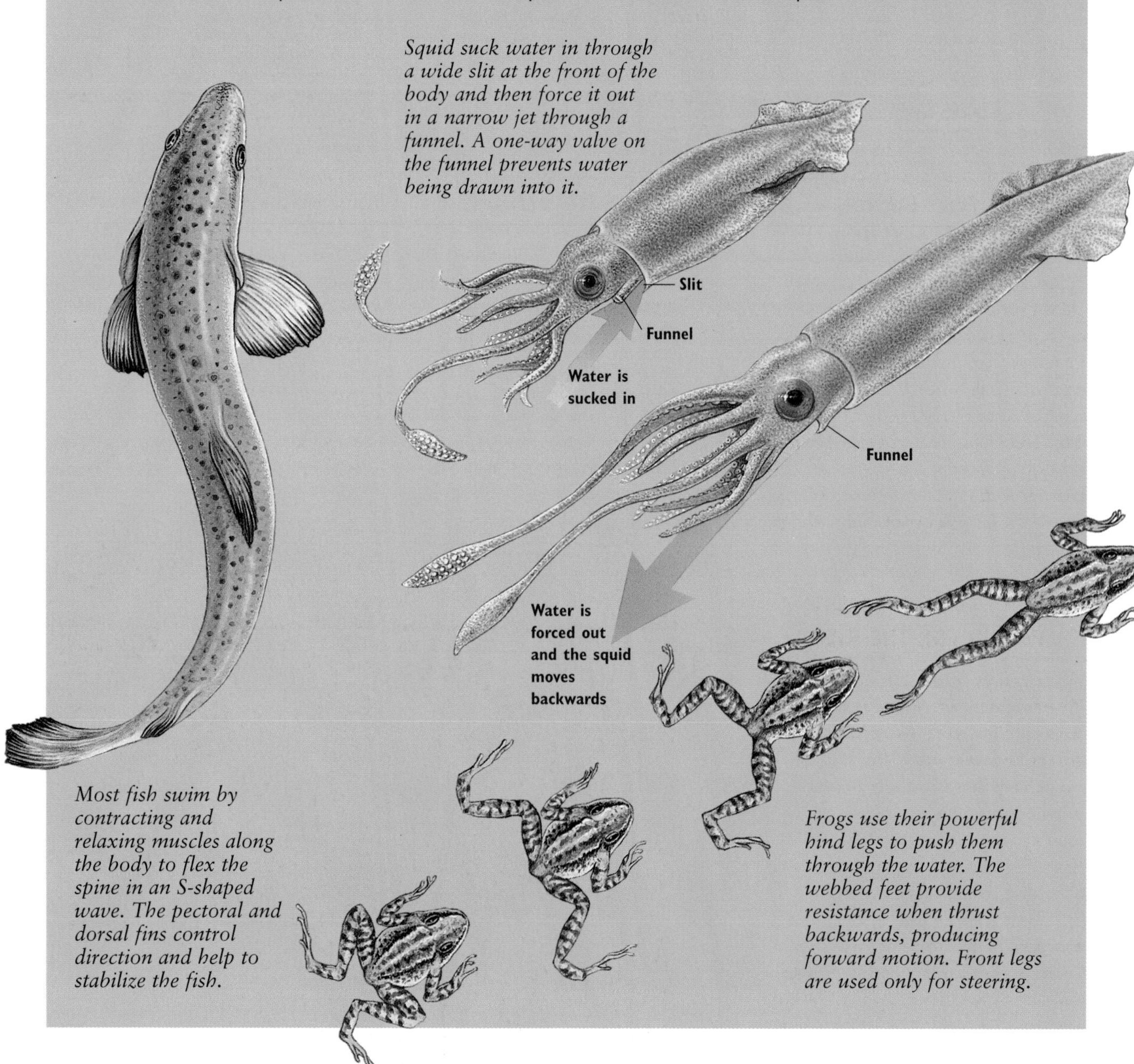

Squid suck water in through a wide slit at the front of the body and then force it out in a narrow jet through a funnel. A one-way valve on the funnel prevents water being drawn into it.

Most fish swim by contracting and relaxing muscles along the body to flex the spine in an S-shaped wave. The pectoral and dorsal fins control direction and help to stabilize the fish.

Frogs use their powerful hind legs to push them through the water. The webbed feet provide resistance when thrust backwards, producing forward motion. Front legs are used only for steering.

ICE HUNTER *Polar bears may travel 1,000 km (620 miles) a year in their search for food. They have been seen to leap 2 m (6 ft) to capture a seal.*

POLAR BEAR IS A STRONG SWIMMER

The polar bear is a true marine mammal that can swim between ice floes for 96 km (60 miles) without pausing for a rest. It can dive down to 4.5 m (15 ft) for up to 2 minutes.

The bear's front paws are almost 30 cm (12 in) in diameter and with partial webbing between the toes it can dog-paddle through the water at about 10 km/h (6 mph). Thick layers of fur and blubber insulate the bear and help to make it buoyant. The bear shuts its nostrils and ears when diving, but the eyes remain open – it can spot a seal from 4.6 m (15 ft) away.

INSECT OARSMAN WAITS POISED FOR THE KILL

A rapacious predator, the backswimmer waits upside-down in the water to attack other insects and tadpoles. With its front two pairs of legs and the tip of its abdomen touching the surface, it responds instantly to the slightest vibration and rows rapidly to the cause of the disturbance using its third pair of legs. These are modified with rows of swimming hairs to resemble oars. It pierces the skin of its prey with its beak-like mouth-parts, injects digestive enzymes, and sucks the resulting soup.

HOW THE HIPPO KEEPS ITS COOL

Africa's hippopotamus is an amphibious mammal. It spends the night on land feeding on vegetation, but returns to the water during the day, where it rests with half of its barrel-shaped body submerged.

Its smooth skin looses water rapidly in dry air so it needs to take refuge in a humid or aquatic environment by day to avoid becoming dehydrated. If caught in the sun, the hippopotamus 'sweats blood', oozing a red anti-sunburn secretion to help it to protect its skin. Although a land mammal, the hippopotamus has many adaptations for a life in water. Its ears, eyes, and nostrils are positioned in a line on top of its head, so it can hear, see, and breathe while the rest of its body is under water.

Its feet are slightly webbed to help it to move through water and on soft mud, and it can close its nostrils and ears when it ducks under, staying below for up to 6 minutes at a time.

The hippopotamus can swim, but more often walks in slow motion along the bottom on its short pillar-shaped legs. If disturbed, it can 'gallop' across the riverbed.

WATER HORSE *The hippopotamus keeps cool in the water by day. It walks delicately across the bottom like an overweight ballet dancer.*

A SNAKE AT HOME IN THE OCEAN

The yellow-bellied sea snake can be seen throughout the Indian and Pacific Oceans. Its streamlined body, up to 1.2 m (4 ft) long, is adapted to life at sea with an underside that is tapered like a boat's keel and a tail that is flattened like a paddle to help it to swim forwards. Valves in its nostrils help it to remain submerged for up to 90 minutes.

A stealthy hunter, the reptile lies motionless in the water like a log. Unwittingly, fish such as mullet and damselfish congregate beneath it. With a rapid lunge the snake catches a meal.

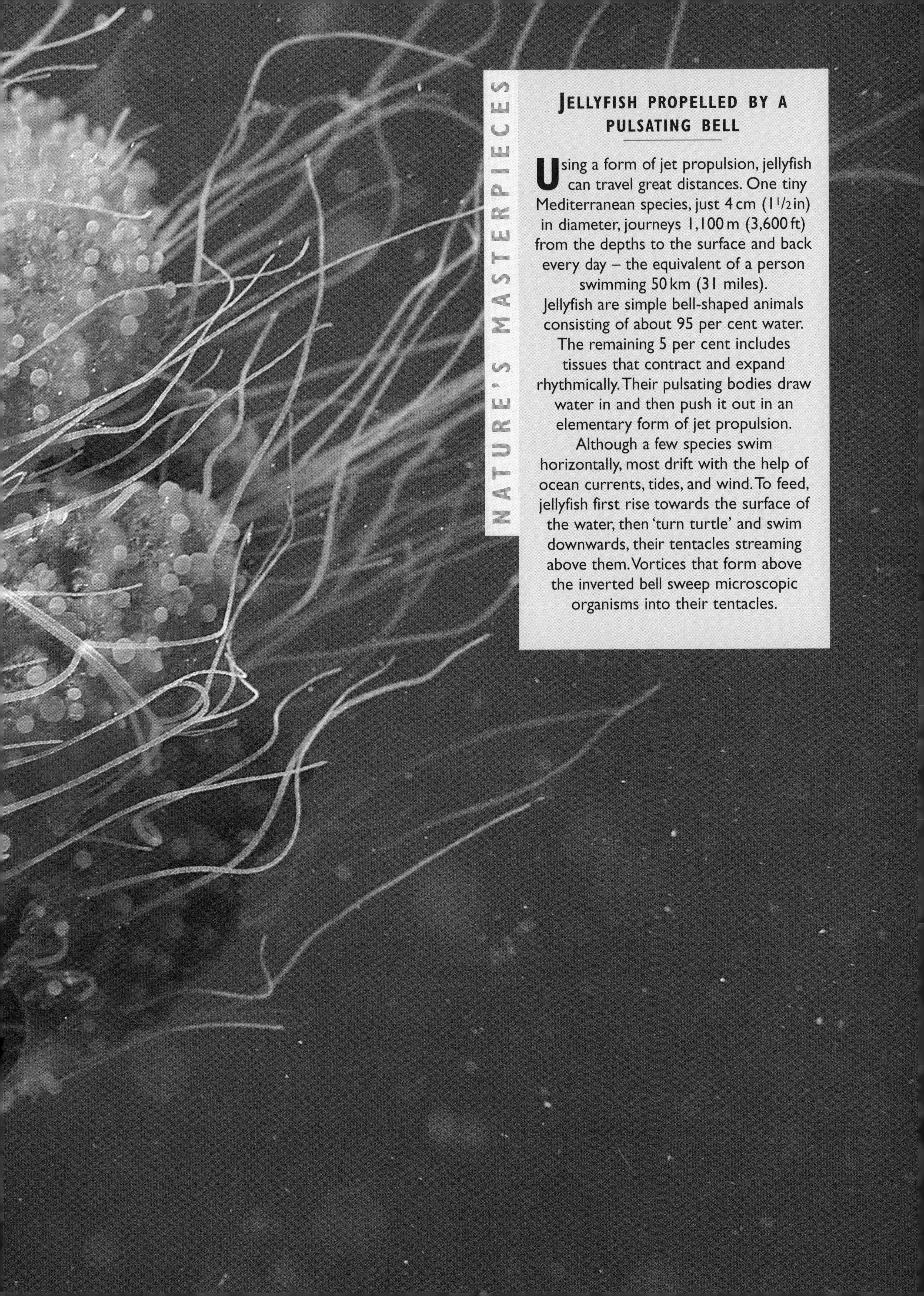

NATURE'S MASTERPIECES

Jellyfish propelled by a pulsating bell

Using a form of jet propulsion, jellyfish can travel great distances. One tiny Mediterranean species, just 4 cm (1½ in) in diameter, journeys 1,100 m (3,600 ft) from the depths to the surface and back every day – the equivalent of a person swimming 50 km (31 miles).

Jellyfish are simple bell-shaped animals consisting of about 95 per cent water. The remaining 5 per cent includes tissues that contract and expand rhythmically. Their pulsating bodies draw water in and then push it out in an elementary form of jet propulsion.

Although a few species swim horizontally, most drift with the help of ocean currents, tides, and wind. To feed, jellyfish first rise towards the surface of the water, then 'turn turtle' and swim downwards, their tentacles streaming above them. Vortices that form above the inverted bell sweep microscopic organisms into their tentacles.

FLYING ACES

The first animals to take to the air were probably the insects, and although some fish, frogs, and reptiles can glide, the only other true powered flyers are birds and bats. These creatures fly to find food or a mate, or to escape predators.

THE FLIGHT MYSTERY OF THE BUMBLEBEE

The bumblebee's wings seem too small to keep its body aloft, yet it is an accomplished flyer. With wings beating at 130–200 times a second, the insect manages to fly at an average speed of 3 m (10 ft) a second. The bee's secret is that its wings are not rigid. They bend and twist, generating lift on both the fore and back strokes, and their irregular cross-section makes for an efficient aerofoil. The four wings are operated by muscles in the thorax: the front pair move together, as do the back pair.

Unlike most insects, the bumblebee can raise its body temperature by shivering its muscles to produce heat. This allows it to fly at the lower temperatures found in more northerly latitudes and high on mountains.

BUSY BEE *A carder bumblebee, in search of nectar and pollen to feed to its colony's offspring, flies from one bluebell to the next.*

DRAGONFLIES HOVER ON AIR DISTURBANCES

A dragonfly keeps itself airborne by creating whirlwinds and vortices in the air. It has four wings, the front pair beating alternately with the pair at the back. Each wing moves in a figure-of-eight pattern causing the air around it to form vortices and eddies which keep the insect aloft.

Using its four-wing configuration, the dragonfly can fly in any direction without turning its body and can hover in one spot. However, when it turns, its fore and hind wings tend to hit each other making a clacking sound. The wings are almost identical to those of ancient dragonflies that flew some 320 million years ago.

LITTLE AUK FLIES ABOVE AND BELOW THE WAVES

The little auk, or dovekie, has wings large enough for flying, yet small enough to propel it through water. In the air, its whirring flight is characterized by rapid wing beats, and it also flaps its wings to 'fly' under water. The auk, which is found in the Arctic and North Atlantic Ocean, resembles a tiny puffin. It pursues prey, such as tiny shrimps, under water, carrying them back to its chick in its throat pouch.

SWIFT'S LIFE ON THE WING

The European swift spends more of its life in the air than any other bird. It eats, drinks, sleeps, preens, collects nest material, and mates on the wing. Youngsters that have just left the nest may be aloft almost nonstop for up to three years. Abandoned by their parents from the moment they fledge, the young birds then migrate south from Europe

HOW INSECTS FLY

Insect wings have adapted over millennia into highly efficient aerofoils. Wings developed from blade-like side extensions of the body that first enabled an insect to glide. Gradually the ability to twist then flap the wings evolved, leading to the powered flight we see today.

A modern insect's wing consists of two membranes supported by a framework of veins carrying blood, nerves, and oxygen. The vein along the leading edge is thicker than the rest, providing the rigidity needed to cut through the air.

The wings are attached by ball-and-socket joints to the thorax (the middle section of the body). Power comes from two sets of large muscles attached to the thorax. One pair pulls vertically on the top and bottom of the thorax; the other pair runs longitudinally. Wing muscles work by changing the shape of the thoracic walls. When the vertical muscles contract, the roof of the thorax is pulled down and the wings flap upwards. Contraction of the longitudinal muscles raises the thorax roof, and the wings beat downwards. The thoracic walls are thick but elastic, ready to spring back to their original position, so only minimal muscle power is needed to work the wings.

The muscles can move remarkably quickly – up to 1,000 times a second in small flies. The wings also have small muscles attached that twist the blade as they move, so each wing is continually changing shape in flight. The leading edge of the wing is tilted downwards during the downstroke and upwards on the upstroke, with lift and power on both strokes. The movement of the flexible wings causes vortices to form around them, and it is these air disturbances that keep the insect aloft.

Most flying insects have four wings, although in some species the front pair are modified. Beetles and earwigs have hardened forewings, adapted as protective wing covers, or elytra. They are propelled by the translucent hind wings. The forewings of flies have become balancing organs, known as halteres, which work in the same way as gyroscopes.

BOUNCY FLIGHT *The tough walls of the insect's thorax are flexible, so that the contracting flight muscles set up a bouncing motion between the sides of the body and the back.*

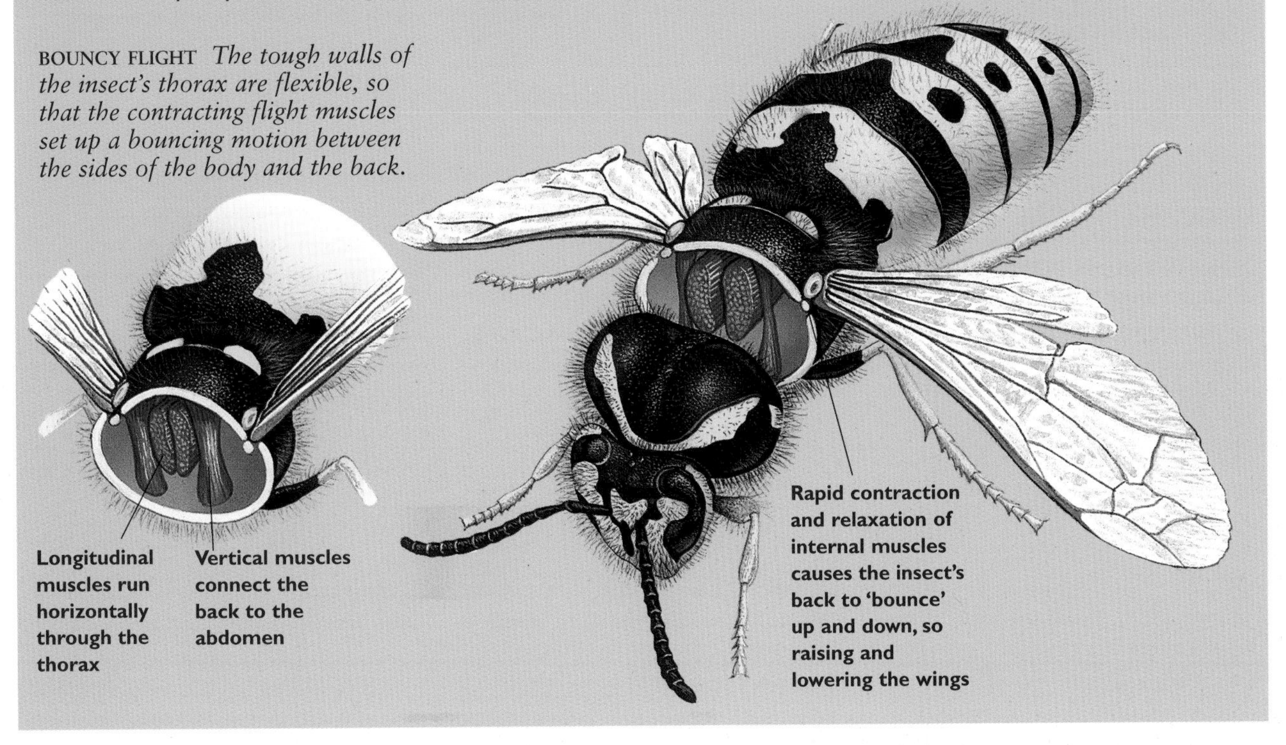

to Africa in loose flocks ahead of them. Swifts fly at about 40 km/h (25 mph), and can be in Central Africa within a few days. Here they spend winter, flying continuously. They might return to Europe that summer, or the following summer. Their first stop will be to look for nest sites, but they land only briefly before setting off on another epic flight. So it may take several seasons of continuous flying before a young swift finds a mate and finally stops to nest and rear its young.

FLIGHT OF THE CONDOR

The Andean condor vies with the maribou stork for the title of 'bird with the largest wings'. Although the wandering albatross has the longest wings, the condor's wings are broad – about 75 cm (30 in) – as well as being up to 3.2 m (10 ft 6 in) long, which gives them a greater area. The bird flies along mountain ridges, using updrafts and thermals to soar without flapping its wings for tens of kilometres. From the air it watches for dead or dying animals to scavenge.

BUTTERFLY BREAKFAST *When the sun rises, the silver-washed fritillary drops from its night-time treetop roost to brambles and thistles below.*

WORLD TRAVELLERS SUNBATHE BEFORE TAKE-OFF

Butterflies rival many birds as long-range world travellers. Migrating butterflies cover vast distances and sometimes fly at great heights. Large cabbage whites have been spotted in the Alps above 3,658 m (12,000 ft), and tortoiseshells have been seen over the Zemu Glacier in the Himalayas at 5,791 m (19,000 ft).

Butterflies have two pairs of thin wings, supported by veins and covered with tiny scales. The front wings are usually larger than the rear pair and overlap them slightly. On each side the front and rear wings work together as a single unit during flight. A medium-sized butterfly's wings flap at 8–12 times a second, propelling it along at about 14 km/h (9 mph).

Each morning, butterflies must raise their body temperature to 30°C (86°F) before they can take off. In order to warm up, they sunbathe with their wings closed, since the dark underscales absorb heat better than pale ones.

SWANS USE WINDS TO TRAVEL LONG-DISTANCE

On December 9, 1967, a flock of some 30 whooper swans set off at dawn from the coast of Iceland. They were heading south to the British Isles to avoid the harsh winter. Over the Inner Hebrides, a civil aircraft pilot spotted them flying at 8,230 m (27,000 ft). Air traffic controllers in Northern Ireland confirmed their altitude and clocked their ground speed at 139 km/h (86 mph). The birds had hitched a ride on the strong winds of a southerly moving jet stream associated with a ridge of high pressure. It is now thought that swans and geese routinely use these winds in the lower stratosphere to fly vast distances with minimum effort.

NATURE'S PRECISION FLYERS

Small, insect-eating bats can be astonishingly accurate flyers. When a Dent's horseshoe bat with a 25 cm (10 in) wingspan was released in a confined space 1 m (3 ft 3 in) wide, 80 cm (32 in) deep, and 30 cm (12 in) high, it flew around without ever touching the walls, ceiling, or floor.

Small bats need precision flying techniques so that they can track and catch fast-flying, highly manoeuvrable insects in midair.

A bat can drop out of the sky when in pursuit: it adjusts the angle of its wings and rolls sideways to lose altitude rapidly in a downward turn. To achieve a tight turn it slips sideways or beats one wing faster than the other.

SPEEDY AERIAL PIRATES

Frigate birds are such speedy and acrobatic flyers they can steal food from other seabirds' mouths. With a wingspan of up to 2.5 m (8 ft 6 in) and a weight of just 1.5 kg (3 lb 5 oz), they are the fastest-flying seabirds, reaching speeds of up to 145 km/h (90 mph).

When hunting, the frigate bird skims the ocean surface at full speed. It can pluck fish or squid from just beneath the surface with its sharply hooked bill, and with an extra burst of speed it can catch flying fish as they leap from the water. On turtle nesting beaches, it swoops down and snatches hatchlings as they cross the sand.

AVIAN AEROBATICS

Hummingbirds can fly forwards, backwards, sideways, straight up or down, and even upside-down. They can hawk for insects at high speed and hover in front of flowers probing for nectar. All this is possible because the hummingbird's shoulder joint, where the wings are attached, can move in all directions.

Most birds' wings move up and down with lift created on the downward stroke, but hummingbirds move their wings forwards and backwards creating lift on both strokes. The wings move exceptionally fast – about 80 beats per second during level flight, and 200 per second during aerobatic courtship displays. Unlike other birds, hummingbirds reach maximum speed the moment they leave their perch.

BLUR OF WINGS *A rufous-breasted hermit hummingbird, attracted by the colour, hovers in front of a hibiscus flower, feeding on its nectar.*

OUT OF THEIR ELEMENT

To gain advantage, some animals leave their normal element and travel through another. Gliding through the air is a favoured way to flee predators, so certain fish, lizards, snakes, and squirrels have developed the means to fly.

FISH THAT FLY OVER THE SEA

The flying fish, found in tropical and subtropical seas, escapes danger by flying above the surface of the water. It flies using greatly elongated pectoral fins. When pursued by predators, such as dolphins, billfish, and sharks, it swims rapidly to the surface, at about 32 km/h (20 mph), keeping its fins flat to its body. It then launches itself out of the water, opens its fins like wings, and glides. As it loses power and drops to the surface, the fish gains a second wind. Vibrating its tail from side to side at about 50 beats per second, it takes off again across the water surface at speeds approaching 65 km/h (40 mph).

Usually the fish glides for about 40–50 m (44–55 yd), but one of the longest flights observed lasted for 42 seconds and covered about 600 m (656 yd). Flying fish have been found on decks of large ships as much as 9 m (30 ft) above the sea surface. They probably took off into a gust of wind and were carried high into the air.

Some predators have worked out how to catch flying fish. Dolphins have been seen to anticipate flight paths, swim rapidly ahead, and grab their unfortunate victims as they re-enter the water. Acrobatic frigate birds swoop down out of the sky and snatch them in midair.

FISH OUT OF WATER *As it begins to leave the water a flying fish propels itself along by trailing the tip of its tail in the sea. Once launched it spreads out its pelvic wings to provide lift as it glides.*

AMPHIBIANS PARAGLIDE BETWEEN TROPICAL TREETOPS

Flying frogs use the webbing between their extra-long toes to glide between trees. To escape predators, such as tree-climbing snakes, the frogs launch themselves into the air with their powerful back legs. They spread their limbs, stretch the webbing between their toes, flatten their body into a saucer shape to maximize surface area, and fly – or, more accurately, glide – travelling up to 45 m (50 yd) between trees.

AIRBORNE SQUIRREL *The northern flying squirrel of North America extends the flaps of skin on either side of its body and glides from tree to tree. The squirrel's overall length is 25–31 cm (10–12 in).*

The frogs live high in tropical rain forest trees in Central America and southern Asia, and their long toes are well adapted for clinging to bark and branches.

GLIDING SQUIRREL LEAPS TO SAFETY

Most squirrels must climb down to the ground in order to reach distant trees. But the flying squirrel has an alternative method: it glides through the air on its flight membrane, the webs of furred skin between its forelimbs and hind limbs. When clambering over branches, the squirrel tucks the membrane away under its belly so it does not trip up.

If pursued through the canopy by a predator, such as a marten, the flying squirrel leaps into space, stretches out its flight membrane and parachutes to the next tree. It can adjust the membrane by changing the position of its limbs, and its bushy tail acts as a rudder, keeping it on course.

As it approaches its landing site, the squirrel turns its body and tail upwards in a braking action before landing on a neighbouring tree trunk. It can then climb the trunk to gain height and take off again. In this way, a flying squirrel can progress through the forest in a series of glides. The giant flying squirrel, with an overall length of 60 cm (2 ft), can glide for 100 m (330 ft), while smaller species, such as the 23–25 cm (9–10 in) southern flying squirrel can manage only much shorter flights.

SNAKES THAT FLY FROM TREE TO TREE

In the tropical forests of South-east Asia, flying snakes take off from a high branch and glide to the next tree. They use muscles to spread their ribs wide, which flattens the body, and undulate through the air with a controlled S-shaped swimming motion, travelling up to 20 m (65 ft) in a single flight. Although the flying snake is an able climber, flying is a more energy-efficient way to travel through the forest than going up and down trees. It is also an effective way to surprise prey, such as lizards, frogs, birds, and bats, and to escape predators including birds of prey, monkeys and other tree-climbing snakes.

LIZARD ALOFT *The colourful folds of skin of the flying dragon lizard enable it to glide and also to impress a potential mate.*

GLIDING DRAGONS DODGE DANGER IN THE TROPICS

Flying lizards are commonplace in the tropical rain forests of Southeast Asia. The flying dragon has a thin membrane of skin on either side of its body. By stretching the membrane taut with its movable, elongated ribs, the lizard can glide 18 m (60 ft) between trees. Likewise, the flying gecko has a fringe of skin running along its sides and webbing between its toes. It launches itself from branches to escape tree-climbing snakes and to hijack insects on adjacent trees.

SQUID USE JET POWER TO ESCAPE THE HUNTERS

When chased by fast-swimming predators, such as dolphins, tuna, or sharks, flying squid escape by taking flight. They fly using jet propulsion. Water is taken into the mantle cavity then squirted out at high pressure through the siphon on the body's underside. The squid accelerate so fast, at speeds up to 32 km/h (20 mph), that they can leave the water altogether and use their lateral 'wings' to glide for several metres.

Found in tropical and temperate waters worldwide, flying squid move closer to shore in summer and offshore in winter. During the day they remain at depth, migrating to the surface to feed at night. It is at the surface that they risk being eaten, for even if they escape into the air there are squid-eating birds, such as albatrosses, ready to pick them off. But because squid swim and fly in large shoals some always get away. They can reach a height of 1.5 m (5 ft) above the ocean's surface, and may occasionally fall into small boats.

AUSTRALIA'S FURRY FOREST FLYERS

Some species of possum glide from tree to tree, to extend their feeding range and to avoid predators on the ground. In the air, flying possums and gliders (their cousins) look like rectangular kites, complete with furry tails. Folds of furred membrane stretched between elongated fore and hind limbs enable them to glide 90 m (100 yd) from the top of one tree to the lower trunk of another. The tail serves as a rudder, although larger species, such as the 50 cm (20 in) greater flying possum, have little control over their flight. The smaller gliders, such as the yellow-bellied glider, can manoeuvre to avoid branches and other snags.

Landing is rather crude. Gliders and flying possums swoop upwards to brake and then collide clumsily with the target tree trunk and hang on tightly. They are active mainly at night.

ATHLETES AND GYMNASTS

Animals are capable of effortless athletics and many species unwittingly chalk up remarkable records while going about their daily lives. In the race for survival, maximum fitness can often mean the difference between life and death.

MAGNIFICENT MARLINS' SURGE OF SPEED

Their superbly streamlined bodies enable marlins to swim faster than 110 km/h (68 mph). With their dorsal and pectoral fins held flush to the body, marlins normally use their speed to catch other fast-swimming fish, such as mackerel and herring. The sharp rapier-like bill slashes at the prey as the marlins streak through the shoal. Then they turn and scoop up casualties. As the marlin starts to attack, its silvery body changes colour, and a pattern of dark stripes appears along its flanks. Nobody knows why.

When hooked by sports anglers these spectacular fish, up to 4.6 m (15 ft) long, will leap 3 m (10 ft) into the air time and again before they tire.

CHOUGHS FLY HIGH ON TOP OF THE WORLD

The record for the highest altitude over land by any bird is held by the chough. This member of the crow family flies regularly at 3,500–6,250 m (11,500–20,500 ft) in the Himalayas, and one flock was seen by a British expedition at 8,235 m (27,017 ft).

The chough is a scavenger, but if carrion is unavailable it will dig for insects. In the European Alps, choughs are often seen at ski resorts, picking over the debris left by human visitors.

HIGHEST FLYERS *A flock of yellow-billed choughs circles at high altitude beneath Mt Lhotse in the Himalayas on the border of Nepal and China.*

BOUNCING ANTELOPE LEAPS TO SAFETY

When predators are about, the springbok of Southern Africa engages in a curious behaviour known as pronking. From a standing start, it jumps into the air and appears to bounce on stiff legs, all four hoofs hitting the ground at the same time. As it jumps, the fan of long white hairs on its back and rump are erect, and the head is bent forward almost to its forefeet. The hoofs are kept together and the back is arched, and the moment the antelope touches the ground it springs back up again. It can leap straight up, forwards or jig to one side. Then, it lowers its fan, raises its head and shoots off at high speed.

Pronking is infectious. One or two animals start and the rest of the herd soon joins in. The jumps signal to a predator, such as a lion, that the springbok is fit and will be able to outrun a pursuer – encouraging the predator to chase after easier prey.

BUSH LEAPERS *Using their pronking tactics, springbok confuse predators, who find it difficult to select an animal that will be an easy kill.*

TENACIOUS HUNTERS' RELAY STRATEGY

African hunting dogs are known for their determination and stamina: when they start the hunt they never give up. The dogs hunt in relay teams during the coolness of dawn and dusk, but may be active throughout an overcast day or on a moonlit night. They are plains hunters and rely on sight to spot their quarry.

First they isolate a target, such as a young zebra, antelope, or adult gazelle, then two dogs give chase. The rest of the pack lopes in the background, fanning out during the pursuit, and as the first pair tires a second pair takes over the hunt. The pattern can be repeated many times, with the prey chased relentlessly at speeds up to 70 km/h (45 mph).

Eventually terror and fatigue weaken the victim and it slows. This is a signal for the dogs to start their final attack. The chaser dogs jump at the flanks and belly, and the rest of the pack follow, ripping chunks of flesh from the victim's body.

Using this method, a large pack can tear apart and devour a medium-sized antelope in 10 minutes.

THE ANTELOPE THAT OUTRUNS ITS RIVALS

The American pronghorn is the champion long-distance runner of all the mammals. It can run with ease at 70 km/h (45 mph) for 6.4 km (4 miles). It can accelerate in bursts up to 86 km/h (55 mph) and has a stride of 8 m (27 ft) when running full tilt.

The pronghorn achieves all this with useful modifications of its body. Its lungs are unusually large. In addition,

its heart pumps more oxygenated blood to the leg muscles, and it has more mitochondria (tiny organs that convert oxygen into energy) in each muscle cell than any other antelope.

Relying on this ability to escape at speed, the pronghorn will inspect anything that moves – including a predator – then flee. Its eyes bulge from its head, giving it 360 degree vision, and long, black eyelashes keep out the sun. Its body is stocky, with pointed hoofs to cushion its long, thin legs.

PRIMITIVE PRIMATE IS A LONG JUMP CHAMPION

The indri, from Madagascar, can leap 10 m (33 ft) from tree to tree in the upright position. It can do this thanks to its powerful legs, forward-facing eyes, long arms, and grasping hands and feet. Resembling a giant teddy bear, the indri, largest of the lemurs, clings to tree trunks and thrusts itself into the air with its powerful hind limbs. As it flies, it

LEAPING LEMUR *The flying leap of the indri is carefully judged. The animal will avoid tangled foliage and land cleanly on the trunk of a tree.*

holds out its arms and tucks its legs under its body, ready for the landing.

The indri is active during the day when it forages for fruit and leaves. It spends most of its life in the trees, coming down to the ground only occasionally to eat earth or bark, thought to be an aid to digestion.

GREAT JOURNEYS

Animals need to maximize their intake of energy-rich foods in order to live and reproduce. Some go to great lengths to find supplies, travelling thousands of kilometres to places where food is plentiful at particular times of the year.

MIGRATING BIRDS' REFUELLING FEASTS

On long migrations, shore birds such as red knots, sandpipers, and turnstones always congregate at traditional places to refuel. They time their arrival to coincide with the availability of seasonal foods. Should they run short of energy on a long migration they can break down some of their body tissues to fuel their flight, but at the first opportunity they set down to feed and rest.

One spring stopover in the United States is at Delaware Bay on the Atlantic coast. Here, 1.5 million waders heading northwards towards the Arctic to breed drop down to feast on the eggs of horseshoe crabs.

The crabs – not true crabs but primitive aquatic relatives of spiders – emerge from the sea in their thousands, each depositing 80,000 eggs in the damp sand. When they leave, flocks of waders and gulls – between 100,000 and 250,000 to a beach – crowd around noisily to pick out the highly nutritious eggs. The birds depend on the eggs to sustain them during their journey. They gorge themselves for up to a fortnight, doubling their body weight in that time. It has been estimated that 50,000 sandpipers alone eat 6 billion eggs weighing a total of 27 tonnes.

A LONG-DISTANCE CRUISE FOR FOOD

Wandering albatross parents travel huge distances across the Southern Ocean to find food for their single chick. On just one foray in search of squid, a bird will journey up to 14,500 km (9,000 miles), covering 900 km (560 miles) in 24 hours, and flying day and night with few stops at speeds of about 80 km/h (50 mph). The albatross is a glider rather than an active flyer, and can cruise for long distances over the ocean while rarely flapping its wings.

Its flight path takes advantage of prevailing winds and the updraughts from waves. When diving, the albatross turns into the wind as it nears the water. This gives it enough lift to glide upwards again.

An albatross chick needs to learn how to fly efficiently, too, as it might spend the first seven or eight years on the wing before returning to its natal home to find a mate.

TURTLES' JOURNEY ALONG HIGHWAYS IN THE SEA

Leatherback turtles are air-breathing reptiles, yet they use deep-sea highways to travel on long and unexpected migrations. Turtles that nest on the tropical beaches of

FEEDING FRENZY *A huge flock of semipalmated sandpipers on North America's east coast continues its spring migration after feasting on the eggs of horseshoe crabs. Local laughing gulls (inset) clean up the leftovers.*

Surinam, in northern South America, often turn up off the coast of Scotland en route to the Arctic, 6,960 km (4,325 miles) away. Satellite tracking has revealed that they make this regular northerly migration in search of meals of jellyfish.

The turtles follow the contours of underwater canyons and mountains, with each individual in a population following almost the same track. They travel at about 60 km (37 miles) a day, and dive often, sometimes to 1,000 m (3,280 ft) or more, holding their breath and remaining submerged for up to 40 minutes. At these icy depths, they keep their muscles warm by maintaining a body temperature several degrees higher than the surrounding sea water.

Many sea serpent stories from the west coast of Scotland could well be the result of sightings of leatherback turtles, which are the world's largest sea turtles with a body length of more than 2 m (7 ft).

THE LONGEST JOURNEYS

● **In the air,** Arctic terns (*see page 182*) are the record holders among bird migrants, flying from one end of the world to the other. But they are not the only long-distance travellers. Golden plovers and sanderlings fly 20,000 km (12,400 miles) between Canada and the south-east coast of South America each year.

Bobolinks migrate from southern Canada to Paraguay and back, island-hopping the 2,000 km (1,240 miles) of open sea through the West Indies from Florida to South America.

The tiny rufous hummingbird makes a round trip of 6,000 km (3,730 miles) to and from the north-west coast of North America, where it breeds, and Mexico, where it overwinters.

● **In the sea,** many species of bony fish, sharks, dolphins, whales, and sea turtles undertake long migrations. In the Atlantic, the sandbar shark ventures as far north as Massachusetts in summer but is found 3,200 km (2,000 miles) away in the Gulf of Mexico at other times.

Bluefin tuna leave the Mediterranean, pass through the Strait of Gibraltar, to reach feeding grounds off the coast of Norway. Green turtles in the equatorial Atlantic journey 2,200 km (1,370 miles) from the Brazilian coast to Ascension Island to breed.

● **On land,** the bulbous-nosed saiga antelope migrates 350 km (220 miles) from winter feeding grounds near the Caspian Sea northwards to traditional birthing sites. From there, it heads 250 km (155 miles) south-west for summer grazing, and then travels east, back to its lowland winter sites.

In northern Norway, the Sami people and their semidomestic reindeer follow a migration route 400 km (250 miles) long between summer and winter feeding sites.

ANNUAL MIGRATION FOR GIANTS OF THE SEA

Each spring, the grey whales of North America's Pacific coast set out on a 20,400 km (12,600 mile) round trip. They leave their breeding sites in the sheltered lagoons of Baja California and travel to their feeding grounds in the Arctic.

The procession is led by the newly pregnant mothers eager to maximize feeding time to nourish their developing offspring. The adult males and nonbreeding females are not far behind, followed by the immature whales and, lastly, mothers and calves.

They take the inshore route, skimming the kelp beds where they can hide from pods of marauding killer whales which prey on their young. They swim at about 8 km/h (5 mph), but they can take off at 20 km/h (12 mph) when pursued. The whales surface every 3–4 minutes and blow a short twin-blast of vapour before lifting their tails and diving below.

NAUTICAL NURSERY SCHOOL FOR WHALE CALVES

Every year the southern right whales of the South Atlantic migrate from Antarctica to their breeding grounds off Patagonia. In summer, they feed on plankton and krill from the surface of the polar sea. At the onset of winter they return to a safe haven at Peninsula Valdez on Argentina's Patagonia coast to mate, give birth, and rear their young, well away from patrolling killer whales.

As the southern summer approaches, playfulness gives way to training. Mother and offspring move back and forth across the bay at high speed, so the calf can build up its muscles for the long voyage to Antarctica.

THE LONG MARCH OF THE CARIBOU

Epic journeys are a way of life for the caribou of Arctic Canada. Hundreds of thousands of caribou travel about 1,000 km (620 miles) in each direction between their summer calving grounds on the tundra and their winter refuges in the great coniferous forests of the taiga.

In early winter, the tundra is covered by a layer of ice that prevents the caribou from brushing the snow aside to feed on the vegetation beneath. So they move into the forests where the snow is powdery. In spring, between February and April, thousands of animals travel back to the rugged tundra together following well-worn trails. A single herd can stretch for 300 km (185 miles).

On reaching the calving grounds in May the pregnant females give birth immediately. The calves grow quickly, and once they can forage for themselves the herds move on a further 200 km (125 miles) to lower, greener pastures.

By July small groups begin to trickle south towards the tree line, arriving in the forests in September, in readiness for winter.

LOCATION FINDER *Young Arctic terns circle their nesting sites to familiarize themselves with the terrain in order to find their way home in the future.*

EPIC FLIGHTS TO THE LAND OF THE MIDNIGHT SUN

Arctic terns embark on some of the longest journeys made by any bird. Those that nest in the Arctic fly to the Antarctic and back each year, a round trip of more than 40,000 km (25,000 miles). This means the bird is flying for eight months of the year, and at either destination it is in perpetual daylight.

Terns take various routes. When heading south in autumn, the Scandinavian and British birds hug the Atlantic coast of Europe and north-west Africa. Some cross the ocean on easterly winds to Brazil and follow the coast of South America, while others either island-hop down the mid-Atlantic ridge or continue along the African seaboard.

Birds from Canada and Greenland ride on westerly winds to Europe and then follow the same route south as European terns.

The terns feed on the way, plunge-diving to capture fish near the surface. The 24 hours of sunlight at each end of the world enables them to maximize their feeding time when they breed in the north and moult in the south.

ON THE ROAD *Vast caribou herds can take a week to pass by a single location as they travel through the inhospitable Arctic tundra.*

MUTTON BIRDS FOLLOW THE PACIFIC WINDS

Short-tailed shearwaters, or mutton birds, cover 32,000 km (20,000 miles) a year on their migration around the Pacific Ocean. They breed on islands in the Bass Strait, between Australia and Tasmania, and when their chicks have fledged they head out into the Pacific and ride prevailing winds in search of food.

They fly east to New Zealand, then north to Japan, and farther north again past Kamchatka to the Bering Sea. Riding westerly winds towards Alaska, the birds follow the west coast of North America as far as California, where they pick up easterly winds that carry them back towards Australia.

THE EEL'S EXTRAORDINARY LONG-TERM VOYAGE

Common eels undertake an unusual migration, swimming thousands of kilometres from river to sea to spawn. Eels live in the rivers and lakes of Europe and North America and when the time comes to reproduce they head for the sea. As they do so, they become streamlined 'silver eels'. Their target is the Sargasso Sea. Here, they spawn and die. The eggs develop into larvae, which are carried on the currents. In the case of European larvae, the Gulf Stream takes them to the Atlantic coasts of Europe.

On entering fresh water, the larvae change into 'glass eels', or elvers. As they swim upstream, their colour deepens and they become known as 'yellow eels'. They feed in the rivers for six or seven years before embarking on the return journey to the Sargasso Sea.

MIGRATING MONARCHS HEAD FOR BUTTERFLY TREES

Monarch butterflies, just 10 cm (4 in) across, travel thousands of kilometres in their short lives of up to eight months. Adult butterflies leave southern Canada and northern USA in July and fly south on a 2,000–3,000 km (1,250–1,860 mile) journey to California and Mexico where they spend the winter on the oyamel 'butterfly trees'.

Monarchs mate in spring, and by March the first butterflies are flying north, laying eggs as they go. When egg-laying is completed the butterflies die. The offspring of the early egg-layers continue the journey north.

NO EASY MEAL *Brightly coloured wings warn predators that the intrepid, long-distance travelling monarch butterflies are unpalatable.*

FINDING THE WAY

Animals use all kinds of cues to find their way across the world and back again. Landmarks, the position of the Sun by day and the stars by night, and even the invisible lines of force in the Earth's magnetic field are their reference points.

DESERT LOCUSTS FOLLOW A STORM

The first sign of a desert locust swarm is a faint cloud on the horizon. As it comes closer, the sky darkens and everything is engulfed in a noisy world of flying and hopping insects. Swarms 4 km ($2^1/_2$ miles) high, covering 250 km^2 (95 sq miles), have been recorded.

Each locust steers by the Sun, aiming to reach a place where it can see that rain is falling. An overriding instinct causes the individual to fly into its swarm rather than out of it, and swarms tend to fly downwind, carrying the locusts into areas of low pressure where rain is concentrated. The wind is important to the locust, but it can be destructive: swarms of locusts blown out to sea have been seen in the Atlantic, 2,200 km (1,370 miles) from the west coast of Africa.

LOCUST SWARM *Desert locusts descend on Mauritania, West Africa, creating havoc and devouring crops as they go.*

AMAZING MEMORY OF THE FISH OUT OF WATER

In the West Indies there is a species of goby, a small rock-pool fish, that has an incredible memory. As the tide recedes, this little fish is sometimes marooned in a rock pool some distance from the sea. It avoids being boiled in the hot sun by jumping from one pool to the next until it reaches the ocean.

The goby never lands on the exposed rocks, but judges the distance to the next pool so accurately that it drops right in. Somehow the goby memorizes the position of the pools as the tide rises and makes a mental map of its territory.

SALMON USE RIVER 'SCENTS' TO FIND THEIR WAY HOME

Some species of salmon rely on chemical memory to find their way home after a long journey. The pink salmon of the northern Pacific travel 4,000 km (2,500 miles) out to sea from the estuary they left as an immature smolt. Yet they find their way back unerringly to the same river several years later to breed. Each river has a 'smell signature' by which a

HOMEWARD BOUND *A school of pink salmon leaves the sea and heads for the spawning grounds in the Chiniak River on Kodiak Island, Alaska.*

salmon can recognize its own stretch of waterway.

It is thought that the Earth's magnetic field controls the salmon's direction-finding in the ocean.

HOW ANTS NAVIGATE IN THE DESERT

Most creatures avoid the scorching desert sunshine, but a tiny ant in the Sahara would be lost without it. It nests underground and makes short excursions in search of food. When it runs, it stops occasionally to turn its head from side to side to take a bearing from the Sun.

The ant has eyes that can detect polarized light in the Sun's rays. The angle at which the light vibrates depends on the position of the Sun. Having found food, the ant does not follow a winding scent trail home. Instead, it uses its map of polarized light to head straight back to the nest entrance, which could be 140 m (150 yd) away, minimising its exposure to the surface heat of 70°C (158°F).

MIGRATING SHARK SIMPLY FOLLOWS ITS NOSE

Female blue sharks use the Earth's magnetic field and geomagnetic information from rocks on the seabed to find their way. Blue sharks mate off the east coast of the USA and cross the Atlantic Ocean to Europe to drop their pups.

While on migration, they descend from the surface to depths of 200 m (660 ft) every 2 to 3 hours. It is thought that they are checking for magnetic bearings from rocks on the sea floor using the pits of Lorenzini in their snouts, which can detect electrical and magnetic activity.

Solidified lava flows from extinct undersea volcanoes are also used by sharks and other marine animals as 'underwater highways'.

HOW BIRDS FIND THEIR WAY

Like NASA space shuttles, birds have several back-up systems for orientation and navigation that they can call upon.

Landmarks are particularly important. Often birds circle a nest or roost site before leaving, as if refreshing their memory of the geography of the place.

They observe the position of the Sun and crosscheck it with their biological clocks. If the Sun is behind clouds, but a little blue sky is visible, they can detect the changing angle of polarized light as the Sun crosses the sky and orientate using that. Night flyers have the stars to guide them, and night migrants flying in the Northern Hemisphere use the fixed position of the Pole Star to find their way.

If the sky is overcast, there is another almost fail-safe system. Thanks to tiny particles of magnetic material in their heads and necks, birds can detect the lines of force that make up the Earth's magnetic field. The lines rise at an angle to the Earth's surface, so, in perceiving the angle at which lines intersect the ground, a bird can pinpoint its position on the planet with considerable accuracy.

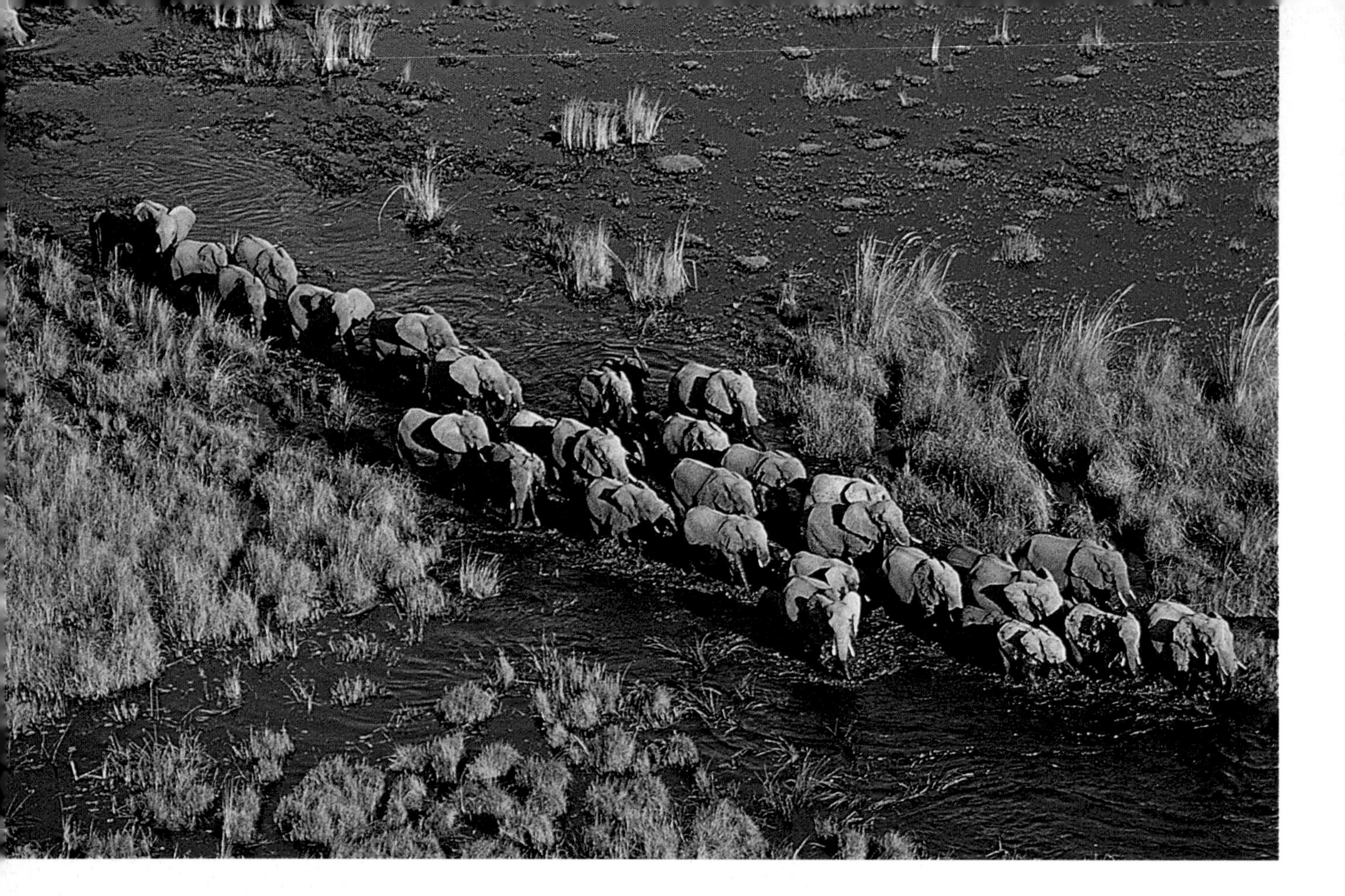

WHY ELEPHANTS NEVER FORGET A LANDMARK

Elephants navigate by remembering the location of landmarks. They embark on long treks, often in search of water, crisscrossing an area of up to 500 km^2 (190 sq miles). In Botswana during the dry season from June to October elephants leave the wilting savannahs and head for water sources that do not dry up completely. Each herd is led by a matriarch, the oldest and wisest female, for it is she who remembers the traditional routes and recognizes the relevant milestones (*see page 225*).

ELEPHANT PROCESSION *A travelling herd of elephants, up to 500 strong, comprises females and offspring. Male elephants journey alone.*

They move at speeds of 4–6 km/h (2 1/2–4 mph), along 'elephant roads' – well-worn paths as straight as any Roman road. The elephants return in the rainy seasons, from October to December and March to June.

HOW WHALES REVISIT HIGHWAYS IN THE OCEAN

Humpback whales can journey up to 6,400 km (4,000 miles) to and from their breeding sites. They breed in tropical waters in winter and return to their feeding grounds in temperate and polar waters for the summer months.

It is thought that the whales find their way by bouncing low-frequency sounds off familiar landmarks on the seabed. They may also obtain visual cues from the Sun, Moon, and stars when they surface to breathe. But the discovery of magnetite (a magnetic material) in humpbacks suggests that they can also detect, orientate to, and navigate by the Earth's magnetic field.

AUSTRALIAN STOPOVER *Humpback whales stop off in the warm waters off Queensland to feed on their migration south to the Antarctic.*

8 KEEPING IN TOUCH

SENSES AND COMMUNICATION

Getting the message across is a priority for many creatures. Whether the aim is to attract a member of the opposite sex, lure a potential meal, or deter or deceive an enemy, the best communicators get what they want by sending clear signals.

WHY HUMPBACK WHALES MAY CHANGE THEIR TUNE

Singing humpback whales switch songs when they hear a catchy new ballad. Each group of whales has its own pattern of squawks and booms that constitute its song, providing information about each individual's location and readiness to mate.

Recordings of whale songs in the Pacific in the 1990s found that over a two-year period one set of harmonies had been replaced by another. The new song was similar to that of Indian Ocean humpbacks. The two groups had mixed and the Pacific whales had adopted the newcomers' song.

SWITCHING ON THE POWER TO FIND THE WAY

Like bats, the Amazon knife-fish and the elephant-trunk fish of Africa use radar to find their way. But instead of emitting pulses of sound, they use electricity. By generating a weak electrical field around their body, these fish can 'read' and interpret any of the pulses that hit a rock, plant, or other obstacle and bounce back. This helps them to build a detailed image of their surroundings.

The Amazon knife-fish has a ribbon-like fin that runs along its belly. In the skin above this is a line of organs that emit a stream of 3–10 volt electrical pulses. The elephant-trunk fish emits pulses from modified muscles at the base of its tail.

ELECTRICAL CHARGE *The elephant-trunk fish from Africa has a well-developed electrical sense, which it uses to recognize friend, foe, or food.*

HARP SEALS PREFER THE GOLDEN OLDIES

The old tunes are the best as far as harp seals are concerned. Unlike humpback whales, killer whales, and bottle-nosed dolphins, which vary their songs, harp seals stick to what they know. They produce a wide variety of sounds and have the richest repertoire of all seals.

Harp seals live on drift ice in the Arctic, travelling south to Norway and Newfoundland in winter. They gather in huge breeding groups of up to 250,000 individuals, all calling under water for mates. There are so many of them that potential partners may ignore males with new or altered songs. Sticking to the same tune may be a way to avoid being 'drowned out'.

MOLE RAT USES CHEMICALS TO KEEP LAW AND ORDER

A queen naked mole rat governs her colony with a cocktail of chemical messages. The moment she becomes queen, she spreads her pheromones, probably in urine, to ensure that no other female can breed.

With a court of nonbreeding workers ruled by a single breeding queen, naked mole rats are the only mammals whose social system resembles that of bees and ants. They live in large underground colonies in eastern Africa. When the queen dies, several females start to show signs of sexual activity. Only one wins to become queen and spread her own hormone suppressant.

THE INTRICATE LANGUAGE OF BIRDSONG

The range and quality of birdsong is phenomenal. The black-capped chickadees of North America have literally dozens of calls, each of which means something different.

Most birdsong proclaims the male singer's hard-earned territory. The simple two-note call of male pigeons makes it clear to other pigeons that he has claimed a patch of woodland and is advertising for a mate.

Birds have a range of relatively simple calls to beg for food, give an alarm, keep in touch, or express fear or pain. Only the passerines, such as nightingales or canaries, have true songs. Females prefer males with the most complex songs – a beautiful song often indicates a healthy singer.

NIGHTINGALE'S SONG *Although the nightingale has one of the most elaborate songs, its message is as simple as that of other birds.*

THE PERFUME WITH A PUNGENT WARNING

Holding his smart, black-and-white banded tail aloft, a ring-tailed lemur wafts a malodorous warning at a rival. Lemurs live only on the island of Madagascar. If two opposing groups of ring-tailed lemurs meet, the males stand up, draw their long tails between their thighs, and wipe them on the scent glands on their wrists, smearing secretions on to the fur. Then they drop back on to all fours, bend their freshly scented tails over their heads and march towards each other, unleashing gusts of war-perfume in an attempt to get the other group to back off.

HONEY-LOVERS' PERFECT PARTNERSHIP

The greater honey guide's favourite food is waxy honeycomb and bee grubs. But the food is locked away inside an impenetrable fortress: a wild bees' nest. The bird cannot break the nest open, so it calls for help from another species – the honey badger, or ratel, a relative of the skunk.

To attract the honey badger's attention, the honey guide launches into an excited chatter, accompanied by a swooping flight. The badger, familiar with the routine, follows the honey guide to the nest. With its thick fur protecting it against bee stings, the badger breaks open the structure. While the animal gorges on the honey the honey guide is able to feast on the spilled honeycomb and grubs.

HOW BABOONS FACE DOWN RIVALS

Among baboons, flashing eyelids tell a rival, in no uncertain terms, to back off. Baboons live in multi-male troops of up to 100 or more throughout sub-Saharan Africa. Males constantly struggle for status, because the dominant ones have most access to females. The complexity of their social relationships demands visual and vocal signals.

TEAMWORK *The greater honey guide in Africa's Kalahari Desert uses a special call to recruit a honey badger (ratel) to open a calorie-rich bees' nest. The badger grunts in reply.*

WARNING SIGNAL *Competition for mates among baboons has led to the evolution of long, sharp canines, displayed here in a 'threat-yawn'.*

If a male wants to express threat or dominance, he fixes his rival with a long, hard stare which he intensifies by raising his eyebrows and retracting his scalp to display contrastingly coloured eyelids. He then blinks rapidly.

To show off his dental weaponry he yawns in an exaggerated fashion. Gelada baboons use a 'lip-flip', in which the mobile upper lip is folded back to reveal formidably sharp teeth.

CAREFUL KRILL PRACTISE SYNCHRONIZED SWIMMING

Krill swim in a curious staggered formation by detecting and avoiding those swimming in front. These tiny, shrimp-like creatures of the ocean congregate in densely packed schools. They propel themselves forwards with paddle-like swimmerets, but this stroke produces a jet stream that makes swimming hard work for those behind.

Sensors on their antennae detect the swimmeret beats of the krill in front through changes in the water pressure, allowing them to adopt a position that reduces their own swimming effort.

HOW ELEPHANTS COMMUNICATE

Elephant language is so rich and complex that so far humans understand only a few of the facial expressions, tummy rumbles, pungent smells, gentle caresses, and trunk calls that elephants use to communicate with each other.

Handshakes and kisses

The elephant's leathery skin is acutely sensitive, with a rich supply of nerves. So physical contact – using trunks, tails, the soles of the feet, and the body's skin – plays an important role in strengthening social bonds.

A mother and her offspring touch the most, the female constantly guiding her calf, reaching behind with her tail to check that it is following, letting it lie against her belly when they rest, wrapping it reassuringly in her trunk, and touching its genitals when it is frightened.

All elephants, whatever their relationship, use touch to communicate: they slide against each other when taking mud baths, or practise sparring as adolescents.

TOUCH AND FEEL *Elephants are social and tactile animals. Strong bonds are formed early in life, helped by close contact during play.*

Rumbles and trumpets

Elephants also talk to each other using low-frequency infrasound that can carry for considerable distances, allowing animals as far apart as 1 km (1,000 yd) to communicate. They have a repertoire of about 30 rumbles, including a greeting rumble for when they meet. If the matriarch is ready to move on, she utters a 'let's go' rumble. There is a contact call and answer, a lost call, a suckle rumble, a reassurance rumble, and a musth or oestrus rumble (from elephants in breeding condition). An elephant's trumpet is audible to humans. It comes from the trunk, which is like a wind instrument, producing a variety of sounds, including the threat trumpet of an aggressive male.

Scents and stinks

Elephants rely heavily on chemical cues, and smell is their most highly developed sense. An elephant's trunk is its 'finger' as well as its nose. It uses the tip of its trunk to investigate another's genitals or mouth for clues about its identity, sex, age, and reproductive status.

To follow a track, an elephant sweeps its trunk over the ground like a metal detector. For long-distance scents, the trunk is swivelled round like a periscope.

Body language

Elephants have good vision at close range and can see some visual signals from 50 m (55 yd). There are also around 100 positions of the body, head, ears, or trunk, each of which indicates something specific to other elephants. An elephant that is annoyed will lower and extend its head, folding its ears back into a V shape. The way an elephant stands in relation to another indicates how it views the other's status.

FRIENDLY TUSSLE *Two bulls test their strength. To show that they are not serious about fighting, each gives the other visual and audible cues.*

WAYS OF SEEING

Although many animals have eyes like humans, this does not mean that they see what we see. Often we can only guess at their view of the world. Some may see little more than light and dark, others see colours we cannot begin to imagine.

THE POP-EYED TARSIER'S ENORMOUS ORBS

No animal has eyes as big, in relation to its body, as the pop-eyed tarsier. Like two huge dinner plates, the eyes of this small primate totally dominate its face.

The eyes are so huge that the volume of the tarsier's eye sockets exceeds that of its brain case. Since the tarsier hunts at night it needs large eyes to gather what little light is available. The eyes have enormous pupils which dilate fully at night and a high concentration of rod receptors in the retinas. Rod receptors capture light and work well in dim conditions.

The bulging eyes are forward-looking and close-set for good binocular vision, allowing the tarsier to judge distance accurately. The eyes are so large that they are almost immobile in their sockets – but the tarsier compensates with its ability to turn its head through 180 degrees.

LOOKING BACK *Tarsiers, which grow to about 13 cm (5 in) long, can turn their heads around so far that they can look behind themselves.*

GIANT SQUID'S SLIDE-PROJECTOR EYE POWER

Giant squid, measuring up to 18 m (60 ft) in length, have eyes as big as footballs. Their eyes are among the largest in the world.

Like other cephalopods, including octopuses and cuttlefish, the giant squid's eyes are more highly developed than those of any other invertebrate, allowing them to see in greater detail and be aware of the slightest movement at a distance.

Squid live and hunt in open water and need good sight to spot prey from afar. Octopuses, however, tuck themselves in crevices and need to be able to spot a potential meal as it passes by. Experiments show that an octopus can discriminate visually between objects as small as 5 mm (1/4in) from a distance of 1 m (3 ft 3 in). It supplements the information from its eyes with its highly developed sense of touch.

Cephalopod vision is similar to that of humans in that an image is focused through a lens on to the retina at the back of the eye. However, in humans the image is focused by changing the shape of the lens while cephalopods zoom the lens forwards and back in order to focus the image, rather like a slide-projector.

SHRIMP'S SOPHISTICATED EYE TO THE MAIN CHANCE

The mantis shrimp has two unique mobile segments in its head, which carry its eyes and horny appendages known as antennules. As carnivorous hunters, mantis shrimps need to judge how far away their prey is and how fast it is moving. To do this, they have the most highly developed compound eyes of all crustaceans.

Not only do the eyes detect moving objects, but, judging from the accuracy with which the shrimps can swim towards their prey and strike with their claw, they can also assess

SHRIMP'S EYE VIEW *The mantis shrimp from Indonesia has good eyesight, but chemoreceptors on its antennules are also useful for detecting prey.*

depth. Mantis shrimps are pugnacious creatures. They wait at the entrance of their burrows on the seabed for small snails, crabs, clams, or fish to pass, then strike rapidly with their large, flexed second limb to spear or smash the victim. Some are so strong that, in captivity, they can crack the glass of their aquarium.

The 350 species of mantis shrimp range in size from 5 cm (2 in) long to giants of more than 36 cm (14 in). Most live in tropical waters.

FLEXIBILITY GIVES HORNED OWL AN ALL-ROUND VIEW

The great horned owl can swivel its head through almost three-quarters of a circle to look behind itself. This ability greatly enhances its chances of spotting prey such as mice, shrews, voles, and other small creatures from a distance.

Owls need to judge exactly how far away and how big their prey is. They get the three-dimensional effect they need by having binocular vision: both eyes face forwards so that the field of view of the left and right eye overlap. This means that owls can see with both eyes simultaneously an object within a central arc of 60–70 degrees, allowing them to swoop down onto their catch with great accuracy.

To keep locked on to an object that has moved from the central three-D field of view the great horned owl will turn its head – something it can do better than any other animal.

Most owls are busy at dusk, but about 50 of the 135 species are truly nocturnal. With only the moon and stars for light, they have evolved enormous eyes to make the most of what little light is available.

Their eyes are so big that they can hardly move in the sockets. The neck-swivelling compensates for this restriction, too.

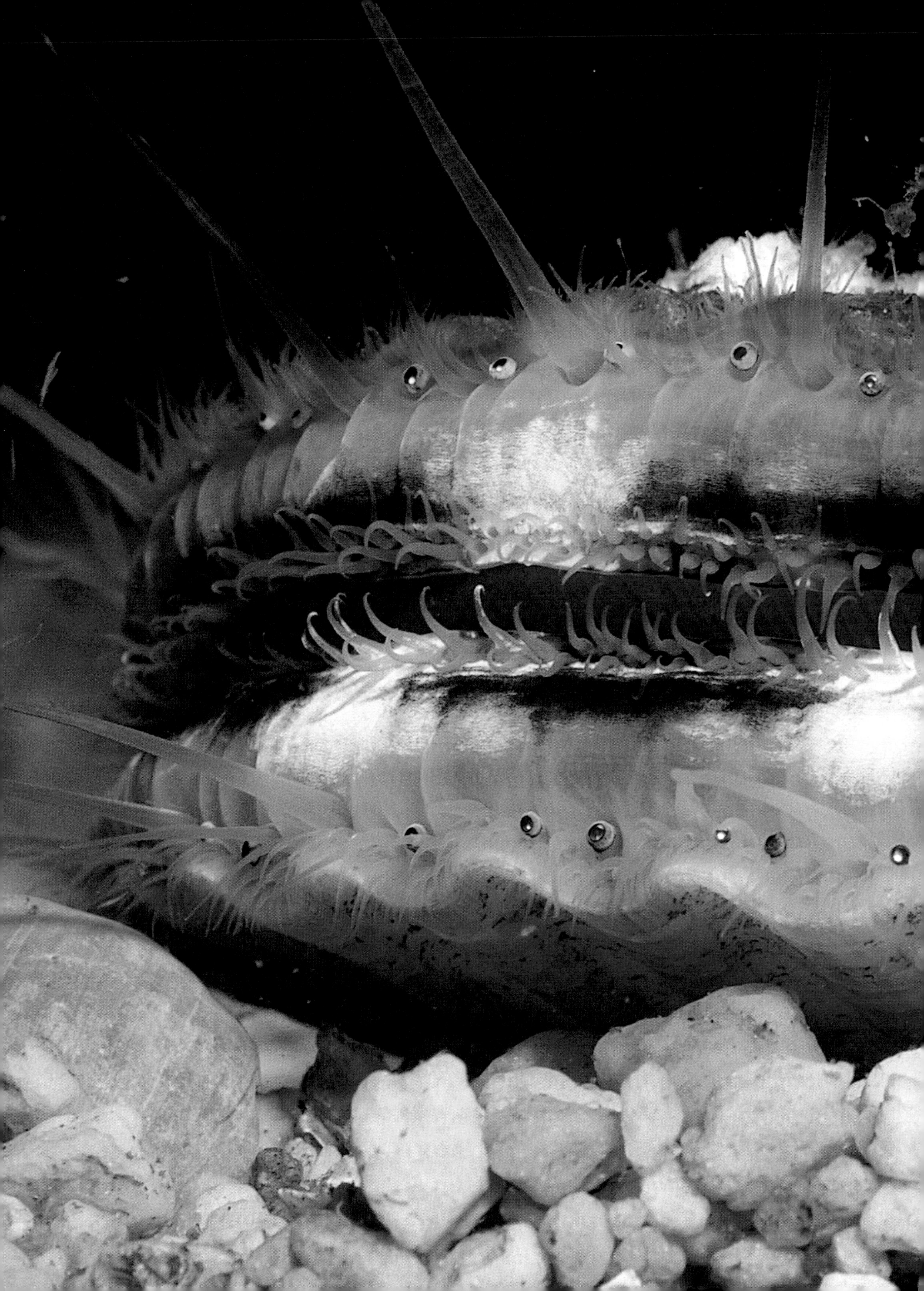

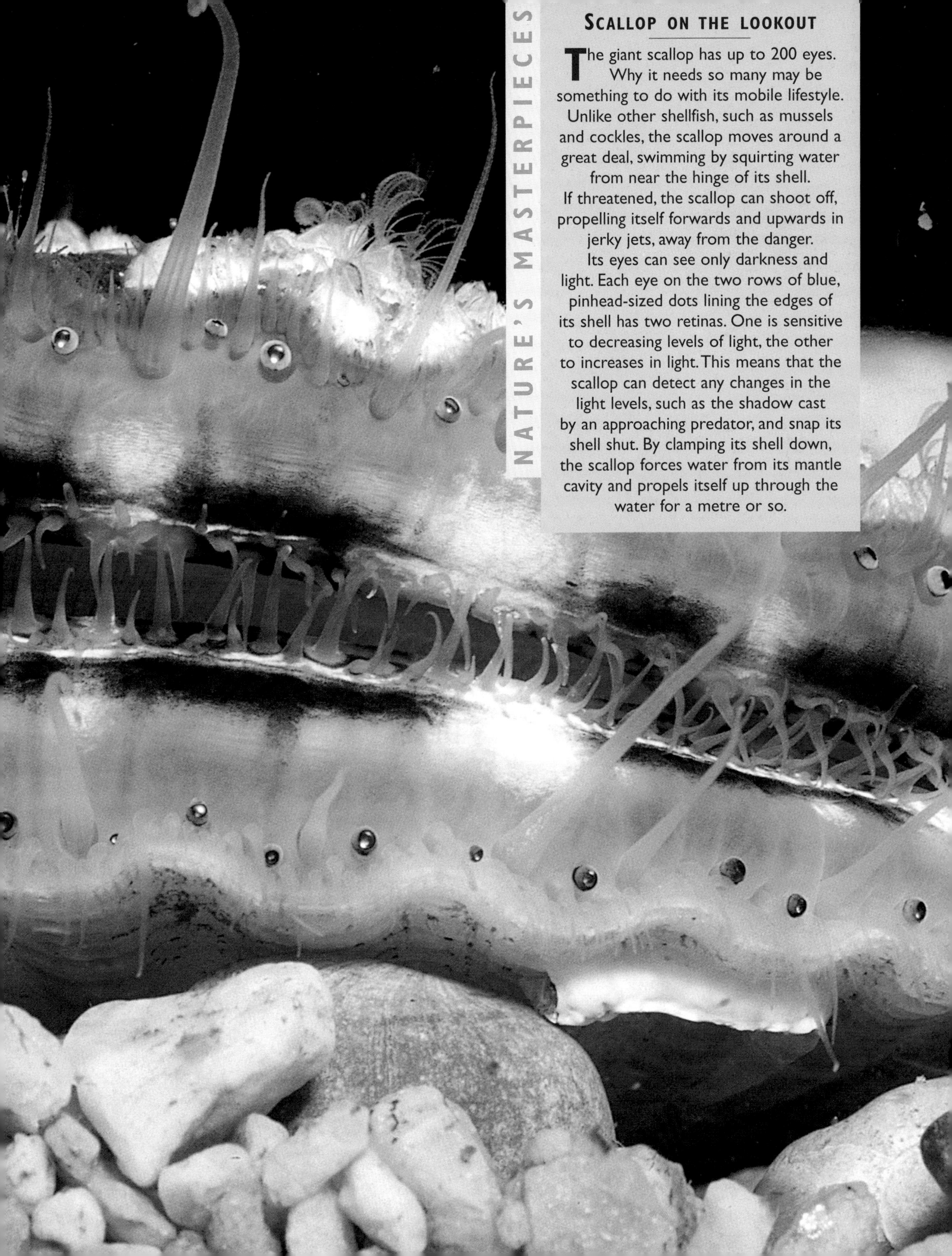

NATURE'S MASTERPIECES

Scallop on the lookout

The giant scallop has up to 200 eyes. Why it needs so many may be something to do with its mobile lifestyle. Unlike other shellfish, such as mussels and cockles, the scallop moves around a great deal, swimming by squirting water from near the hinge of its shell.

If threatened, the scallop can shoot off, propelling itself forwards and upwards in jerky jets, away from the danger.

Its eyes can see only darkness and light. Each eye on the two rows of blue, pinhead-sized dots lining the edges of its shell has two retinas. One is sensitive to decreasing levels of light, the other to increases in light. This means that the scallop can detect any changes in the light levels, such as the shadow cast by an approaching predator, and snap its shell shut. By clamping its shell down, the scallop forces water from its mantle cavity and propels itself up through the water for a metre or so.

ALL EYES *Huge compound eyes enable the southern hawker dragonfly to detect and catch its elusive aerial prey on the wing.*

DRAGONFLY'S SUPERIOR FLICKER-VISION

A dragonfly's huge, bulging eyes cover more or less its entire head. Thanks to these, the dragonfly can see nearly all the way round, scanning for prey while keeping a watch for hungry birds that may be interested in a meal themselves.

Its compound eyes are composed of more than 20,000 tiny, six-sided facets, each with its own tiny lens and retina. The minuscule facets are stimulated whenever something moves through their field of vision, providing the dragonfly with a sort of flicker-vision that enables it to detect even the slightest possible movement.

The number of facets in an insect's eye is an indicator of the sharpness of vision that can be achieved. Compared with a dragonfly, an ant, which has only nine facets, has very poor sight indeed.

INSECT RUNWAYS LINED WITH UV LANDING LIGHTS

To humans, the alpine arnica flower looks like a splash of sunshine yellow. A honeybee, though, sees purple petals with crimson lines leading to a crimson centre. This is because, unlike humans, the bee sees the world through the ultraviolet end of the spectrum.

Many plants depend on insects for pollination. They entice them with a rich reward – nectar. To reach the nectar, insects must brush past the plant's sexual organs, dusting themselves with pollen and rubbing it off on the next plant they visit.

Plants have evolved colourful blooms to attract pollinators, but the signals are often in ultraviolet and invisible to humans. These signposts serve to direct pollinators towards the centre of the flower. Foxgloves have 'runways' of stripes while oxeye daisies direct bees with clear 'arrows' on a white background.

which produces the hormone melatonin and is important in controlling sleep, mating, and hibernation cycles. It may also monitor changes in natural light levels and tell the tuatara when it is time to hibernate, for example.

SHADES HELP HYRAXES TO KEEP AN EYE ON THE SKY

To cut out the glare from the fierce African sun, hyraxes have in-built sunglasses. This increases their chances of spotting aerial attackers such as Verreaux eagles. By day, the rabbit-sized rock hyraxes like to sunbathe on rocky outcrops. In order to keep watch for attackers without being blinded by the bright light, hyraxes have evolved spade-like shields (umbraculae) in the irises of their eyes to shade their pupils.

SPARKLING BRILLIANCE OF A BIRD'S-EYE VIEW

The world as seen by birds is bursting with many more detailed, sharper images than can be seen by us. The retina of a human eye contains 120,000 light-sensitive cells, known as rods,which transfer information received to the brain. The more rods, the sharper the image. A sparrow has 400,000 rods and a peregrine up to a million, giving it extraordinary visual acuity.

Birds also have superb colour vision thanks to a profusion of colour-sensitive cone cells, and photoreceptors that are sensitive to ultraviolet light. They can see hues we cannot even begin to imagine.

In addition, birds' eyes have a higher 'flicker-fusion' rate, allowing them to see minuscule details such as the movements of a butterfly's wings that are invisible to us.

WHY SOME MONKEYS SEE RED

Among mammals, only African and Asian monkeys can see reds and greens in addition to the blues and yellows seen by most other mammals. Being able to see reds and greens helps monkeys to find ripening fruit, so it seems strange that some South American monkeys are colour-blind.

Research with Geoffroys marmoset, which lives in Brazil, has shown that the X chromosome in its cells carries the gene for one of three possible pigments in its retina. Each pigment is sensitive to a different colour, so individual Geoffroys marmosets see quite different colours and some cannot see reds and greens.

THE THREE-EYED REPTILIAN RELIC

The tuatara, a primitive reptile which lives on islands near New Zealand, has a rare 'third eye'. From the outside, this appears as no more than a tiny spot at the top of the tuatara's head, but just under a small hole in the skull there is an internal 'eye' complete with a lens, retina, and nerves connecting it to the brain. The tuatara cannot see with this eye, but the eye is linked to the pineal gland,

LIVING FOSSIL *The tuatara has remained unchanged for 140 million years. Its 'third eye' was a common feature in many ancient reptiles.*

DAZZLING BEACONS *The scales on a butterfly's wings that create colour, such as on this birdwing butterfly from Australia, are modified hairs.*

MESSAGES ON BUTTERFLY WINGS

Butterfly wings are the most elaborate visual signal there is. The colours and designs on wings come either from 'proper' pigments or from the effects of microscopic physical structures in the scales, which split the light falling on them and reflect back only a part of it as colour.

Some butterflies have ultraviolet markings on their wings, which act like an invisible ink since most vertebrates cannot see UV. So while a butterfly can advertise itself to mates – its wings revealing, at a glance, its sex, physical condition, and readiness to mate – it can avoid attracting the attention of a number of predators.

When the already visually stunning, metallic-blue morpho butterfly is filmed in flight using a camera sensitive to UV light, its wings are transformed into dazzling, flashing beacons. It can talk, in code, to other butterflies, while remaining invisible to many of its enemies.

HOW KING PENGUINS SEE IN THE DARK

Thanks to its highly specialized eyes, the king penguin hunts with ease in the ink-black waters of Antarctica. Unlike other diurnal (day-living) birds, the king penguin has superb vision in low light. It is able to shrink its pupils to tiny pinholes while still on land, thus minimizing the amount of light that enters its eyes. This way, it gets accustomed to the dark even before it sets off hunting.

When the king penguin dives into the water its pupils expand 300 times, reaching more than 1 cm (1/2 in) across. Most birds can expand their pupils only 16-fold. At speeds of 1.4 m (4 ft 6 in) a second, the penguin dives down to 300 m (985 ft), passing, in terrestrial terms, from bright sunlight to dusk in just 70 seconds. In spite of hunting in darkness, it can, in dives lasting just 4–5 minutes, find and eat around 2,000 lantern fish (which are barely 8 mm (3/8 in) long) in a 24-hour period. Lantern fish have light-producing cells, so the bigger the king penguin's pupil, the easier it is to spot the telltale specks of light.

HOW TWO EYES SERVE AS FOUR

The '*quatro ojos*' (four eyes) fish of the Amazon can look through both air and water at the same time. The fish has two eyes that function as four.

Stuck right on the top of its head, each eye is split into an upper and a lower section. Light from the world above the water passes through the width of the egg-shaped lens, which gives good long-distance vision. While swimming just below the water's surface, the fish can simultaneously check the world above for predators, such as birds, while viewing the surface and depths below for food.

SHARP HEARING

There are sounds we cannot hear which are essential to animals. Their sensory world is extraordinary and they often explore it in strange ways, from listening with their knees and stomachs to breathing in sounds.

BIRDS' SECRET WORLD OF INFRASOUND

Some birds are able to detect sounds with ultralow frequencies, beyond the range of the human ear. Humans can hear frequencies of only 20–30 Hz or above, while birds such as pigeons can respond to frequencies as low as 0.05–10 Hz. These low-frequency noises, known as infrasound, travel through the atmosphere for hundreds of kilometres and may help birds to find their way at night or in cloud.

The capercaillie, a large grouse found in northern Europe, broadcasts part of its song on a frequency which is clearly audible to humans, but loud though it is, the call travels only 500 m (550 yd). To compensate, there is an infrasound component to the bird's song which can be heard by other potential mates and rivals over twice as far away.

SOUND SYSTEM *A capercaillie's song can be heard by humans, but by inflating its large windpipe the bird can produce a low frequency boom too deep for the human ear.*

TOAD WITH AN UNUSUAL INNER EAR

The fire-bellied toad is able to listen with its lungs. Most frogs and toads have an internal 'middle ear' to receive and transmit sound to the sensory organ of the inner ear, which connects with the brain, but some species have other means of detecting sound. In the case of the fire-bellied toad, sound waves travel through the toad's mouth and resonate in its lungs, which lie just below the skin. Then the waves are transmitted through soft tissues to the toad's inner ear.

This unusual arrangement would have allowed early amphibians to hear sounds under water as well as on land.

BEWARE THE TOAD *The fire-bellied toad listens through its lungs, while its brightly coloured belly warns predators of its poisonous skin.*

CRICKETS GO OUT ON A LIMB TO LISTEN

Crickets have a membrane of stretched skin on their front legs that acts as an ear. Grasshoppers have a similar membrane, known as a tympanum, but in most species this is on the underside of the abdomen.

The membrane vibrates in response to sound and, in turn, stimulates receptors which send a sound message to the insect's brain. This way, crickets and grasshoppers can hear calls made by potential mates hidden in the undergrowth.

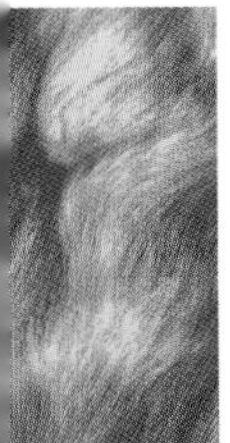

STRANGE SOUNDS

From songs and screams to whirrs and whines, the natural world is never truly silent. Some noises that sound irritating to humans are music to an animal's intended mate. Other songs are more beautiful than anything we could compose.

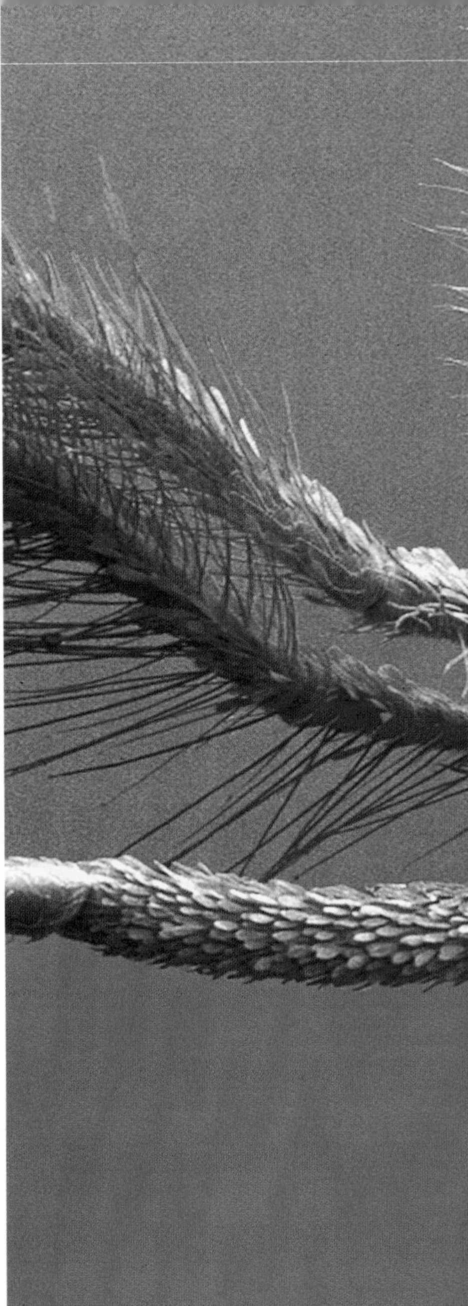

THE GIBBONS' ACROBATIC DAWN CHORUS

Every morning, the forests of South-east Asia resonate to the musical duets of gibbon pairs. The climax is a narrow-frequency wail from the female that is designed to carry through dense forest (humans use a broader frequency which does not carry voices so well).

As the sun rises over the horizon, the air above the canopy becomes slightly warmer than the air below and any sounds that filter up through the canopy are reflected back down by the warm layer of air. So, by calling at dawn from the tallest trees, gibbons ensure that they can be heard up to 3 km (2 miles) away. The calls are often accompanied by spectacular acrobatic displays as the gibbons gracefully swing their gangly bodies along branches on long arms, covering up to 12 m (40 ft) in one leap.

The duet announces that this part of the forest is occupied. It probably also serves to reinforce the pair bonds between adults.

CALL OF THE WILD *Singing from the treetops, the female white-handed gibbon from Thailand provides clues as to her size, age, sex, and identity.*

THE SONG OF THE HUMPBACK WHALE

The haunting songs produced by male humpback whales are more complex than any other sounds in nature. Booming echoes follow bass grunts and piercing high notes as the tune reverberates across the world's oceans. Sonographs show that humpback songs are highly structured, with distinct notes and lyrical themes that are repeated and varied over about 12 minutes, and in some cases up to 35 minutes – the longest known song. The whale may repeat the entire song for hours on end.

Female humpbacks listening to the opera may use the songs to assess a male's physical condition. Certain sections of songs, or calls, also help individuals and populations to identify each other and coordinate travel during the annual migration to and from breeding grounds. When sound-transmission conditions are favourable, the deep notes carry for many thousands of kilometres.

LISTENING DEVICE *A scanning electron micrograph of the head of a male mosquito shows its acutely sensitive feathery antennae.*

TUNING INTO A SUITABLE MATE

The male mosquito's antennae are the most sensitive known among insects. They enable him to pick out a female of the same species from among dozens of mosquitoes and insects flying nearby (there are more than 3,000 species of mosquito).

Long bushy hairs attached to the male's antennae vibrate in response to sound waves, transmitting the information along each antenna to the Johnson's organ at its base. The sound frequency that causes the hairs to vibrate the most vigorously is identical to the drone made by the wings of a female of the same species. With his antennae vibrating at maximum frequency, the male knows that he has tuned in to a mosquito of the right species and sex.

To avoid confusion, males have a different flight tone, and immature females have a flight tone outside the hearing range of males so as not to attract suitors. Things do go wrong: male mosquitoes have been known to fly straight into the mouths of opera singers droning at a crucial note.

WATER BOATMAN SINGS THROUGH THE WAVES

The water boatman harnesses wave power to enhance the range of its love messages. This tiny aquatic bug that swims just below the surface of ponds, rubs its forelegs together to generate a noise of about 40 decibels that can travel through water for about 40 m (44 yd). If a receptive female calls back, the male swims towards her and launches into a loud courtship serenade. Nearby males burst into song, too, sometimes coming closer to fight for the female. The male that succeeds in mounting her produces a shrill victory call.

MONKEYS WITH A LANGUAGE OF THEIR OWN

The way vervet monkeys warn each other of danger is one of the closest examples we have of animals using words like humans. Vervet monkeys have specific 'words' for different sorts of danger. Their alarm vocabulary includes words for leopard, bird of prey, and snake.

Vervets are widespread throughout the African savannah, where they live in large troops. Should one of them spot danger, it gives the alarm.

But different predators attack in different ways so, in order to prompt its peers into appropriate evasive action, the alarm-giver 'describes' the danger. In response to a loud bark, the monkeys run for the very tops of trees – a leopard has been spotted. If they hear a chuckle alarm, they look up,

WORDS OF WARNING *A vervet monkey has taken to a tree to warn of danger. Vervets have a large vocabulary of alarm calls.*

check for hunting eagles and dive into the nearest bush. At the sound of a high-pitched chattering, the vervets immediately stand up on their hind legs and search the grass for a snake.

PERCUSSIONISTS OF THE INSECT WORLD

Cicadas are the loudest insects of all, and when many call at once, the sound can be deafening. Abundant throughout the tropics, male cicadas sit high in the branches of trees to call for mates.

Each insect produces his call with a pair of built-in tymbals (hardened plates on either side of his thorax), which he flicks rapidly in and out. The tymbals click as they buckle in and click again as they buckle out, like a tin lid. Most of the abdomen behind the plates is hollow, forming a resonating chamber which amplifies the sound so that it can be heard up to 0.8 km (1/2 mile) away.

Each of the 3,000 species of cicada has its own particular click rate. Most call at between 200 and 600 clicks a second. The clicks run together into one long hum which often reaches 112 decibels. As soon as one cicada starts to sing, males in the vicinity join in. Then each male synchronizes his clicks with those of his neighbours, producing an ear-splitting chorus.

CICADA CHORUS *The resonating chamber of this bladder cicada contains a gossamer membrane, thought to amplify the insect's song.*

THE DEAFENING BOOM OF THE BLUE WHALE

The sound emitted by blue whales is louder than the sound of heavy gunfire at close range or the roar of a rocket. Blue whales, found throughout the world's oceans, are the biggest creatures ever to have lived on Earth, growing to 20–33 m (65–110 ft) in length.

This vast body contains a massive larynx and huge set of lungs, which allow it to produce a deep, intense booming call that lasts for 30 seconds or more. The low-frequency pulses measure up to 188 decibels and can be detected at least 850 km (530 miles) away, probably much more. No one knows for certain, but the call is probably involved in courtship.

THE RATTLESNAKE'S DEADLY WARNING

The distinctive sound of a rattlesnake warns of the proximity of one of the deadliest reptiles on the North American continent. The venom from a diamondback bite can kill a 100 kg (220 lb) man within an hour. Every time the rattlesnake sheds its skin, the horny sheath at the end of the tail remains, adding another segment to its rattle. When threatened, the rattlesnake rubs each segment against the next, creating a sinister, rattling buzz.

The rattle carries information about its owner's condition – the faster the rattle, the livelier the snake.

RAUCOUS RATTLE *The diamondback rattlesnake's rattle betrays its owner's age – the more segments a rattle has, the older the snake is.*

WAKE-UP CALLS *Mantled howler monkeys, found in Costa Rica, do most of their calling soon after dawn and just before dusk.*

THE LONG-DISTANCE CALL OF THE HOWLER MONKEY

When howling at full strength, a howler monkey can shout to other monkeys as far as 16 km (10 miles) away. From a distance, the eerie wails of a troop of howler monkeys – South America's noisiest land animals – sound like the roar of a busy motorway. Closer up, the racket is deafening.

Howler monkeys are leaf-eating primates. They live in groups of about 20 individuals. Because the home ranges of different troops overlap, the howlers must make their presence known to avoid confronting another group. They do this by shouting.

Both males and females howl, but the male is louder, thanks to a specially adapted, cage-shaped bone in his throat, the hyoid, which, along with his thick neck and drooping double chin, acts as a resonating chamber to amplify the sound. Other howlers answer back. If there is a territorial dispute, the two groups will engage in a prolonged vocal battle and the issue is usually settled without the howlers having to come into physical contact.

SHRIMPS THAT GO SNAP, CRACKLE, AND POP

Shallow, subtropical waters all over the world fizz with the constant crackles and pops made by shrimps. Sometimes the cacophony is so intense that submarines use it to escape sonar detection. The shrimps, each no thicker than a forefinger, produce the noise with a series of tiny explosions.

Sea water contains tiny air bubbles. When the snapping shrimp shuts its claw, it creates a jet of bubble-filled water travelling at 110 km/h (68 mph). As the micro-bubbles zoom out from the claw, they swell and then collapse violently. This process of cavitation is what makes the fizzing sound. No other animals are known to create cavitation bubbles. The crackles can stun or even kill small prey, as well as being useful for communication.

THE BIRD THAT GOES BOOM IN THE NIGHT

Male kakapos living in the mountains of New Zealand may call for seven hours each night, hence their nickname 'owl parrot'. Kakapos are rare, secretive, ground-living parrots. They dig a series of bowl-shaped arenas in prominent places on hillsides. Their aim is to attract females by projecting their calls far and wide. A booming male will inflate his entire body and produce one boom every 3 seconds for up to 45 seconds.

An inflatable air sac in the male's thorax amplifies the sound which can reach females 5 km (3 miles) away.

THE SENSE OF TOUCH

All the kissing, hugging, stroking, and tickling that goes on in the natural world has a purpose. Touch is loaded with information. It is used to communicate, to establish identity, to reinforce bonds, and even to 'see' and 'hear'.

KISS-ME-QUICK PRAIRIE DOGS

Black-tailed prairie dogs greet each other with a kiss: a quick peck for strangers and a longer embrace for those they know better. Then they inspect each other's anal glands for an identity check.

If the two prairie dogs do not know each other, they will either move on or fight. If they are from the same coterie, a close-knit group of about 30 dogs, they will greet each other enthusiastically with an open-mouthed kiss and then graze amicably side by side. They live in the grasslands of the western USA.

FAMILY GREETING *Two black-tailed prairie dogs check each other out with a kiss. In this way they can tell whether they are rivals or friends.*

HOW FROGS AND TOADS HANG ON TO THEIR MATES

Male frogs and toads are so keen to keep hold of fertile females that they clasp them in a tight hug known as amplexus. A male, having found a female, grabs on to her from behind and there he remains until she spawns, producing her eggs for him to fertilize. Only then will he release his amorous grasp.

The duration of amplexus depends on the temperature: when it is cold, bodily reactions tend to slow down. In Europe, common frogs and toads usually complete amplexus in a day. The arrow-poison frogs of tropical South America finish much more quickly, but female frogs in the cold Andes may have to carry their partners around for more than a month.

INESCAPABLE HUG *The common European toad grips the female firmly under the armpits and waits for her to spawn.*

HOW THE AYE-AYE 'HEARS' WITH ITS FINGER

The aye-aye uses its long, wizened middle finger like an antenna, listening for echoes. The nocturnal aye-aye lives only on Madagascar (*see page 306*). With this bizarre finger, it raps on wood, listening with its enormous ears pressed forwards. On detecting a hollow and hearing the faint rustles made by a hidden grub, the aye-aye rips off the bark and opens up the hollow, into which it inserts its twig-like middle finger and hooks out the grub.

WHISKERS GIVE ANIMALS SECOND SIGHT

A cat's whiskers are so sensitive that the animal can use them to 'see' a solid object in total darkness. These long, tough hairs are arranged in four rows along each side of a cat's nose, and more whiskers poke out from above the eyes. Cats can move their whiskers at will to check how big a gap is by detecting how air currents flow around the object.

The walrus also has a fine moustache which it uses to search the muddy depths for clams and mussels. It can even hold its catch with its whiskers.

WHY GROOMING HELPS MONKEYS TO RELAX

Grooming is the social glue that binds individuals together and reinforces existing relationships. It is so important to primates that some species spend up to a fifth of each day doing it. All monkeys and apes groom one another, flicking through each other's hair, plucking out skin parasites such as lice and ticks and removing flakes of dead skin and vegetation.

When an animal is being groomed it relaxes. It is thought that the mild pain caused by tugging during grooming releases endorphins (the body's natural painkiller), which suppress the production of stress hormones and make the animal feel calm.

Grooming helps to relieve the stress that builds up from competition in groups. If, for example, a subordinate female monkey is attacked by a dominant one, her body releases stress hormones. She may rush to a friend who soothes her by grooming. But grooming does more than help a stressed individual to feel better. High levels of stress reduce a female's fertility, so grooming is also important for survival.

STRESS THERAPY *Thailand's long-tailed macaques are compulsive groomers. Grooming is often used in exchange for favours, such as food.*

TOUCHING MOMENTS *Dolphins, including these Atlantic spotted individuals, have a skin rich in nerve endings that is sensitive to touch.*

DOLPHINS USE TOUCH TACTICS TO WARN OF DANGER

Dolphins are highly tactile creatures, forever pushing, nudging, prodding, stroking, and smacking one another. Spotted dolphins, faced with potential threats such as attack from killer whales, work together to nudge their calves out of the way, ready to fend off an attack. After feeding, dusky dolphins touch and caress each other with their flippers, sometimes swimming belly to belly or poking at each other with their noses. Touching happens most when when two subgroups meet having been apart for several hours.

JACKALS' GENTLE NIBBLES

For golden jackals, like all canines, almost every aspect of social interaction involves touching. Golden jackals, which live in Asia and Africa, are constantly licking, sniffing, and grooming one another to reinforce social bonds. In courtship, reciprocal nibbling sessions can last up to half an hour. A male, for example, may begin by nibbling a female on the front leg, then proceed to groom her waist, throat, and underside.

Parents also groom their pups regularly and intensively, while pups groom each other and their parents.

THE SENSE OF SMELL

The great thing about smells is that they linger. Using scent, an animal can leave important chemical messages for other animals to decipher at a later time. This is why a sense of smell is so significant in the animal kingdom.

THE ORIBI'S BATTERY OF SCENT MESSAGES

With a battery of six different sets of scent glands, the oribi has more odours oozing from its body than any other mammal. To the oribi, one of Africa's smaller antelopes, smell communicates a wealth of important information. Even when individuals are dwarfed by long grasses, they are always in olfactory contact with each other.

The biggest of the six scent glands in the male is the size of an eye socket and is situated below the ear. There are also glands in a groove in front of the eye, in the groin, beneath brushes of long hair on the front legs and shorter brushes on the hocks, and in the split in the hoofs.

Males mark their large territories by biting a grass stem and then wiping their eye gland over it. It is thought that the groin gland is used for mating cues and alarm signals, while secretions from around the ears produce a scent 'aura' that identifies each individual. Glands on the feet and knees probably leave a scent trail that helps to mark out the territory.

TONGUE TIED TO SMELL AS WELL AS TASTE

Snakes do not 'sniff' out their prey; instead they taste the air with their tongues. Snakes have a personal chemistry set in the roof of their mouths called the Jacobson's organ, which helps them to analyse airborne molecules and track prey. A snake collects airborne molecules by flicking its thin, moist, forked tongue in and out. The snake then transmits the chemicals to the Jacobson's organ by pressing the tip of its tongue into pits in the roof of its mouth. The pits are lined with fine hairs that pick up the chemical messages and send them to the snake's brain for interpretation.

The ability to 'smell-taste' the air is so well developed that a snake can accurately track a rat or mouse to eat, recognize a predator, or find a mate, even in complete darkness.

HOW PLANKTON DRAW SEA BIRDS TO THEIR PREY

Storm petrels use their extraordinary sense of smell to home in on a feast of tiny crustaceans known as krill. These shrimplike animals eat plankton (tiny, floating organisms), which release a chemical called dimethyl sulphide (DMS) when grazed by krill.

Research around the sub-Antarctic island of South Georgia shows that several birds, including Wilson's storm petrels and black-bellied storm petrels, follow the telltale smell of DMS to track down huge masses of krill. These birds hunt by night, so they cannot use vision to find their prey. Sniffing out DMS may also help birds to navigate, by detecting varying patterns of the chemical.

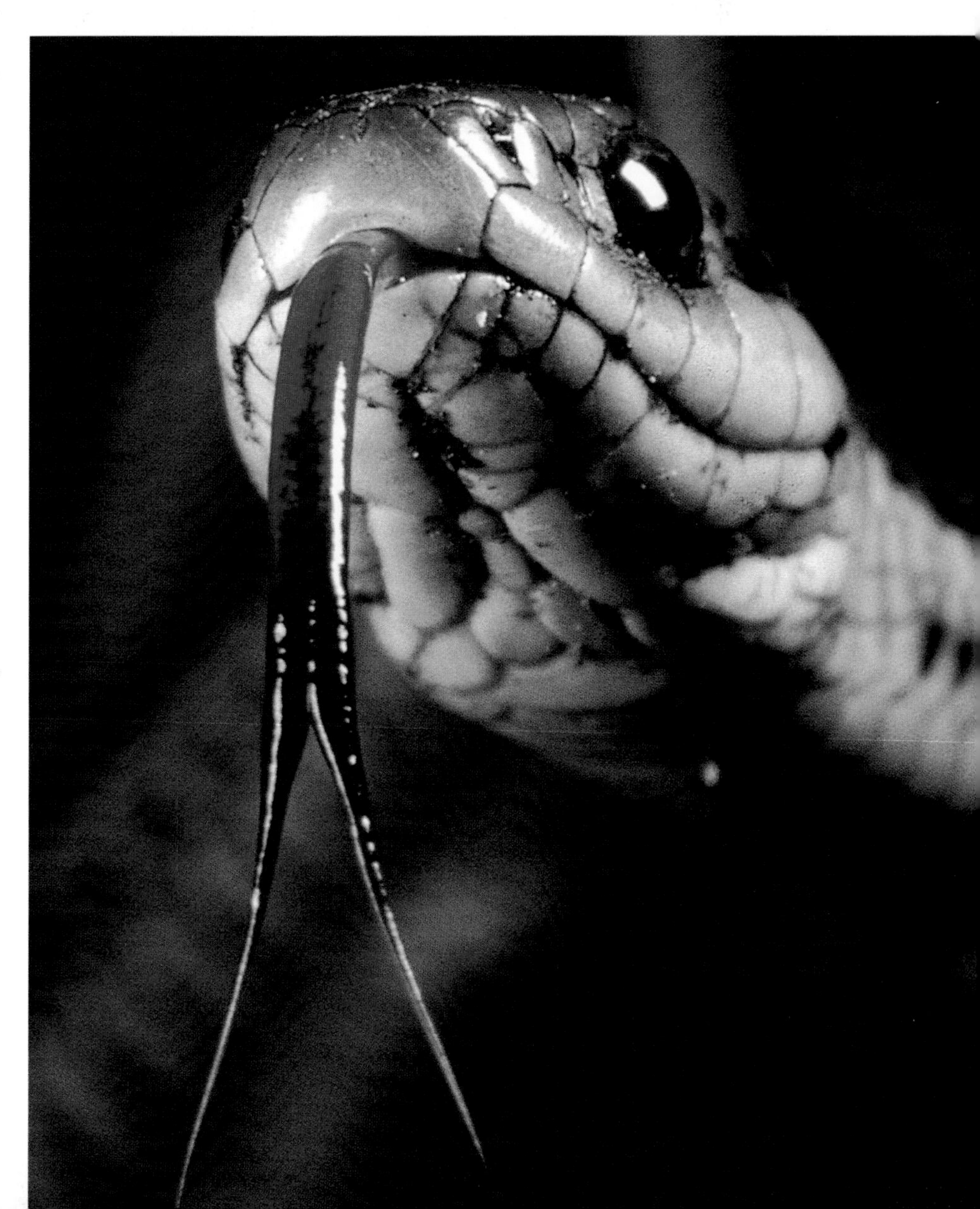

FORKED TONGUE *Snakes such as the common garter snake have very poor hearing but can taste their prey on the air from some distance away.*

SNIFF AND TELL *A submissive spotted hyena in Kenya's Masai Mara presents its rump and anal glands to a more dominant animal.*

HYENAS MAKE A MARK WITH PASTE

By wiping their backsides on grass stalks all over their territories, hyenas leave others in no doubt about who they are. The scent comes from a pouch between the anus and the base of the tail that can, remarkably, be turned inside out.

Glands lining the pouch produce copious amounts of a greasy, stinking paste that is chemically unique to each individual. A single sniff of another animal's paste conveys vast amounts of information about its identity, rank, and even what it has been doing recently. All of this is vital for hyenas, whose social system is among the most complex known in mammals.

The scent pouch is particularly big in brown hyenas, which secrete two distinct pastes from the glands lining it: a white secretion followed by a black paste. Apart from scent-marking, the pouch is also turned inside out during social interactions. To show their submission, low-ranking striped and brown hyenas present the everted gland to their superiors. Spotted hyenas evert the pouch when they are excited or aggressive.

MILLIPEDE SCENT REPELS LEMUR PARASITES

Black lemurs in Madagascar have discovered an excellent insect-repellent: millipede secretions. When wounded or threatened, millipedes release a cocktail of toxic chemicals, including chlorine, iodine, and hydrogen cyanide. Black lemurs have been observed biting into these millipedes, then rubbing the wounded body all over their fur to protect themselves from parasites.

ASIAN DEER WITH A POWERFUL PERFUME

A single grain of musk from a musk deer will scent more than 50,000 m³ (1.7 million cu ft) of air. The waxy, sludge-coloured secretion that male musk deer produce from a gland on their bellies is a powerful natural perfume and has a retail value three to five times higher (per gram) than that of gold. Its prime function is probably as a signal to attract females. For thousands of years it has been highly prized in Asia for its medicinal properties.

Musk deer live in high forests in central and north-eastern Asia. The small pouch in which the male stores his musk can hold no more than 28 g (1 oz). The continuing high demand for this special product is endangering the wild populations of musk deer.

WINGS SPREAD WAFTS OF COURTSHIP ODOUR

The more urine, saliva, and genital secretions a male sac-winged bat can collect, the better his chances of finding a mate. Males have large, white openings on their forearms – the 'sacs' that give the bats their name. Until recently, it was thought that glands within these sacs secrete the odours that females find so alluring. A Costa Rican colony revealed the true answer. At the same time every day the males fill their sacs with their bodily excretions. The male will then seek out a female and hover in front of her, wafting clouds of his courtship perfume into her face until she succumbs to his smelly charms.

SCENT-MARKING

A scent-mark is like an up-to-date curriculum vitae. It reveals to a passing animal the owner's identity and sex, and probably its age, its movements, and sometimes even its physical condition (whether it is healthy, ready to mate, or whether it is feeling territorial).

Dung and urine are frequently recycled into scent-marking materials. Rhinoceroses defecate in the same spot (a midden) time and time again to reinforce their borders, scattering the dung by kicking backwards. Hippopotamuses whirl their tiny tails as they defecate, spraying a shower of fresh dung at a rival on a border.

Other creatures use urine. Members of the dog family spray a little urine on trees or rocks. Pottos in West Africa urinate on their hands and feet before setting off to patrol their home range. Every smelly step leaves a 'footprint' long after the potto has passed by. In Madagascar, mouse lemurs use urine as a method of sexual control. The urine of the dominant male contains chemicals that make other males sterile, by causing testosterone levels to fall.

Many animals have specialized scent glands with which they manufacture their own chemical signature. The glands occur all over the body, particularly on the head, chest, abdomen, and around the anus and genitals. The ring-tailed lemur has a scent gland on each wrist, the badger has one next to its anus, and the tiger has glands between its claws.

As well as marking territories, animals scent-mark each other to establish a group identity. Domestic cats rub themselves against people in their homes to deposit a trace of their own perfume.

DUNG SHOWER *Scent-marking with dung and urine is more conspicuously frequent among the more dominant animals.*

BOUNDARY MARKER *A South African territorial white male rhinoceros defecates in 20 to 30 dung middens. He scatters the pile with slow kicks.*

COLOUR SIGNALS

Animals use colour like a code to attract mates, scare enemies and identify individuals. Colour can help creatures to disappear. From fast flashes to lengthier transformations, some animals are able to change their looks completely.

LIVING KALEIDOSCOPES

Cephalopods, including octopuses and cuttlefish, communicate by flashing vivid, changing colours across their soft bodies. A male octopus, for example, may flush crimson if he sees a female or if he is angry or frightened.

An octopus can supplement a colour change with bizarre textural changes, raising small bumps all over its skin. If it is trying to hide, it can use colour and texture as camouflage, blending in perfectly with its environment.

Cephalopods control the colour of their skin in response to nerve signals. Elastic pigment cells in the skin, linked to muscle fibres, can expand (to dilute the colour) and contract (to concentrate the colour), allowing the animal to produce complex patterns of stripes, bands, or dots over the surface, of its body. In this way it can interact visually with potential mates, rivals, and even other animals.

RED IS THE TARGET FOR HUNGRY CANARY CHICKS

When a canary chick is hungry, the lining of its mouth flushes crimson, providing its parents with a target in which to deposit food. The mouths of ravenous chicks flush a deeper red than those of chicks which have recently been fed, so the parents know which chicks to feed first.

Canaries are found in the Azores, Canary Islands, and Madeira. A red gape is a 'truthful' signal: the chicks do not cheat to get more food. If a chick is digesting food its blood supply will be directed towards its gut, hence a lighter coloured mouth, but a chick with an empty stomach has more blood available to colour its mouth.

COLOURFUL CHAT *Giant cuttlefish males in the Andaman Sea exchange information by flashing changing colours at one another.*

THE CHAMELEON HAS A COLOUR FOR EVERY OCCASION

Chameleons are experts at camouflage, but they also change the colour of their skin to threaten rivals and court mates. During an aggressive encounter, panther chameleons in Madagascar, for example, signal their rage by flushing their bodies with violent, intimidating reds and yellows.

Another species advertises its sexual availability with colour. In the mating season, bright-green males seek out grey-brown females. But if a female turns very dark, with black on her back and green lower parts,

COAT OF MANY COLOURS *Chameleons are slow movers and as a result do not have to change colour as fast as other animals. This panther chameleon, found in Madagascar, took half a minute to alter the colour of its skin.*

she is not willing to mate. The male ignores this rejection at his peril, for she can deliver a powerful bite if he persists.

A female who has already mated makes it clear that she is pregnant and no longer in search of a mate: about two or three days after copulation, she changes colour, developing an array of yellow stripes and black spots on a turquoise background.

False eyes fool predators

The sudden appearance of a huge pair of 'eyes' can deceive hungry birds into believing that they have not disturbed a tasty butterfly, but some other, much larger and more dangerous creature. Eyespots occur on the wings of a wide range of insects. Some are simply dark, round blotches, but the more sophisticated eyespots comprise concentric circles which mimic a real eye's iris and pupil. The best eyespots, such as those of the owl butterfly, include small, pale, off-centre patches – just as though light were glinting from the moist surface of a real eye. The zigzag emperor moth has an eccentric pair of 'eyes' on its underwings, each with a black 'pupil' surrounded in concentric rings of orange, pink and brown.

The farther apart the two eyespots, the better: widely spaced eyes are indicative of a much bigger animal than those close together. Eyespots often occur on insects that have good camouflage and so can utilize the startle effect: if a predator comes too close, the insect suddenly opens its wings, exposing a pair of terrifying, staring 'eyes'.

EASY TO SEE *The distinctive markings of the killer whale help individuals to follow the movements of other members of their group under water.*

ORCAS AIDED BY DISTINCTIVE PATTERNING

The black-and-white livery of killer whales is not just an ornamentation; it helps individuals to see each other. There is a white patch above the eye, a thin white patch extending up into the flank, and a grey saddle behind the fin. Killer whales hunt cooperatively (*see pages 86 and 99*) and need to be able to see each other so that they can coordinate their movements to herd shoals of fish. The striking black-and white pattern is a key to their foraging success.

THE GRASSHOPPER THAT ESCAPES IN A FLASH

Many insects buy themselves time to escape from a predator by using a 'startle effect' – a sudden burst of colour. The European grasshopper *Oedipoda miniat*, for example, looks like a mottled stone when sitting still, but if a predator comes too close it exposes bright pink hind wings in a sudden flash display. While the predator is recovering from this surprise, the grasshopper makes a quick getaway.

GAUDY OUTFITS BETRAY SEA SLUGS' UNUSUAL TASTE

With their vivid colours and spectacular patterns, sea slugs are among the most beautiful creatures in the world's oceans. Their soft bodies are painted in dazzling shades of red, yellow, orange, blue, and green. These fantastic colours may serve to warn predators that a sea slug's flesh is foul-tasting and even poisonous, for many sea slugs have special glands in the skin that produce sulphuric acid.

Some sea slugs even hijack the defences of their own prey. When a slug feeds on an anemone, some of its meal's stinging cells pass into the sea slug's body unchanged, ending up in the mollusc's mantle, ready to fire.

ALL DRESSED UP *Soft-bodied sea slugs, such as Kunie's chromodorid nudibranch, deter predators with an alarming array of bright colours.*

BODY LANGUAGE

From complex displays and dance routines to semaphore and even sex, body language helps to attract the attention of others. Animals have evolved a vast range of signs and signals, which they use to get the message across.

FIDDLER CRABS WAVE THE FEMALES OVER

At low tide, a male fiddler crab will stand at the entrance to his sandy burrow and madly wave his enlarged front claw in the air. He is trying to attract a passing female. This massive claw is solely to arouse a mate's interest: it is useless for feeding.

Before he catches sight of any specific female, he performs what is called the lateral display. He holds the claw in front of his face. Then he swings it back to an open position, rotating it up and then back down to the starting position.

While waving his claw, the fiddler crab also lifts his body up and down. The lateral display is not very intense, so the male can keep waving for a long time without getting tired. If a female comes within sight, his waving becomes much more energetic. The crab frantically bobs up and down, raising and lowering his claw in a simple but vigorous gesture.

THE WOLF SPIDER'S CAREFUL APPROACH

The male wolf spider ensures that the female knows he wants to mate by rearing up as high as possible on his legs. By doing this and waving wildly at her with his palps (appendages on his mouth-parts), he can avoid becoming her next meal.

The palps are specially patterned in black and white to help to identify him. If the female gets the message, she responds by juddering her front legs. As soon as he sees these favourable movements, the male knows it is safe to approach her.

THE SEXUAL LANGUAGE OF BONOBOS

To bonobos, sex means more than reproduction: it is also an important means of communication. While other primates communicate through grooming, vocalizations and gestures, bonobos use sex to express themselves. These primates, sometimes called 'pygmy chimpanzees' (since they are slightly smaller than

CRAB COURTSHIP *The male fiddler crab impresses females with the vigour of his wave. In order to attract a mate, a bigger, fitter male, will wave more when it is later in the season. It is possible that he may be saving his energy for when unmated females are less choosy.*

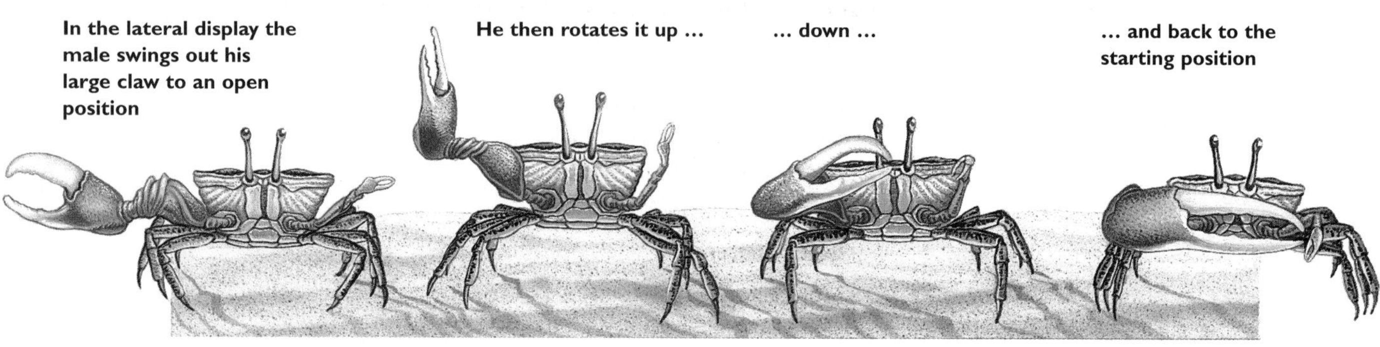

In the lateral display the male swings out his large claw to an open position

He then rotates it up ...

... down ...

... and back to the starting position

THE HONEYBEE'S DINNER DANCE

In the hive honeybees use dance as a means of communication. Every wriggle, waggle, spin, and step is loaded with valuable information. The dance that a returning forager performs to give details of a food source is the most sophisticated and complex in nature.

Honeybees have three distinct dances. The 'round dance' is used for food found within 25 m (80 ft) of the hive. The honeybee walks round in a tight circle, regularly changing direction. Watching closely, her nestmates touch her constantly with their antennae. How often she changes direction and the overall vigour of the dance indicate the quality of the food.

When the bees leave the nest they fly in ever-increasing circles around the hive and quickly discover the food for themselves. If the food is more than 100 m (330 ft) away, the dancer describes its location precisely with the 'waggle dance'.

The choreography of this dance is a figure-of-eight movement flattened top to bottom. The longer the straight walk between the two ovals, and the more she waggles her abdomen, the farther away the food is. But which way? To send the bees off on the right bearing, the dancer turns herself into a compass. The angle of her straight run from the vertical (she is dancing on a vertical hanging slat) is the same as the angle between the Sun (as seen from the hive's entrance) and the food. Using her internal clock, she is able to compensate for the Sun's movement. Also, the more high-frequency buzzes she includes, the more enthusiastic her actions, the more nutritious the food.

The third type of dance, an intermediate between the round dance and the waggle dance, is used for food located at distances of 25–100 m (27–110 yd).

COMPASS POINTS *A figure-of-eight dance with a waggle across the middle is used to indicate a distant food source. The angle of the waggle indicates to the other bees the bearing they should follow.*

common chimpanzees), live only in the dense forests of the Congo Basin, in Central Africa.

All bonobos, young and old, male and female, make sexual contact frequently, with every possible combination of partners, including infants. Sexual interactions range from a brief, casual contact of the genitals to sexual intercourse. Sex is used in play, to resolve conflict, to greet, to express dominance and submission, to reinforce friendships, to diffuse tension, to comfort, and to relax. It is even used in negotiation and trade: a male, for example, might decide to share food with a female only after she has agreed to have sex. Bonobos share one curious feature with humans: they often have sex face-to-face. The reason for this is not clear, but it may have something to do with the fact that bonobo relationships can be intense and intimate. By facing each other during sexual contact, partners can monitor each other's faces for cues about their receptivity.

DIVERSIONARY TACTICS *A north European ringed plover suddenly runs from its camouflaged nest. Its aim is to distract a predator.*

THE RINGED PLOVER'S CONVINCING ACT

The ringed plover uses its acting talent to feign a broken wing in order to protect its ground nest from an approaching predator. At a safe distance from the predator, the bird launches into an extraordinary performance, holding its wing out awkwardly and dragging it pitifully across the ground while uttering a plaintive distress cry.

This ploy is simple but effective. A predator will look up to see what the fuss is about, thus losing sight of the camouflaged nest. It will decide that an injured bird is an easy meal and follow the plover farther and farther away from the nest. When the plover judges its nest to be safe, it takes off, the 'broken' wing healed, leaving the predator staring in astonishment.

LIZARDS FLY THE FLAG

When they are ready to breed, many male lizards develop brightly coloured facial flags. Anolis lizards in South and Central America grow large dewlaps, flaps of brightly coloured skin that hang below their throats. To display these, the males nod vigorously, flashing their splendid neck-flags at rivals. Some males inflate these special throat sacs to show off the brilliant hues.

SPLASH OF COLOUR *A male blue-eyed anolis lizard in Costa Rica displays his lurid, expanded dewlap to challenge a rival.*

THE GREY REEF SHARK'S STIFF THREATS

The grey reef shark uses its entire body to spell out a clear, aggressive message if it senses that its territory is being invaded. Holding its pectoral (side) fins stiffly pointing downwards, the 2 m (7 ft) long shark, common on Australia's Great Barrier Reef, hunches its back up, lifts its snout, and starts to swim in an exaggerated manner, flexing its tail sideways. The more threatened the shark feels, the more intense the display becomes.

If the threat moves away, the shark will seize the opportunity to escape. But those who ignore the shark's message can be sure that a lightning-fast, slashing attack will follow.

CHIMPS' FEELINGS WRITTEN ALL OVER THEIR FACES

With subtle changes in their facial expressions, chimpanzees can communicate a wealth of moods and feelings. To call to one another, they use a 'pant-hoot' sound, accompanied by one of two 'calling' faces. The tight-lipped, hard-stare 'display face' is a sign that the chimpanzee is aggressive. This is the expression seen when a chimpanzee attacks another individual.

Subordinate or highly excited chimpanzees respond with full 'fear' grimaces – their lips pulled back above the gums and their teeth wide apart. A

PULLING FACES *Chimpanzees have many facial expressions. They call other groups using a pant-hoot (top). A submissive animal uses a fear-grin with teeth clenched (above), while relaxation is shown by a closed mouth or drooping lip (right).*

similar expression, but with the teeth clenched, also indicates submission and is used by a low-ranking animal approaching a superior one.

In the 'play face' adopted by youngsters, the top teeth are carefully covered, probably to show that no harm is intended. 'Pouting' with a whimpering sound communicates dissatisfaction – the animal may want to be fed or groomed.

BABY MINDER *To calm a potentially aggressive situation, male Barbary macaques will sometimes hand over a baby as a peace offering.*

BARBARY MACAQUES MAKE PERFECT FATHERS

Male Barbary macaques play an extensive role in the care of their young. A male will readily baby-sit an infant that is not his own, often picking out a 'favourite' baby and carrying it around with him. He will groom it, play with it, and show it off to other males.

These monkeys live in mixed sex groups of 10–30 individuals in Algeria, Morocco, Tunisia, and Gibraltar. When it comes to breeding, a female macaque will look for a mate who appears to have good parenting skills.

MALE MUDSKIPPERS SHOW OFF WITH PRESS-UPS

When the male mudskipper is ready to display his bright breeding colours, he does press-ups. The mudskipper – a fish, about 15 cm (6 in) in length, with long, large fins – lives in tropical estuaries around the Indian and Pacific oceans. When the male is ready to breed, the colours on his body brighten up and he develops a golden chin and throat.

He shows off his colours by using his strong pectoral fins to raise himself up in front of a female. To catch her eye, he will leap into the air, spreading his bright dorsal fins.

GRANT'S GAZELLES FLEX THEIR MUSCLES

The threat and dominance displays of East African Grant's gazelles are more highly developed than most other gazelles. They involve strutting, prancing, and ritualistic flipping of their heads in a series of swaggering postures.

The most distinctive gesture, head-flagging, is unique among gazelles. Two rivals stand broadside on to one another, facing in opposite directions. As they approach, they face away, chins and heads lifted. Then, at the same instant, they whip their heads around to face each other, flaunting their magnificent horns.

Head-flagging may help two rivals to work out which of them is the more powerful. They can spend 15 minutes in one of these confrontation duels which may include up to 20 head-flagging displays. Usually, the dispute is settled without any physical contact. But if a confrontation escalates, they lower their heads and interlock their horns in a sparring bout.

HEAD TO HEAD *When ritual display fails to establish a victor, male Grant's gazelles lock horns and fight. Injuries are extremely rare.*

SUGAR-CANE WEEVILS WEED OUT THE WIMPS

Female sugar-cane root-borer weevils have an unusual method of ensuring they are mated by the strongest males: they pretend to be males themselves. When a female mounts another female, male weevils swarm over them. Large male weevils often interrupt a 'mating' couple, attempting to dislodge the 'male' so as to mate with the female on the bottom themselves. The males that hurry over to the paired females tend to be the biggest weevils. The small ones do not bother.

It seems that by acting like a male and mounting one another, females summon over the best mates.

LEARNING AND MEMORY

Being able to remember incidents, places, or outcomes gives some animals a distinct advantage. If a creature can learn from these memories and adapt its behaviour on the outcome of an earlier trial, the benefits are enormous.

ELEPHANT SHREW CARRIES A ROAD MAP IN ITS BRAIN

The African elephant shrew memorizes every inch of its territory, so that if it senses danger it can make a quick escape. Every morning, the elephant shrew trots around its territory, removing twigs, leaves, and other debris from the paths with a neat side-kick of its forelegs. The purpose of this exercise is to refresh its mental map of its home area, including the spatial relationships of paths and bolt holes, as well as to keep the trails clear.

The mental map is much bigger than the route taken on the regular morning walk and if it must, the elephant shrew will deviate from the most familiar path and take a short cut, which it has memorized at some time in the past.

KEEPING IN STEP *An elephant shrew in Africa's Kalahari Desert knows its territory so well that it follows in its exact footsteps when on patrol.*

CREATIVE MONKEYS WATCH AND LEARN

Japanese macaques learn from each other and have been known to invent new, useful behaviours. Macaques on a beach on the island of Koshima, in southern Honshu, were regularly thrown sweet potatoes to encourage them into the open, where they could be studied.

The food was covered in sand, but the monkeys ate it. Then, one day, a young female carried her sweet potato to the edge of the water, dipped it in with one hand while rinsing sand and mud off with the other, and ate it. It must have tasted better, because she did this with every sweet potato she picked up from then on.

After a month, another young monkey copied her. Within a few months nearly every member of the troop had learnt to wash sweet potatoes. The ones who never picked up this new behaviour were the oldest macaques – they seemed to be too set in their ways to accept innovation.

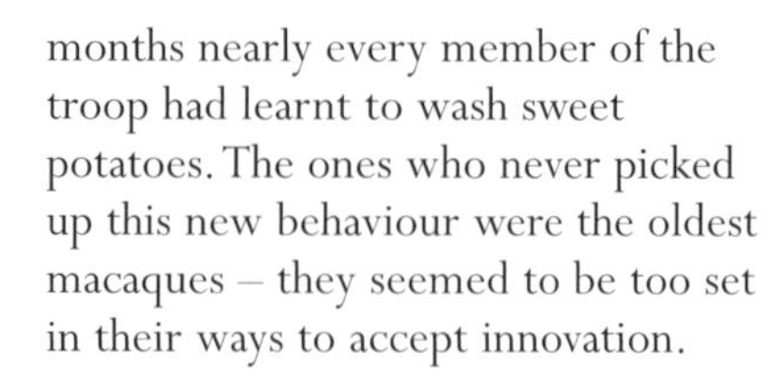

NORTHERN FUR SEALS REMEMBER LONG-LOST KIN

A northern fur seal can remember other individuals, even if it has not seen them for several years. While they are nursing, female northern fur seals living on Alaska's Pribilof Islands leave their pups to go to sea for up to a week at a time in search of fish. On their return, they use a mixture of sight, smell, and sound cues to track down their young among hundreds of others in the breeding colony.

Voice recognition lasts way beyond the breeding season. Mothers and pups have been found to respond to each other's voices even if they have not seen each other for four years.

THE OCTOPUS THAT WEIGHS UP ITS OPTIONS

One of the cleverest brains in the ocean is found in the huge head of the octopus. If presented with two different shapes, one leading to a food reward, the other to a mild shock, an octopus can learn to select only the 'right' shape.

An octopus can also learn to tell objects apart just by how they feel, using the sensitive tactile receptors in its suckers to distinguish not only between a rough and smooth surface, but also between surfaces with different degrees of roughness.

Behaving independently, one octopus learned to tug the stopper out of a bottle to get at a shrimp inside, and others in laboratories have been known to wait until night to raid fish from nearby tanks and return to their own tanks by the next morning.

BRAIN POWER *Octopuses can remember events for several weeks and are remarkable in their ability to solve problems using insight.*

PRUDENT BIRDS THAT SAVE FOR THE FUTURE

Some birds save for hard times ahead by stashing away emergency rations. Each autumn, the Clark's nutcracker, which breeds in the mountains of the western US and south-western Canada, buries up to 33,000 pine-cone seeds or hazelnuts in the ground. Months later, when the forests are in the grip of the bitter winter weather, the nutcracker breaks into these hidden larders, each of which holds just four or five seeds. In spring, its young chicks also benefit since the remaining caches of food help to supplement their diets.

Other birds also stash food. Willow tits hold the record for the number of seeds that an individual bird will store in a single day – more than 1,000. These careful caterers memorize landmarks (trees, fallen logs, boulders, and fence posts, for example) in order to recover their buried treasure, even when the ground is covered in snow. Some put their own markers, such as small pebbles, nearby.

The birds never remember all of their hidden foodstores. But this can be good for the future, too: some of the forgotten seeds germinate and grow into saplings, providing food for the next generation of birds.

NUTCRACKER SWEET *The Clark's nutcracker can carry as many as 90 pine-cone seeds in its pouch without impairing its ability to call and eat.*

RHESUS MONKEYS LEARN TO PLAY DUMB

Among rhesus monkeys, which live in Asia, underdogs are not stupid: they only pretend to be so in order to keep the peace. Dominant monkeys appear to be more competent at learning new tasks, while lower-ranking troop members seem to be slow to catch on. Recent research shows that subordinate monkeys learn just as quickly, but they have to keep a low profile.

Captive rhesus monkeys were given tests (learning which of two differently coloured boxes contained food) to see how low-ranking individuals were affected by the presence of dominant individuals. When the whole troop was faced with the task together, dominant monkeys seemed quick to catch on; subordinates appeared slow to learn. But when subordinate monkeys did the task alone, they performed much better than when the dominants were present. The presence of high-ranking monkeys does not affect the subordinates' ability to learn, but it does inhibit them from revealing what they know. The low-rankers, to avoid aggression, learn to play dumb.

SOLITARY WASPS FIND THEIR OWN WAY HOME

A female sandwasp is able to find her way back to her nest with the help of local landmarks. These might include a church steeple, mountain peak, clump of trees, or even pebbles on the ground. Before

she sets off in search of juicy caterpillars or other tasty morsels for her grubs, the wasp makes a few short, looping flights around her burrow, memorizing the position of visual cues such as sticks and stones in relation to the entrance hole.

Then she departs, using the position of the Sun to keep note of the direction. Even if it is cloudy, the solitary wasp can detect polarized light and so use the Sun as a compass.

Later, carrying her paralyzed caterpillar or other prey with her legs, she flies back to the burrow. The Sun may have moved, but that is not a problem: thanks to an internal clock, she knows how to compensate for that. Using first the distant landmarks, then the smaller ones, she navigates unhesitatingly back to her tiny burrow.

HOMEWARD BOUND *A female sandwasp will lay her eggs on this caterpillar providing a ready meal for her grubs when they hatch.*

AGED ELEPHANTS ARE A GUIDING FORCE

An old cow elephant may not be able to breed, but she can still make a difference to her family's survival, thanks to her memory. The matriarch's accumulated experience and thorough knowledge of her home range mean that she can guide her family to seasonal water sources and lead them away from areas of danger.

She may remember a spot where, after an isolated shower of rain, for example, a lush patch of grass will grow, leading her family up to 30 km (18 miles) to reach it, following familiar landmarks along the way. If there is a drought, the matriarch may recall a distant riverbed where water is only a few tusk-digs below the dried-out mud.

Full-grown elephants need to drink 70–90 litres (15–20 gallons) of water every day, so the ability to remember the location of water sources is vital for survival.

The home range of African elephants is about 750 km^2 (290 sq miles) where there is lots of food and water, and 1,600 km^2 (620 sq miles) in drier places. Elephants, both African and Asian, often use the same route to and from sources of supplies for generations, wearing out tracks known as 'elephant roads' through even dense forest.

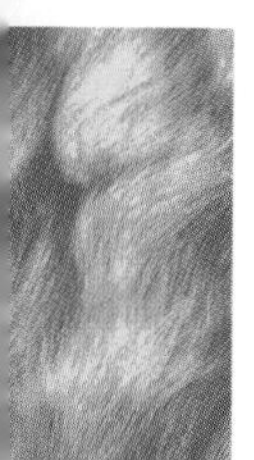

PROBLEM-SOLVING

Intelligence is one of the greatest survival qualities of all. Some animals have a limited capacity for problem-solving but have evolved behaviours that mimic intelligence. Others such as apes have found clever solutions to a range of challenges.

SHORE BIRDS THAT MUST MASTER A TRICKY SKILL

Oystercatchers have devised several feeding methods, some of which are more difficult to learn than others. At low tide, flocks of oystercatchers swarm on to the mud flats to feed. Some families concentrate on hunting for worms. Their youngsters watch and within six weeks learn to sprint up to a wriggling head and stab at it.

Other families feed on shelled molluscs: they amble across the mud flats, probing with their beaks to tug out mussels. But eating something from a shell is a tricky skill to learn, and it may take adult birds up to a year of patient demonstration before their chicks pick up the technique. Some oystercatchers simply lay the mussel on the mud and hammer it with their beaks until the shell smashes. Birds with particularly sharp bills learn to insert their beaks in between the two halves of the shell and prise it open, saving themselves a lot of energy.

FEEDING SKILLS *While turnstones delicately pick among pebbles an oystercatcher (right) employs the 'smash and grab' technique.*

CROWS CRACK THE PROBLEM OF WALNUTS

Walnut-eating American crows seem to have acquired a grip on basic physics. They have learned that they need to drop nuts with thicker shells from greater heights in order to smash them open.

Several species of birds tackle nuts and other hard-to-get-at food by dropping them from the air to break them open, but the walnut-eating urban crows in North America seem to have a deeper understanding of the issue. As well as assessing how thick the nutshell is, they have learned that different surfaces have different smashing abilities. Nuts released on to asphalt and concrete do not need to be dropped from very high, but if the crow is flying above soft earth, it will fly higher before letting go of its nut. And if there are nut thieves around, the crows are smart enough to fly low, so that a dropped nut is not stolen before they have had a chance to eat it.

MOUNTAIN GORILLAS GRASP THE NETTLE

Nettles are covered in stinging hairs, but mountain gorillas have worked out how to eat these plants without getting hurt. The stings grow mainly on the leafstalk and leaf edges of nettles, but this is not a problem for mountain gorillas as long as they handle the plant using delicate, precise, and well-coordinated hand movements.

First, the gorilla grabs the stalk of the nettle plant near the base, then strips several leaves off in one smooth movement, sweeping its fist upwards. Holding these plucked leaves in its lower fingers, the gorilla repeats the gesture. To eat the nettles, the gorilla holds them in a bunch in one hand and uses the other hand to twist off all the remaining stalks (where most of the stings are) in one go.

Then the gorilla folds the bunch of leaves into a neat parcel. This tucks the stings safely into the middle of the leaf sandwich, where they will not affect the animal's sensitive lips.

STRICT VEGETARIAN *Mountain gorillas live almost entirely on vegetable matter. They rest during the lengthy process of digestion.*

WEAVER ANTS' SILKEN STITCHES

Weaver ants, found in Africa, are experts at making nests out of leaves held together by home-grown silken thread. Nest-building is a family affair: the larger weaver ants curl up the leaves, bring the edges together, and hold them in place with their jaws and feet, while smaller ants collect some of the colony's grubs, which produce a glue-like silken thread. To secure a seam, the ants squeeze the grubs' abdomens like tubes of glue, moving the grub to and fro across the leaf edges.

ANTING SOLVES AN ITCHY ISSUE

Many birds have discovered that they can obtain a natural insect repellent from ants. The repellent is formic acid, which the birds use to disinfect themselves of the irritating parasites that live in their feathers. Some ants squirt the acid

A FIRM BOND *The larval silk used by weaver ants to bond leaves together is the same substance that their grubs produce to make cocoons.*

from their abdomens when disturbed. Grabbing an ant in its bill, a starling will stroke the insect across its wing and tail feathers, distributing the formic acid that the distressed ant is producing. This behaviour is known as 'anting'. Other birds grab beakfuls of ants in one go, reaching what appears to be a state of ecstasy as the angry ants rampage through their feathers.

CLEVER CROWS HAVE TWIGGED HOW TO CATCH GRUBS

On the Pacific island of New Caledonia, crows use three different tools to hunt for fat grubs in the crowns of palm trees. The first tool consists of a sharp leaf stem. The crow uses this to probe in the leaf litter, sometimes impaling a grub on the end of the spike. But it has also learned that the grub, if provoked by enough sharp prods, will bite on the spike and can then be pulled out. A twig with a curved end is the crow's second, more elaborate probing tool.

The crow's third implement is a harpoon made from long, stiff leaves of the pandanus plant, which bear saw-like spines. The crow uses this to jab at a grub until the teeth snag in its flesh.

SMART SLIME FINDS ITS WAY OUT OF THE MAZE

Lowly slime mould can figure out the way through a maze. At one stage in its life slime mould exists as a slimy, creeping plasmodium that feeds on decaying vegetation. When a plasmodium was put in a maze with oat flakes at each end, it sent out strands which spread to fill the maze.

On touching the oat flakes, the slime mould sent chemical messages back along its length. Strands of plasmodium that had crept into 'dead ends' shrank back. Within eight hours, the plasmodium consisted of a single thread, linking both food sources.

9

THE SECRET LIFE OF PLANTS

LONG AND SHORT LIVES

Some plants stretch life to its absolute limits. A number live for thousands of years, but grow so slowly that they hardly seem to be alive at all. Others have a far busier schedule, packing their whole life span into just a few weeks.

BARREN DESERT BECOMES A CARPET OF FLOWERS

After rainfall in the desert, the normally barren floor becomes a carpet of colour. The interval between rain can be a year or more, and moisture is like a starting gun, triggering a race to reproduce. Within days small plants germinate in their millions. Their seeds may have been lying on the ground for years waiting for a rare desert storm to bring them back to life. Soon the ground is covered in exotic blooms which, fertilized by insects, form seeds of their own. But the spectacle is short-lived: as the sun burns away the moisture, the plants wither and die. All that is left are shrivelled leaves and scattered seeds – the seeds that will wait for the next rainfall, when the frenetic cycle of life will begin again.

FLORAL FLOURISH *In the deserts of the American south-west, California poppies burst into bloom after winter rain.*

WORLD'S OLDEST POT PLANT

Most pot plants survive for a few years, but one specimen at London's Kew Gardens is well into its third century. The cycad known as *Encephalartos altensteinii* was taken to London from South Africa in 1775.

Cycads are tropical or subtropical plants that were around even before the dinosaurs roamed the Earth. Found in woodland, or in open dry places, they look like palms, with a single trunk and a single tuft of leaves, but they produce male and female cones instead of flowers and seeds.

IMMORTAL CREOSOTE BUSH

In the deserts of North America grows a shrub that theoretically can live for ever. The creosote bush, named for its pungent smell, grows very slowly. It flowers and produces seeds like most plants, but it also sprouts new plants from its roots. These 'clones' spread out around the parent plant in a ring. They then sprout new plants from their roots, so the ring steadily increases in diameter, by just 1 m (3 ft 3 in) every 500 years.

The oldest known specimen, 'King Clone' in California's Mojave Desert, has existed for more than 11,000 years. The original plant died long ago, but its clones, which all share the same genes, live on, making the plant potentially immortal.

TIME TRAVELLING SEEDS

When a 3,400-year-old bean was found in the tomb of King Tutankhamun of Egypt, few guessed that it would germinate and grow. Yet for seeds, even 3,000 years is not a record. The world's oldest living seeds were found in 1954 by a miner in the Canadian Arctic, who came across a hoard of deep-frozen Arctic lupin seeds in a network of lemming burrows. The 10,000-year-old seeds germinated after they were thawed, producing perfect lupin plants.

WINDSWEPT PINES OLDER THAN THE PYRAMIDS

The world's oldest individual plants – as opposed to clones – can live for five millennia. High up in Canada's Rocky Mountains and California's Sierra Nevada, twisted and gnarled bristlecone pines cling to slopes of frost-shattered rock.

Slow growth is the secret of their long lives. When young, they resemble Christmas trees, with straight trunks and needle-like foliage. But after their thousandth birthday, icy winds and searing mountain sun take their toll, often pruning back branches as fast as they grow. Once a tree is into its third or fourth millennium, most of its wood may be dead. But the hard, resinous wood is resistant to decay. So these ancient pines stay standing long after they die. Their growth rings provide a detailed picture of the local climate stretching back 8,000 years.

ANCIENT MONUMENT *The twisted shape of this long-dead bristlecone pine bears witness to a lifetime's exposure to powerful winds.*

LIVING ON LIGHT

Unlike animals, plants do not derive their energy from food, but take energy directly from the Sun, converting it using photosynthesis. They are experts in self-sufficiency, surviving on little more than sunlight, water, and air.

WHY PLANTS CAN LIVE IN A BOTTLE

If you put a plant into a bottle garden and fasten the stopper, it will grow for months or even years, as long as it has enough light. Yet if an animal runs short of fresh air, it suffocates. How do the plants survive?

The answer is to do with the way in which plants process sunlight. During the day, when they carry out photosynthesis (*see opposite*), they use up carbon dioxide from the air and give out oxygen. At night the process goes into reverse, and they take in oxygen and produce carbon dioxide as waste. In this way the air inside the bottle is constantly recycled. On a far larger scale, this is how the trees of the planet help to make the whole of the Earth's atmosphere fit to breathe.

PLANTS' TIRELESS WATER CIRCULATION

Plants keep enormous volumes of water on the move in a process that is invisible and involves no moving parts. They absorb water through microscopic hairs on their roots, then send it on a journey via their stems to their leaves, where it is used in photosynthesis or evaporates into the air. The flow of water is continuous and a big tree can suck up more than 2,500 litres (550 gallons) of

COMPETING FOR LIGHT *In an Australian rain forest, trees and climbers struggle with each other to get the sunlight they need to survive.*

water a day. The evaporation of water from the surface of the leaves is called transpiration and forms part of the global water distribution cycle.

LIGHT-CATCHING SECRET OF DEEP-SEA SEAWEED

Seaweeds need light to survive, yet off the Bahamas one species of red algae grows at a record-breaking depth of 268 m (879 ft). Here sunlight is a thousand times weaker than it is at the surface.

Samples of the seaweed have been brought to the surface to discover the secret of its survival. Research reveals that red algae has light-trapping pigments that are particularly good at intercepting the blue-green light that filters down to these depths, so the seaweed is over 100 times better at using light energy than seaweeds near the surface or plants that grow on land. Because it wastes so little of what it collects, it can live in perpetual gloom far below the waves.

PHOTOSYNTHESIS

Plants obtain their energy directly from sunlight, using a substance called chlorophyll to capture the energy. Chlorophyll is bright green, and it is packed away inside leaves. When sunlight shines on a leaf, chlorophyll soaks it up and then passes on the energy, so that it can be used to combine water and carbon dioxide to make glucose (oxygen is produced as a by-product and is released through the leaves).

Plants use glucose as a fuel and they can also turn it into hundreds of other substances, including sticky sugars, floury starch, and building materials that make some kinds of wood, such as ebony and greenheart, almost as hard as metal. Over many years, some plants grow 140 m (460 ft) tall, and end up weighing more than 1,000 tonnes – thanks to energy collected from the Sun.

Animals cannot carry out photosynthesis, but they depend on it for survival. This is because photosynthesis allows plants to grow to become food for plant-eating animals, which in turn are food for hunters. Without photosynthesis, the food chain would break down.

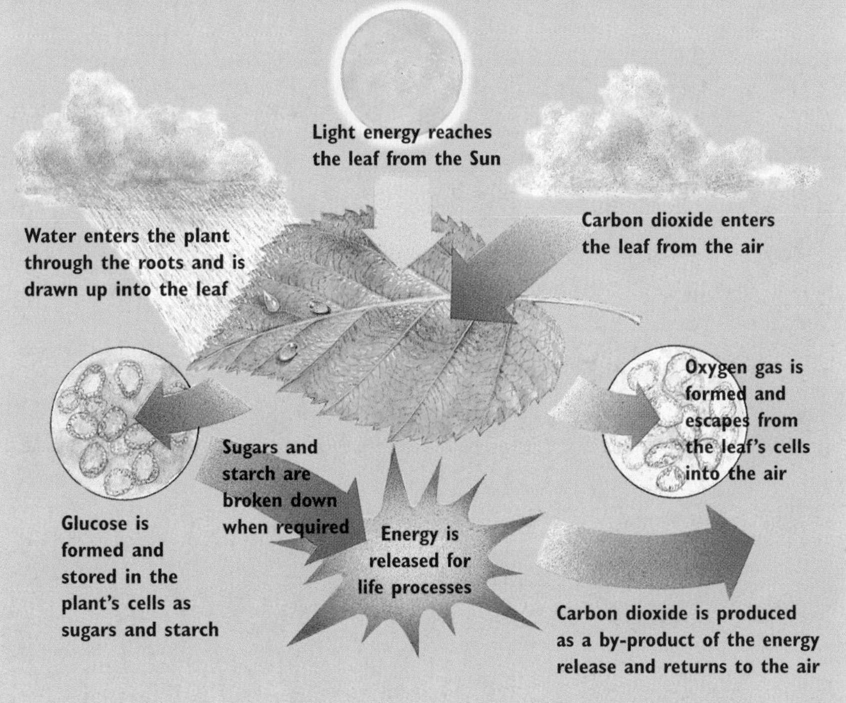

WHY PLANTS HEAD FOR THE SUNNIEST SPOT

Plants always grow towards the sunlight, using a chemical called growth hormone to guide them. Growth hormone always gathers on the side of the plant that is in the shade. For example, if a plant is growing at an angle, instead of straight up, the growth hormone gathers on its underside. This makes the cells on the underside grow faster, which bends the plant until it is growing upright once more.

Armed with this simple but ingenious system, plants slowly search out the light. Although they cannot 'see' sunlight, they always know exactly where to grow.

GOING FOR GROWTH

Plants do not move around like animals, but they are very active. Some grow far faster than animals, rushing upwards in their quest for light. Others are so efficient at growing and spreading that they become troublesome weeds.

A STRANGLEHOLD ON RIVERS AND LAKES

One of the most problematic of weeds, water hyacinth has choked rivers in the Far East and some of Africa's lakes. The plant reproduces by budding off small parts that become independent plants. It spreads across the water surface, doubling the area it covers every two to three months, smothering the life below and clogging boat propellers.

Water hyacinth originally came from South America, where it grows in quiet backwaters and lakes, kept in check by plant-eating insects. During the 19th century plant collectors noticed its pretty violet flowers and began to use it as an ornamental pond plant in places as far away as Africa and the southern USA. It spread into the wild, and has run riot ever since.

Ways of controlling the water hyacinth are being worked on, including releasing one of its natural enemies, a root-eating beetle.

LIVING RAFTS *Water hyacinth has air-filled floats in its leafstalks, which help it to spread across the surface of rivers and lakes.*

SMALL-SCALE ORIGINS OF GREAT UNDERWATER FOREST

Off the coast of California, beds of giant kelp form one of the world's largest underwater 'forests'. Giant kelp is the biggest and fastest growing seaweed, sometimes putting on 40 m (130 ft) in a year. At full speed, it can grow 60 cm (2 ft) a day. The plant anchors itself to the seabed and grows towards the surface, buoyed up by gas-filled 'floats' in its fronds.

Though it reaches huge proportions, giant kelp starts life as a microscopic spore, which grows into a tiny plant called a gametophyte. Gametophytes produce male and female cells, and when two of these meet, a tiny adult plant starts to form.

Giant kelp will only grow under certain conditions. The water depth has to be between 5 m and 40 m (16 ft and 130 ft) deep, and its temperature must not exceed 20°C (68°F). But in the right conditions, it grows so thickly that it can easily be seen from the air.

SUPPORTING ROLE *Giant kelp beds create a friendly environment for marine creatures, the mass of green fronds providing a welcome refuge.*

GIANT BAMBOO'S RAPID GROWTH SPURT

Young shoots of a giant bamboo grow 90 cm (3 ft) in 24 hours. This growth rate is hundreds of times faster than many other plants, and about 3,000 times faster than the teenage growth spurt in humans. Found in southern and South-east Asia, this huge member of the grass family towers up to 30 m (100 ft) tall.

The giant bamboo has hollow stems, which look like fat green spears when they first erupt out of the ground. Their headlong rush upwards is powered by billions of cells storing energy-rich chemicals, which constantly grow and divide. As the stem reaches about 15 m (50 ft) things start to slow down. The new stem sprouts side-branches with feathery leaves and gradually turns hard and golden brown.

These mature stems are as tough as tree trunks, but are hollow and surprisingly light. They make an ideal building material, and, in the Far East, they are tied together and used for scaffolding.

HOW LICHENS MAKE A LASTING EPITAPH

Lichens are slow growers: they spread 1 mm (3/64 in) a year. They can survive on bare rock, colonizing it in raised patches. In order to determine their growth rate, a study was made of the lichens found on gravestones. The lichens could not be older than the stone itself, and the biggest specimens are likely to have started life soon after the stone was set in the ground. So by measuring the diameter of the largest patches, the growth rate could be calculated.

MARKING TIME *Slow-growing lichens are the plant-like product of the interactions between fungi and algae. They thrive in cool, damp climates.*

MEXICAN FERN PALM TAKES LIFE SLOWLY

The Mexican fern palm grows 50 times slower than human fingernails. This works out at 1 mm (3/64 in) a year, one of the slowest growth rates of any plant on Earth. Despite its name, the Mexican fern palm is not a palm at all, but a cycad that lives in dry tropical oak forests in Central America.

Perhaps it is their slow growth that makes cycads generally long lived. One Mexican fern palm raised in a garden managed to grow just 10 cm (4 in) during its first 120 years. At this rate, a fully grown plant 1.5 m (5 ft) tall would be nearly 2,000 years old.

Even reproduction is a lengthy process for the plant. It produces male and female cones, which can take 50 years to develop. The female cones then form seeds, which take a further year before they are ready to be dispersed.

USING THE SUN'S ENERGY

Leaves are like solar panels for plants, trapping as much energy in sunlight as possible. In windy or cold places, leaves can be smaller than a fingernail. But in warm climates, some leaves reach gigantic sizes.

TINY PLANTS THAT STEAL LIGHT IN THE RAIN FOREST

In tropical rain forests, tiny plants called epiphylls attach themselves to the damp leaves of trees and plants so they can steal their light. But some plants have a way of fighting back: their leaves have pointed tips, so that rainwater soon drains away. Because their leaves stay dry, they cannot be used as a home by epiphylls.

Epiphylls are not alone in the ruthless pursuit of light. All plants to some extent compete with their neighbours for sunlight. Trees rely on their height, while climbing plants scramble up other plants around them.

LIVING ON LEAVES *Tiny epiphylls live on other plants but do not bring any benefits to their host. In this sense they are true parasites.*

THE SAIL-LIKE LEAVES OF THE RAFFIA PALM

The raffia palm from Madagascar holds the record for the world's longest leaves. The 20 m (70 ft) leaves grow in a gigantic tuft at the top of its trunk, each one divided into dozens of narrow, spiky leaflets, which stick out sideways from the central stalk. The leaflets alone can be more than 1 m (3 ft 3 in) long.

Giant leaves are very efficient at collecting light, but they take a lot of energy to make. They also need strong stalks to hold them in place, because even a gentle breeze can blow them about like giant sails. To combat these problems the raffia palm has only about 12 fully grown leaves at any one time. Trees with much smaller leaves – such as oak trees – often have a quarter of a million or more.

TARO'S ENORMOUS LEAVES ARE NATURE'S UMBRELLAS

The leaves of the wild taro (or elephant's ears) are so large that people use them as umbrellas in tropical downpours. The biggest wild taro leaf ever measured had a total area of about 3 m^2 (32 sq ft) – which is about the same size as an average double bed. The plant, which has the largest undivided leaves in the world, grows in swampy forests and streamsides in South-east Asia.

FLOATING FOLIAGE *The Amazon water lily can grow so big because it has unlimited access to direct sunlight and is supported by water.*

THE SILK TREE SHUTS UP FOR THE NIGHT

At dusk, the fern-like leaves of the silk tree appear to be on the move. Each of its leaves is split into 12 leaflets and each leaflet is divided again into 30. As the sun sets, the tree begins its nightly shutdown as each pair of leaves folds like clasped hands. Other low-growing plants also fold up their leaves, to protect them and conserve moisture, but none can rival the silk tree's leaf-closure display.

FLOATING WATERBEDS ON THE AMAZON RIVER

Looking like huge green dinner plates lying on the water, the Amazon water lily's leaves measure 2 m (7 ft) across. Each 'plate' has a rim about 15 cm (6 in) high, cut with deep notches that let rainwater escape, so that the leaves do not sink under the weight of the water.

These monster leaves stay afloat with the aid of spongy ridges on their undersides, which contain air pockets. The ribs are covered with vicious spines that prevent fish and water snails from chewing them. The leaves' upper surface is smooth and rubbery.

When these gigantic water plants were discovered by plant-hunters in 1801, they caused a sensation: the leaves can take the weight of an adult, but only if lying down.

FEEDING ON ANIMALS

Plants need nutrients – just as humans need vitamins and minerals from food. Most plants obtain these from the soil. But some rely on a different source: they catch live animals, and then soak up nutrients from the bodies of their prey.

CHEMICAL FACTORY FLIES FIND IRRESISTIBLE

To many flies the scent of rotting flesh acts rather like a signpost, indicating a good place to lay eggs. This is why North American pitcher plants smell so bad. Like their oriental counterparts, they use fluid-filled traps to catch their prey, but these traps grow up from the ground, instead of at the ends of their leaves.

The pitchers can be up to 90 cm (3 ft) tall, and look like slender horns with frilly lids. Each one is a complete chemical factory, turning out substances designed to make the trap work. As well as powerful scents, these substances include slippery waxes, chemicals that make insects drowsy, and ones that digest their bodies once they have fallen in. After several weeks of operation, a pitcher can contain hundreds of corpses, piled up in its tank of fluid.

DICING WITH DEATH *Attracted by its scent, a butterfly investigates a North American pitcher plant. If it falls into the trap, downward-pointing hairs will stop it climbing out.*

WHERE INSECTS MEET A STICKY END

Glistening on the dark surface of a peat bog, the fleshy leaves of a sundew plant look as if they are covered with raindrops. The droplets are actually globules of sticky glue, waiting to trap unwitting insects. The glue is produced by hundreds of tiny hairs on the surface of the leaves. When a small insect lands on one of these hairs, the glue immediately holds it fast. As the insect struggles to escape, the entire leaf starts to react, as nearby hairs bend over towards the victim. Soon the insect is hopelessly trapped, and is flooded with digestive juices from the sundew. The insect's body starts to break down, allowing the plant to absorb the liquefied nutrients.

Sundews need this kind of dietary supplement because they live in boggy places where the acidic, permanently wet, soil contains very few nutrients. There are dozens of species of sundew, found in places as far apart as Alaska and New Zealand.

STUCK FAST *This fly has made the mistake of landing on a sundew leaf. Trapped by the leaf's sticky hairs, it has no chance of escape.*

THE VENUS FLYTRAP'S DEADLY JAWS

The Venus flytrap has a startling response when an insect lands on one of its leaves. The leaf snaps shut along a central hinge, closing in less than half a second. As the two parts of the leaf come together, a set of overlapping teeth close like a cage, ensuring that the insect cannot escape. The leaf takes about ten days to digest its meal, and once it has finished, it opens for business once more.

The flytrap's spring-loaded leaves are

extremely sensitive to touch, but the plant has a way of ensuring that they do not go off by accident. Each half of the leaf has three touch-sensitive hairs, which trigger the leaf into action. If one or two hairs are touched, the leaf stays open, but if all three are touched, even lightly, it suddenly shuts. Venus flytraps originated in the eastern USA, but are now grown as pot plants all over the world.

THE UNDERWATER TRAPS OF THE BLADDERWORT

Living in the world's ponds and pools, buoyed up by a mat of stems, bladderworts catch prey using scores of underwater traps. Each trap, or bladder, is about the size of a pinhead, and has a valve equipped with minute hairs. Normally the valve is closed, but if a water flea or other minute animal touches the hairs, it springs open, and the animal is sucked inside. The valve quickly closes again, so the victim cannot escape. When the bladder has digested its catch, the underwater trap is automatically reset.

PITCHER PLANT'S SLIPPERY SLOPE TO DISASTER

In the forests of South-east Asia, oriental pitcher plants use a well-designed trap to capture insects. The plants grow elaborate pitchers at the very tips of their leaves, which are filled with a sweet-smelling liquid.

Attracted by the smell, insects land on the rim of the plant where they may find droplets of a honey-like liquid that the plant has secreted to encourage them to stay. But once an insect begins to explore, it soon loses its footing and plunges into the liquid. The steep, slippery walls of the pitcher plant's leaves ensure that escape is impossible. The insect is drowned and digested in the pitcher's pool. Frogs sometimes follow, in pursuit of the insects, and share the same fate.

FATAL ATTRACTION *Some oriental pitchers are on the ends of branches, but in this plant, the pitchers grow enticingly on the forest floor.*

TREES

Masterpieces of natural engineering, trees are also the ultimate answer in the plant world's struggle to reach the light. There are tens of thousands of species, some so rare that they have only recently been discovered and named.

NATURE'S HEAVYWEIGHTS

Giant sequoias weigh in at 5,000 tonnes and grow to more than 90 m (300 ft) tall. They are probably the biggest and heaviest single living things that have ever existed on Earth. Some are more than 3,000 years old, and their trunks can be 12 m (40 ft) wide at the base, their roots covering more ground than a football pitch.

Today, these mighty conifers are confined to 75 groves in the Sierra Nevada mountains of northern California, but millions of years ago, they and their relatives were common throughout the Northern Hemisphere.

In the frontier days, it took a team of four lumberjacks 22 days to cut through one of these trees, which could provide enough timber to build a small town. Today, giant sequoias are protected within National Park and forest land across their limited range.

TOP TREE *A giant sequoia called 'General Sherman', in California's Sequoia National Park, weighs an estimated 6,000 tonnes.*

THE TALLEST TREES ON EARTH

Coast redwoods are the tallest trees alive today. The biggest are 110 m (360 ft) tall. These redwoods live in the misty seaboard of northern California and southern Oregon, where constant moisture helps them to grow. But while they are currently the world's tallest trees, they have probably not held the title for long. In Australia and Tasmania, mountain ashes can grow to even greater heights. In the 19th century a fallen tree was estimated to have been 144 m (470 ft) tall – one-and-a-half times the height of the Statue of Liberty. Few of today's mountain ashes exceed 100 m (330ft), because the biggest trees have all fallen victim to the axe and saw.

THE SPREADING BANYAN'S MULTIPLE TRUNKS

With more than 1,000 trunks, it is no wonder that banyan trees grow the biggest crowns of any trees on Earth. Some cover an area of more than 1 hectare ($2^1/_2$ acres) – about as much as 30 tennis courts. A single specimen looks like a group of trees with its many trunks standing close together.

Banyans become like this because of the remarkable way they grow. The trees lower special rope-like roots from their branches. When the roots reach the ground, they dig their way down into the earth. Meanwhile, the hanging part of the root gradually thickens, slowly turning itself into an extra 'trunk'.

In their native India and Sri Lanka, banyans are considered to be sacred, and are often found in the grounds of temples and holy places.

SPECIAL SUPPORT *Banyan trees often grow outwards more than upwards, thanks to the support from their extra trunks.*

REACHING FOR THE SKY *Despite their slender proportions, fully grown coast redwoods are about as tall as a 20-storey building.*

NORTHERLY FORESTS THAT MOVE WITH THE SEASONS

In Alaska, where the winters are bitterly cold, there are forests with trees leaning in all directions. The cause of this phenomenon is permafrost – the layer of permanently frozen ground just a metre or so below the surface of the soil.

The roots of trees such as larches and spruces cannot penetrate this layer, so they have only a shallow grasp on the ground. In winter, the whole ground is frozen solid and the trees have no difficulty standing up. But in summer, when the surface layer thaws, the ground often slips, tipping the trees over as it moves. The annual cycle of freeze and thaw makes it difficult for trees to grow up straight.

OVERWEIGHT AND UPSIDE-DOWN

Africa's baobabs are probably the fattest trees in the world. Some have a circumference or 'waist size' of more than 50 m (165 ft), which is much bigger than their height. According to an old African tradition, the baobab is a tree that grows upside-down, and it is easy to see why: its branches are short and stumpy, ending in twigs that look like roots, and its trunk is often fatter in the middle than at the ground.

Baobabs use their enormous trunks to store water. Although they look solid, their trunks are surprisingly soft and in dry weather elephants sometimes gouge into the wood with their tusks to get at the moisture contained within. These strangely shaped trees also attract other animal visitors: bats, for example, flock to baobabs that are in flower, to feed on their sugary nectar. Baobabs often shed their leaves in the dry season, and can grow to a great age – some are more than 2,000 years old.

ANCIENT TREE THAT THRIVES IN THE CITY

The history of the maidenhair tree, or ginkgo, goes back more than 50 million years. Its closest living relatives are conifers, but unlike them, it has leaves shaped like fluttering fans, which turn bright yellow and are shed in autumn.

Ginkgoes come from the Far East, and are extremely rare in the wild. They have long been revered in China and Japan, and temple gardens have been their refuge, helping the species to survive. Westerners discovered the ginkgo about 200 years ago and specimens were planted in European parks. Remarkably, this survivor from the long distant past is very good at coping with the modern world. It will grow in polluted city air, and is rarely affected by disease. As a result, it can now be seen in London, New York, and busy streets all over the world.

STILL GOING STRONG *The ginkgo's lime-green leaves have a unique shape that has not altered in millions of years, yet the plant still thrives.*

RAREST TREE HAS LONELY EXISTENCE

***Dendroseris neriifolia* is as rare as a species can be: only one mature specimen exists.** It grows on remote Robinson Crusoe Island in the South Pacific. With long, drooping leaves and tiny pale-cream flowers, it is hardly eye-catching, and it has never been collected and grown in parks and gardens. But like other endangered trees, it could have uses waiting to be discovered, so it is being protected, in the hope that the species will survive.

TREE RINGS – A WINDOW ON THE PAST

In parts of the world that have changing seasons, trees build up a hidden record of their lives. Each spring and summer, their trunks grow outwards as well as upwards, adding extra wood in telltale rings. When a tree is cut down, these rings show how old the tree is, and how much it grew in every year.

Experts who study growth rings, known as dendrochronologists, use them as a window on the world's past climate. This is because each year's growth depends on the weather, and on events that affect it. In good years, with plenty of warmth and moisture during the growing season, trees grow well and their rings are wide, but in bad years, they grow very little, so their rings are particularly thin.

By studying the pattern of rings in long-lived trees, such as bristlecone pines, and the semifossilized remains of oaks, dendrochronologists have built up a record dating back 7,000 years. This enables the calculation of accurate dates for events that happened far back in human history, such as mini ice ages.

CIRCLES OF TIME *Trees usually grow fastest when young, so the widest rings are near the centre.*

DAWN REDWOOD IS ROOTED IN PREHISTORY

In 1944, some leaves and cones collected in a remote village in China were found to match multi-million-year-old fossils. They belonged to a tree, the dawn redwood, thought to be long extinct.

Dawn redwoods grow up to 40 m (130 ft) tall and are close relatives of California's coast redwoods. Unusually for conifers, they shed their small needle-like leaves every year. About 90 million years ago, dawn redwoods were common in Europe, America, and China, but climate changes forced them into a retreat. Today, only a few thousand still exist in the wild, but many more are growing in parks and gardens around the world.

GIANT PINES COME OUT OF HIDING

When an Australian biologist clambered into a deep mountain gorge in 1994, he found trees that no botanist had seen before. Remarkably, the gorge was only 150 km (93 miles) from the outskirts of Sydney. The trees had spiky leaves and knobbly bark, and some were nearly 40 m (130 ft) tall, making it even more astonishing that they had remained hidden for so long.

The trees were named Wollemi pines, after the national park in which they were found. They belong to the same family as the Chile pine or monkey puzzle (*Araucaria auracana*), which also has spiky leaves. But while Chile pines are fairly common, the Wollemi pine is incredibly rare. Only about 40 trees have been discovered, growing in the gorge. With so few in existence, the race is on to protect them against man-made disasters such as accidental fires. Seeds have been gathered and planted, and the trees' exact location is being kept secret – as it always was until a few years ago.

HIDDEN TREASURE *Before it was found growing in Australia the Wollemi pine was known only from 2-million-year-old fossils.*

MYSTERIOUS SENSES

Plants show a range of reactions to the world around them. They can respond to touch, to being moved or blown about by the wind, and to changes in the weather. They can even sense changes in daylength, which tell them when to flower.

THE PLANT THAT FEELS TROUBLE AHEAD

The aptly named sensitive plant reacts swiftly to the threat of being eaten. If it is touched, it immediately folds its leaves, making it look as though they have disappeared. The leaves can take hours to recover, by which time the animal that touched them will have moved on in its search for food.

This sudden movement takes less than a second, and is triggered by small swellings that are highly responsive to touch. One swelling is at the base of each leafstalk and there are others at the base of every leaflet (the tiny segments that make up each leaf). A light touch makes just a few leaves fold up, but if the plant is given a sharp tug – for example, by a hungry cow – all its leaves react.

The sensitive plant's rapid reactions are very unusual in the plant world, and have probably helped it to become widespread. Originally from Central and South America, it has spread throughout the tropics, and in some places is now a troublesome weed.

HOW PLANTS LEARN WIND SENSE

Plants can adapt their growth pattern to suit their environment. For example, in windy places a tall plant would risk being damaged or even blown over, so those growing in exposed positions tend to be short and stocky. Experiments show this 'wind sense' at work. In one, thale cress plants were gently touched every day to imitate the wind. After several weeks, the plants had become much shorter than usual – a change that would make it easier for them to withstand the breeze. In another experiment, maize plants were grown in trays and given a periodic shaking to make them rock and sway. By the time they were fully grown, their stems were extra thick, but their seeds were unusually small.

This 'wind sense' explains why crops grow best in sheltered places.

FLOWERING AND DETECTING DAYLENGTH

A 'darkness detector' that senses the changing length of the night is crucial to a plant's survival. Plants need to flower at the right time of year to ensure that their flowers are pollinated. Chemical reactions involving a pigment called phytochrome, which occur in a plant's leaves, ensure that they can detect the appropriate moment.

For example, chrysanthemums flower in autumn, when the nights are getting longer. If they are grown in a greenhouse and given light around the clock, they never flower at all. Plants that need long nights can be prevented from flowering by a single minute of light in the middle of the night.

Other plants are exactly the opposite, flowering in spring, when the nights are getting shorter. Keeping these plants lit at night actually encourages them to bloom.

RAPID REACTIONS *The sensitive plant's leaves collapse if touched. If it is given a harder knock, the entire leafstalk folds back, making the plant look as though it has wilted.*

SIGN OF SPRING *Once these horse chestnut buds have burst, they cannot go into reverse. As a result, it is vital to the tree's survival that they do not open too soon.*

BUDS KNOW WHEN TO BURST

Trees in temperate regions never start to grow too early, because a special 'sense' tells them when winter is past. Instead of responding to warmth, this sense works by detecting cold. The sticky buds of a horse chestnut, for example, form in late summer and become dormant as autumn wears on. They will not open until they have had several weeks of cold weather – even if they are picked and brought into the warmth indoors. This sense is vital to horse chestnuts and other trees, because it prevents them growing in winter if the weather turns unusually mild.

HOW ROOTS GROW DOWN

Plants have a sense of gravity, and they use it to grow their roots downwards, to have the best chance of finding water. No one knows how plants perceive gravity's pull. One explanation involves tiny starch particles that are often present in their cells. Although these particles weigh very little, they slowly sink to the lowermost side of root cells. This could make roots head downwards.

To see this 'gravity sense' in action, bring home an acorn or conker that is sprouting a root, and turn it upside-down. Within hours the root will bend back on itself and head downwards.

CLIMBERS AND STRANGLERS

In the struggle for survival, some plants climb up others in their search to reach the light, saving time and energy by not growing strong stems or trunks of their own. Most do little harm to their hosts, but some can have deadly effects.

LIANAS, THE ORIGINAL 'HANGERS-ON'

Lianas, the rope-like plants that hang down from jungle canopies, are often centuries old. They start life by climbing into saplings, and then keep in step as the young trees grow towards the light. When a liana reaches the treetops, it spreads throughout the canopy, forming a tangled mass of stems, high up above the forest floor.

By the time a liana is mature, the trees that it originally climbed up have often died and fallen down. As a result, the liana's stem is left hanging in the air. It may also send down aerial roots, which eventually dig into the ground. These stems and roots can be more than 40 m (45 yd) long, and some are as thick as a man's thigh.

ROPE TRICK *In a rain forest in Costa Rica, lianas hang from the treetops. Their leaves are hidden away, high up in the canopy.*

SWISS CHEESE PLANT IS ALWAYS ON THE MOVE

The Swiss cheese plant found in homes and offices is a captive version of a plant that snakes its way through tropical rain forests. In Central and South America, the plant starts life sprouting from a small seed on the forest floor. The slender seedling with tiny leaves then grows into the shadows, in search of a tree to climb. As it climbs, it slowly changes shape. The stem thickens, and it

ROVING CLIMBER *Heading up a tree trunk, a Swiss cheese plant grows towards the light. It may have been up and down several trees already.*

sprouts coin-shaped leaves which press themselves against the tree. As the plant grows higher, its leaves get bigger and develop long stalks. By the time it reaches the treetop, the leaves are the size of dinner plates, with holes like a Swiss cheese.

Soon the plant descends again. Its leaves grow smaller, and when it gets to the ground it finds another tree. The cycle is then repeated with the oldest part of the plant dying off while the young part continues to grow.

THE STRANGLER FIG'S DEADLY EMBRACE

The strangler fig clasps its host to death, denying it light and nutrients. Strangler fig seeds are scattered in tropical treetops via bird droppings. When a seed germinates, it produces a cluster of leafy branches and a clutch of slender roots. At first, the roots often dangle, but soon they clasp the host tree's trunk, and set off on a long journey to the ground.

When the roots reach the forest floor, the lethal part of the strangler's life cycle comes into operation. Drawing water from the ground, it starts to grow much faster than its host. Its branches shade its victim's leaves, and its ever-thickening roots clasp the host's trunk. Starved of nutrients and light, the tree dies. The strangler stands tall on its victim's corpse, which slowly rots away.

EMPTY SHROUD *From the ground, this strangler fig's root network looks like a woven tube. The roots once clasped a tree which has rotted away.*

CLIMBING HIGH WITH SPRING-LOADED TENDRILS

Passionflowers, like many other climbers, rely on thread-like tendrils to support their weight. Tendrils are the plant equivalent of fingers: they are slender but remarkably strong, and have an uncannily good sense of touch.

As a passionflower grows, it sprouts tendrils from its stem which reach out and feel for solid support. Within an hour of touching a suitable object, the tendril starts to react. Cells on the outside grow faster than ones on the inside, making the tip of the tendril coil round. Later, the middle of the tendril starts to wind up like a spring, pulling the passionflower closer to its anchorage. By the time the plant is fully grown, thousands of tendrils keep it safely in place.

COILING UP *Spring-like tendrils allow passionflowers to grow up through other plants. Because the tendrils are coiled they stretch without breaking.*

RATTAN PALM LEANS ON SUPPORTERS

With trunks measuring 180 m (200 yd), rattan palms are the longest trees in the world. But they are not the tallest, because they 'flop' under their weight as they grow. Rattans, from South-east Asia, have slender, flexible trunks. Their leaves have large backward-pointing spines that hook over the branches of other trees, helping the rattan to stay in place. Otherwise these snake-like palms would flop to the ground.

LIFE OFF THE GROUND

Perching plants, or epiphytes, have adapted in various ways to enable them to survive without soil. Instead of growing on the ground, they live high above it, in trees, on rooftops, and even on electricity or telephone wires.

THE PLANTS THAT LIVE ON THIN AIR

Air plants are remarkable in that some appear to live on nothing. Native to Central and South America, many grow on trees or large cacti, but others perch on telephone wires, clinging on with their small roots.

Air plants' most important survival aids are their narrow, scaly leaves. Each is covered in microscopic winged scales and hairs, which absorb water. The water is channelled into miniature reservoirs, where it is gradually absorbed, allowing the plants to survive in heat and strong sun. Some species, including the Medusa's head air plant, also use their leaves to stay in place: they curl them around branches and wires, becoming hard to dislodge.

BALANCING ACT *Held in place by its roots, an air plant perches on a wire. Like other bromeliads, it collects all of the water it needs with its leaves.*

ROOFTOP SHRUB THAT SPREADS ACROSS CITIES

Buddleia can sprout on walls and rooftops, thanks to its wind-blown seeds. The purple-flowered shrub was discovered more than a century ago in China and taken to Europe and North America, where it became popular with gardeners. It also proved good at growing on derelict ground, railtracks and buildings, and has since spread across many of the world's cities. Its tiny seeds sprout in damp cracks in old concrete and mortar, and send long roots deep into walls.

OASES IN THE TREETOPS

The largest perching plants are giant bromeliads, from rain forests in the American tropics. They have a rosette of spiky leaves, and look like huge shuttlecocks high up in the trees. Some can be more than 1 m (3 ft 3 in) across, and weigh as much as an adult man. Bromeliads need lots of water, and the structure of their leaves ensures that they never run dry. They work like gutters, feeding rainwater into the centre of the plant, which forms a watertight tank. The plants top up their tanks during downpours, then slowly absorb the water. The tanks, containing up to 5 litres (1 gallon) of water, are used as drinking and breeding pools by tree frogs and insects. The plants benefit from their nutrient-rich droppings.

TREETOP GARDEN *In rain forests, bromeliads can be so plentiful that branches break under their weight. These specimens are from Venezuela.*

ORCHIDS THAT GROW IN THE SKY

Half of the 18,000 species of wild tropical orchids live clamped to tree trunks and branches high up above the forest floor. Some of these exotic plants are cultivated by orchid experts, but many are little known, and most do not have common names.

Orchids survive high off the ground by collecting rainwater with their

roots, which spread out over branches like swollen strands of spaghetti. The roots have absorbent surfaces which soak up water and nutrients that the rain washes out of tree bark and rotting leaves. As long as an orchid has enough light, this supply of water and nutrients is all it needs to survive.

THE FERN WITH ITS VERY OWN COMPOST HEAP

The staghorn fern solves the problem of having no soil by making its own. It lives high off the ground on tree trunks in the tropical rain forests of northern Australia and New Guinea, and gets its name from its long antler-like leaves. But it has a second set of leaves that spread out around the trunk. These hold the fern in place, and also form a cup, which collects leaves and other debris that fall from the canopy. This debris builds up inside the cup, turning into nutrient-rich compost.

GENTLE GRIP *Many epiphytes need a level perch, but the staghorn fern can cling onto tree trunks with its specially adapted leaves, or fronds.*

PARASITES AND HOSTS

Parasitic plants steal what they need to ensure their survival from other plants, giving their hosts nothing in return. Most parasites break into their hosts from outside, but a few lurk inside them, and are visible only when they flower.

SPAGHETTI PLANT THAT STEALS ITS SUSTENANCE

Dodder lives above ground but it has no leaves and, once it is fully grown, no roots. Instead, it winds its way around other plants like strands of living spaghetti, stealing everything that it needs to survive.

Dodder clings to its host plant with tiny sucker-like swellings, which grow at intervals along its stems. These suckers slowly force their way into the host plant's tissues, until they make contact with the pipelines that carry water, mineral nutrients, and food. They then siphon off some of these supplies. As the dodder plant grows, it develops into a sprawling mass that may tap into the stems of many hosts. Common dodder is often found on heather, gorse, or nettles, but the many species of dodder worldwide attack a wide variety of hosts.

Dodder grows clusters of small pink flowers, which drop their seeds on the ground. The seedlings grow roots when they sprout, but these shrivel up once the dodder plant has found a host and taken up a parasitic way of life.

VAMPIRE PLANT *Dodder sends out its strangling stems, which literally suck the goodness out of its host plant. It then moves on to another victim.*

UNDERGROUND THEFT KEEPS ROOT PARASITES ALIVE

Broomrapes live underground, tapping into plant roots for all their nutrient and water needs. Found worldwide, they break cover only when ready to flower. There are more than 100 species of broomrape. Some attack weeds while others infest clover and alfalfa, causing problems on farms. Like many parasites, they are particular about their hosts: each species often attacks only one type of plant, so experts can identify them from the infested host plants.

INTO THE LIGHT *Broomrape flowers are often dark yellow or rust-brown, fitting colours for plants that spend most of their lives hidden in the soil.*

HOW BIRDS SPREAD MISTLETOE'S STICKY SEEDS

Common mistletoe steals all its water and minerals by growing into its host tree's water transport system, just under the bark. Found throughout Europe and northern Asia, often on poplar trees, mistletoe relies

SNEAK THIEF *If the plants around it were cleared away the Australian Christmas tree would soon die, revealing its parasitic lifestyle.*

on its own evergreen leaves to use sunlight, but it could not survive without the help of thrushes and other birds. Birds eat mistletoe berries, which are very sticky, so the birds often need to clean their beaks. As they wipe their beaks against a branch of a tree, some of the seeds become 'planted', sticking firmly to the bark. In this way, mistletoe is able to spread. The seed sprouts, and an extra-thick taproot tunnels into its new host.

AUSTRALIA'S CHRISTMAS TREE STEALS UNDERGROUND WATER

From above ground the Australian Christmas tree looks like a self-reliant plant. But beneath the arid soil its roots break into the roots of nearby plants to steal their water. They tap into plants up to 150 m (490 ft) away and thanks to this subterranean robbery, the tree rarely runs dry.

Out of more than 1,000 species in the mistletoe family, this plant is one of very few that grows into a tree, reaching 10 m (33 ft) tall. It produces golden-yellow blooms during early summer, Australia's Christmas time.

PARASITIC PLANTS THAT ARE RARELY SEEN

Rafflesias live inside their hosts for months or years, invisibly feeding on their victims. When the time comes to reproduce, the parasite's flower buds burst out of its host, like an alien exploding into life.

Rafflesias, from South-east Asia, are not the only plants that live this way: in the Mediterranean region of Europe, *Cytinus hypocistis* attacks wild rock roses, producing yellow flowers surrounded by bright red scales.

PLANT INVADER *These bright clusters are the flowers of* Cytinus hypocistis *– a parasite that spends nearly all its life inside its host plant.*

PLANT DEFENCES

*Because plants are not mobile, they need in-built protection against their enemies. Some of these defences work by using poisons (*see pages 258–9*), but many more make it difficult or painful for hungry animals to feed on flowers or leaves.*

THE PRICKLY PEAR'S DOUBLE DEFENCES

The fleshy stems of the prickly pear cactus are covered with clusters of sharp spines to ward off hungry animals. But it also has a second line of defence that is less easy to see: tufts of tiny golden-brown hairs at the base of the spines. These hairs have barbed tips, and easily rub off if touched. They work their way slowly into a creature's skin – an uncomfortable experience that lasts for many days. After tangling with these spines and hairs, browsing mammals, such as deer, soon learn to leave the prickly pear alone.

Originally from Central and South America, the cactus is now widely grown because it makes good hedging.

IMPENETRABLE BARRIER *The oval 'pads' of prickly pear plants break off easily and take root. Over years, a single plant can become a thicket.*

THE WAIT-A-BIT THORN'S BARBED DEFENCE

Africa's wait-a-bit thorn is well named: any animal, or person, tangled up with it will find that it takes a long time to get free again. The tree has backward-pointing thorns and whip-like branches. If a browsing antelope or a buffalo tries to reach the nutritious leaves, the thorns quickly catch in its fur. As the animal struggles to free itself, it pulls against the thorns, making them dig even deeper. At the same time, its struggles cause other branches to wrap around it, making its predicament even worse.

The wait-a-bit thorn is native to Africa's grassy plains, a part of the world teeming with plant-eating mammals, so it needs its formidable defences to survive.

INSECT SOUP *The teasel grows on poor, dry soils. Its 'moats' help to catch and conserve rainwater, while also trapping insects that decompose to release valuable nutrients.*

THE TEASEL'S WATERY TRAPS

The prickly teasel uses moats to protect its flowers and leaves from insects. Teasels have large, spear-shaped leaves that grow in pairs up their prickly stems. Each pair of leaves is joined together at the base, forming a watertight hollow the size of a small coffee cup. When it rains, water trickles down the leaves and fills the hollow, surrounding the plant's stem like a moat. When insects climb up the plant, they meet these moats, and either turn back or drown.

MULLEINS ARE A HAIRY MEAL FOR INSECTS

Roadside plants called mulleins deter insects with a thick layer of hairs. Mulleins are widespread in Europe and Asia. From a distance, their leaves and stems appear to be covered in white felt, but under the microscope the 'felt' is revealed as a forest of branching hairs. To reach the leaves, an insect first has to penetrate these hairs. If it then tries to feed, the hairs stick inside its mouth, making a very indigestible meal.

NO WAY THROUGH FOR NECTAR THIEVES

The red-flowered sticky catchfly, found in central and southern Europe, has hairs on its flowering stems tipped with blobs of glue. If climbing insects try to reach the flowers, intent on raiding nectar without spreading pollen in return, they meet the hairs and get stuck fast.

Meanwhile, above this sticky barrier, pollinating bees and butterflies are free to come and go. Fruit farmers have learned from this example. By putting sticky bands around the trunks of fruit trees, they prevent crawling insects from reaching blossom and fruit.

TOUGH ON THE OUTSIDE TO KEEP LEAF-EATERS AT BAY

As well as being prickly, the holly tree's leaves have thick edges, which deter caterpillars. Caterpillars like to start at the edge of a leaf and nibble their way inwards, but because holly leaves have such toughened edges, most caterpillars find them difficult to chew and leave the plant alone. Its prickly leaves also deter deer and other large animals. The holly tree is native to western Europe.

ALLIANCES WITH ANTS

Plants and insects often depend upon each other, which explains why insects visit flowers. Ants in particular engage in many special relationships with plants, sometimes acting as security guards while the plant offers them a home.

THE ANT PLANT'S HIDDEN HELPERS

The ant plant from South-east Asia looks like a warty football with a single leafy stalk growing on the side of a tree trunk. Because the plant lives off the ground, it has difficulty collecting nutrients, and it depends on ants to survive.

The ant plant's swollen stem is riddled with cavities, each connected by a narrow tunnel to the outside. These cavities house a complete ant colony, from the single queen, who lays eggs in her royal chamber, to the young grubs, which develop in special nurseries. The ants live on sugary nectar made by the plant, which the worker ants collect and feed to the queen and young.

For the plant, the payback comes in the form of ant droppings, which build up in its hollow chambers. These contain valuable nutrients that the plant can absorb, instead of having to get them from the ground.

SAFE INSIDE *A cross-section through this ant plant reveals the ants' nesting chambers within the protective walls of the stem.*

WHISTLING THORNS ON THE AFRICAN PLAINS

One species of African acacia uses both vicious thorns and biting ants to deter animals from eating its tasty foliage. If, despite the whistling thorn tree's protective prickles, a giraffe tries to eat the leaves, ants rush out to attack the animal's sensitive mouth-parts.

The whistling thorn provides the guardian ants with food and purpose-built homes. At the base of the thorns are hollow growths, each about the size of a large grape. The ants chew narrow openings in these growths, then hollow them out, turning them into nurseries for their young. When the wind blows across the ants' 'front door', it makes a whistling noise, which is how the tree gets its name.

WHISTLING IN THE WIND *The growths formed by whistling thorn trees become as hard as wood, making them an ideal habitat for ants.*

In tropical South America, some species of acacia are also protected by ants. But in Australia, where there are few very big mammals, acacias do not need this kind of defence.

A HOME HIDDEN UNDER A WAX PLANT LEAF

The wax plant, a climber from the Philippines and Indonesia, has an ingenious way of housing its insect allies. Its stems are narrow, and it does not have thorns – instead, it uses its leaves.

Wax plants climb up trees. Their leaves initially grow in pairs, but then one of them withers away while the other lies flat over the trunk. The leaf is convex, like an upturned plate,

EATING THE DEAD *Swimming through the digestive fluid of a pitcher plant, this ant is preparing to haul out a grasshopper corpse.*

which means there is a space between it and the bark. Ants live in this space, ready to leap to the wax plant's defence at the slightest provocation.

THE CECROPIA TREE'S LIVING SHIELD

If the cecropia tree is brushed, even slightly, hundreds of attacking azteca ants suddenly swarm over its leaves. Found in Central and South America the cecropia provides the ants with a home and a steady supply of food. In return, the ants protect the tree against intruders. A horde of soldier ants can often overpower or repel small mammals and birds.

The ants live in the tree's hollow branches and trunk, which they enter by boring tiny holes. They feed on a nutritious secretion that oozes from the base of the leaves, harvesting it just as other insects harvest nectar from flowers. The cecropia's guards are so vigilant that nothing stays in contact with it for long. Browsing animals, such as deer and tapirs, are quickly seen off, particularly if they try to eat the tree's tender tips, where the ants get most of their food. Climbing plants like passionflowers and lianas are literally cut away by the ants using their jaws like secateurs.

As a result, cecropia trees usually have a competitive edge, and quickly spring up in areas that have been cleared. But if their ants leave, the trees often become sickly and die.

ANTS THRIVE WHERE OTHER INSECTS PERISH

Living inside a deadly trap might seem a dangerous thing to do. But for some species of ants, in the jungles of Borneo, it is part of a partnership for survival. These ants live aboard a carnivorous pitcher plant, and they behave like robbers at a grave. Like other pitcher plants (*see pages 238-9*), the ants' host plant catches insects, using jug-like traps to drown and then digest them. But if the plant catches an extra-large insect, it sometimes has difficulty dealing with the remains of its prey and the corpse slowly decomposes, fouling the pitcher until it can no longer work.

This is where the ants come in. In a trap that spells doom for most insects, the ants are completely at home and have no problem climbing down into the juices, and hauling out anything that is too big for the plant to digest. They pull their booty to the top of the pitcher, where they break it up and eat it, dropping the occasional fragment back in for the plant to digest.

TOXIC DETERRENTS

Chemical warfare is the first line of defence for many plants. Some plants are so deadly that they are able to kill anything that tries to eat them. Others use poisons to teach their enemies that they are simply too dangerous to touch.

IVY'S STICKY, LONG-LASTING POISON

Poison ivy is North America's most renowned toxic plant. Its three-part leaves are covered in a toxic oil. If the oil comes into contact with bare skin, or the inside of an animal's mouth, it can trigger a severe and very painful reaction.

Poisons that work by touch are normally quick to lose their effect, because they evaporate into the air. But with poison ivy this is not the case. Its poison, called urushiol, is sticky, and it is easily carried about on clothes, or even in the tiny specks of ash that make up smoke. If clothes are not washed, they can trigger a reaction in the wearer months after they originally brushed the plant.

DANGER: KEEP OFF *Poison ivy's toxic oil has a rapid and painful effect on skin, preventing deer and other forest animals from eating its leaves.*

BATTERIES OF NEEDLES KEEP ANIMALS AT BAY

Nettles are armed with spiky hairs that work like miniature hypodermic needles. If anything touches one of these hairs, the tip breaks off, injecting formic acid into the intruder's skin. The acid is the same as the one made by ants, and its aftereffects can last for several hours.

A brush with nettles is never pleasant, but an encounter with the tree nettle, or ongaonga, from New Zealand, can be extremely dangerous. This nettle grows into a shrub or small tree up to twice as tall as an adult man, and its stings deliver large amounts of toxin. Tangling with a tree nettle can have fatal results, so it is not surprising that cattle and other plant-eating mammals leave it well alone.

THE DUMB CANE'S TOXIC CRYSTALS

An animal may sample the dumb cane's leaves once, but it will never go near it again. The tropical plant is often grown for its showy green and yellow leaves. But in its natural habitat it lives on the forest floor of Central and South America, where it is wide open to attack. For protection, its leaves and stem are filled with razor-sharp crystals of calcium oxalate, a fast-acting poison.

If a creature such as a deer tries to eat the leaves, the effects are almost immediate. The crystals pierce the animal's mouth, and the poison starts to work. It inflames the inside of the mouth and the tongue, so that the animal finds it difficult to breathe. Having once endured this experience, it gives the plant a wide berth. People, too, make the mistake of sampling the dumb cane's leaves. The unfortunate victim may be unable to speak for many hours, hence the plant's name.

MIXED RECEPTION *Foxgloves welcome pollinating bees, but their poisonous leaves ensure that mammals keep well clear.*

FOXGLOVES GO FOR THE HEART IN SELF-DEFENCE

Some plants keep animals away by giving them a heart attack. Foxgloves use this form of self-defence, and so do many milkweeds. Their heart-stopping poisons, called cardiac glycosides, are stored in all parts of the plant, but particularly the leaves.

These poisons make sure that foxgloves and milkweeds escape being eaten by browsing mammals, such as cattle and deer. But their defence system does have some loopholes. Insects are very good at overcoming plant poisons, and some even 'borrow' them, so that they can protect themselves. One of the best known of these borrowers is the monarch butterfly, which grows up on milkweed leaves. Monarch butterflies and their caterpillars contain high concentrations of heart-stoppers, so birds have learnt to leave them alone.

RUBBER TREE'S POISONOUS LATEX

If an animal bites into a rubber tree, it gets an unpleasant surprise. Instead of oozing sap, the tree produces latex – a milky fluid that is packed with poisons, giving it a fiery taste. Like real milk, latex contains billions of microscopic droplets that are suspended in a runny fluid. When latex is exposed to the air, the fluid part evaporates, but the droplets stay behind, creating sticky blobs that clog up an animal's mouth-parts, feathers, and fur.

Rubber trees are not the only plants with this method of defence. In East Africa, the candelabra spurge has succulent branches that look as if they would make an ideal meal. But with latex to protect it, this and other members of the spurge family are almost immune from animal attack.

CONIFERS' ANCIENT WEAPON

Long before the first flowering plants appeared on Earth, conifers were using chemical warfare to keep insects at bay. Their weapon is resin – a sticky, aromatic fluid that seeps through special pipelines in their wood. Resin protects conifers from attack by fungi and beetle grubs. It also turns the leaves into an indigestible food.

If a conifer's bark is damaged, resin flows out, eventually sealing the wound. The resin turns hard and glassy, forming a translucent lump on the side of the tree. If buried when the tree dies, these lumps can turn into amber – a clear, yellow, glass-like substance that can be more than 60 million years old.

STUCK FAST *Trapped in resin oozing down a branch of a conifer tree, this unfortunate ant is doomed to a slow and sticky end.*

PLANT CAMOUFLAGE

Animals use camouflage to hide from predators, but this kind of trickery is rare in the plant world. The best places to find camouflaged plants are deserts, where specimens can be so well hidden they are almost impossible to see.

LIVING STONES' BURST OF BRILLIANT BLOOMS

Living stones look just like the rocks and pebbles of the surrounding desert landscape. These plants from Southern Africa are often less than 5 cm (2 in) high, with a squat, flat-topped 'trunk' formed by a pair of fleshy leaves. The leaves are brown or greenish grey and on their upper surface they have speckles and marbling that imitate the patterns seen in rocks. Between the two leaves is a narrow cleft, where the plant's flower buds are hidden.

Their heavy camouflage helps living stones to avoid becoming a meal for a desert animal. But after the rainy period at the end of the long desert summer, the masquerade is put aside for a few weeks as the living stones burst into bloom.

SECRET STORES *Living stones' strangely shaped leaves act as water stores – one reason why they need to be protected from desert animals.*

THE INCREDIBLE SHRINKING CACTUS

Some cacti avoid detection by literally shrinking out of sight. The sea-urchin cactus, from Texas and Mexico, is easy to spot after it rains, because it looks like a spineless sea urchin sitting on the desert floor. But after several months of drought, the cactus becomes a flattened disc – hard for prying eyes to spot.

The rare Winkler cactus, from Utah, shrinks into the ground once it has finished flowering, and often becomes covered by dust and grit. Such camouflage reduces the chances that animals will find it and nibble away at its flesh.

HIDDEN FRUITS *These tiny objects are the fruits of a living baseball plant, perched on top of its firm, but slightly rubbery fleshy stem.*

SOUTHERN AFRICA'S LIVING BASEBALLS

The leafless stem of the living baseball plant is exactly the same size and shape as a baseball: only its colour, a shade of bluish green, shows that it is actually a plant. Found in South Africa, the plant's unusual shape helps it to avoid unwanted attention. Any animal that tries to eat it is in for a shock because the living baseball is packed with a milky latex, that tastes very unpleasant indeed.

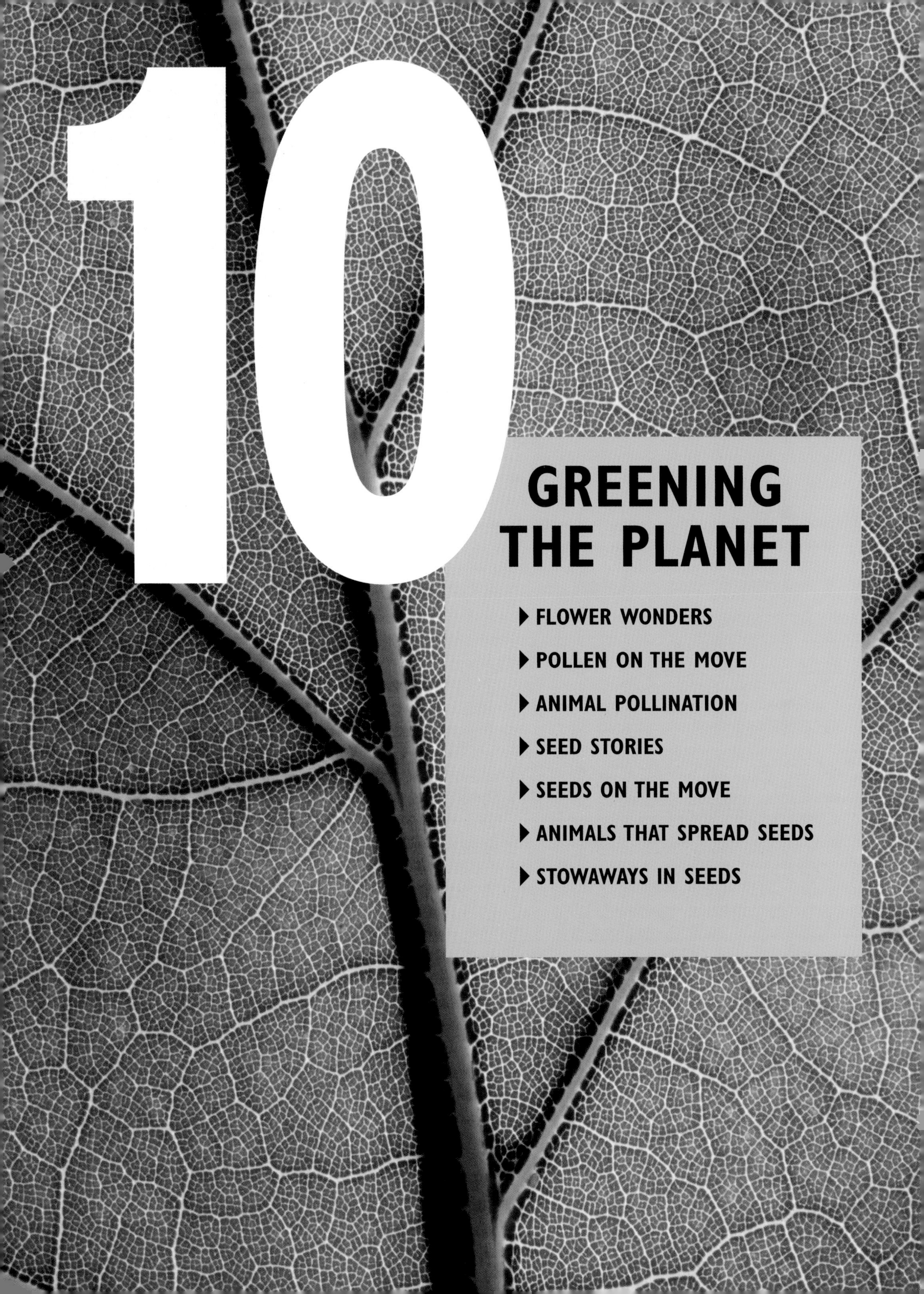

10 GREENING THE PLANET

- FLOWER WONDERS
- POLLEN ON THE MOVE
- ANIMAL POLLINATION
- SEED STORIES
- SEEDS ON THE MOVE
- ANIMALS THAT SPREAD SEEDS
- STOWAWAYS IN SEEDS

FLOWER WONDERS

For a quarter of a million different plants, reproducing involves growing flowers. Flowers can be as big as a coffee table or smaller than a grain of sand, but they all carry out the same vital job: enabling plants to make seeds.

THE BIGGEST FLOWER IN THE WORLD

The world's biggest single flower comes from a plant that spends most of its life unseen. This is because the plant, called rafflesia, is a parasite. It lives in the jungles of Malaysia, inside the roots of tropical vines, and is visible only when it blooms once a year. The first sign of a rafflesia flower comes in the form of a fleshy cabbage-like bud, which slowly erupts out of the ground. This gradually opens to form an immense reddish brown bowl, ringed by rubbery white and purple petals. The flower can be up to 1 m (3 ft 3 in) across and weigh 7 kg (15 lb). It has a powerful odour of rotting flesh, which attracts flies in their thousands.

Once pollinated, the rafflesia flower forms a large squashy fruit filled with seeds. This must then await the heavy tread of a large animal, such as a rhinoceros or elephant. When the fruit is trodden on, it bursts, and the seeds stick to the animal's feet.

As the animal moves through the jungle, the rafflesia seeds get pushed into the soil. In this way they stand a good chance of being planted near the roots of another host vine, so that the process of parasitic life can start again.

THE TALIPOT'S SPECTACULAR FINAL FLOURISH

The talipot palm from South-east Asia produces a single gigantic burst of flowers, after which it dies. It takes the tree 60 or 70 years to reach the point when it is ready to bloom. Its creamy white flower head sprouts from the tip of its trunk and can be up to 6 m (20 ft) long, about a quarter of the tree's original height. It looks like a huge multibranched feather boa rising above the foliage

THREATENED FLOWER *In South-east Asia's vanishing rain forests, rafflesia flowers are a rare sight – each one opens for just a week, once a year.*

and contains up to 60 million individual flowers – more than any other single flower head in the world.

This stupendous display lasts for several weeks, until finally all the flowers have withered and the tree is ready to set seed. Once this is done, the talipot's reserves are exhausted. It slowly dies, having completed its cycle of reproduction in a spectacular way.

MICROFLOWERERS WITH PETAL-FREE BLOOMS

The world's smallest flowering plants are duckweeds, which look like tiny green specks floating on the surface of ponds. They have a leafless, ball-shaped stem, and usually a single root, which trails in the water below. There are many kinds of duckweed, and some are so small that they can fit through the eye of a needle. The world's tiniest species comes from Australia. It has no roots, and when fully grown, measures 0.33 mm (1/100 in) across – only just big enough to be visible with the naked eye.

Duckweeds produce the tiniest 'blooms' on Earth. Their flowers form inside microscopic pouches, and are extremely simple, without any petals. They are so small that a bouquet of a dozen could easily rest on the sharp end of a pin.

IRRESISTIBLE STINK OF THE DEVIL'S TONGUE

The devil's tongue has a huge and unusual flower head with an intense aroma of rotting fish. The smell is so repulsive that it has been known to make people pass out. But flies react to it very differently.

The plant grows in the tropical forests of Sumatra. Its flower head can be taller than a human, and consists of a yellow spire, resembling a folded parasol, rising vertically from inside a huge dark mauve cup. Although the smell is appalling to humans, it appeals to its pollinators – flies that feed on rotting meat. Dozens may be drawn towards the flower head by the apparent promise of a meal.

INSIDE A FLOWER

Many flowers have brightly coloured petals, which help to attract insects. But the most important part of any flower is in the middle, where its male and female organs are found. The male organs, called stamens, produce pollen. This yellow dust-like substance contains male cells. The female organs, called carpels, are designed to collect pollen, so that seeds can be formed. Each has a landing platform (stigma), which gathers the pollen, and a chamber (ovary) which contains a female ovule. Most flowers do not use their own pollen to make seeds. Instead, pollen is carried from flower to flower by insects, or by wind or water. This ensures that plants breed with each other, which encourages a good mix of genes and gives the seeds the best chances of survival.

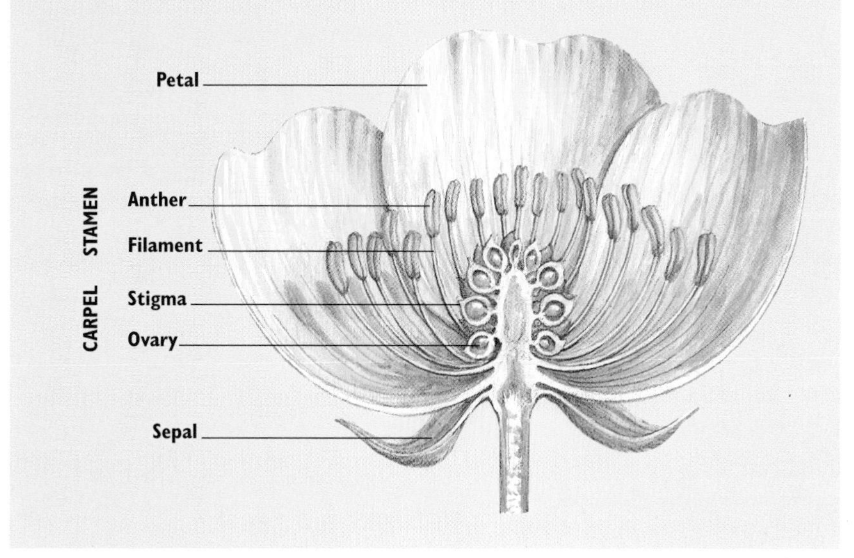

FLOWERS THAT MELT A ROUTE THROUGH THE SNOW

The eastern skunk cabbage, from North America, has a remarkable warm-up routine. The plant flowers in late winter or early spring, and it often sprouts when snow is still covering the ground.

To break their way through this icy carpet, skunk cabbage flowers generate their own heat in a series of chemical reactions that 'burn' a sugary fuel. They become up to 25°C (45°F) warmer than their surroundings – easily enough to melt their way to the surface. The skunk cabbage also uses heat to spread its smell, attracting insects in the cold spring air.

SIGN OF SPRING *The skunk cabbage is initially maroon in colour, but when its fuel is spent the plant turns green and the noxious smell fades.*

POLLEN ON THE MOVE

Before a plant can produce seeds, it needs to exchange pollen with other plants of its kind. Plants have evolved some remarkable ways of ensuring that their pollen reaches its goal by using animals, wind, or water as transporters.

PINE POLLEN FLIES HIGH ON TINY WINGS

Pine pollen has a special pair of air-filled wings which help it to travel through the air. Grass pollen is also airborne, but its grains are round and smooth. The difference in structure is to do with the way the pollen is shed. Pine pollen is released from the top of trees, instead of near the ground (like grass), so its wings help it to 'glide' down.

Pines release pollen when the weather is dry, but if it then becomes thundery, interesting things can occur. Thunderclouds form around columns of warm, rising air, which suck in more air from near the ground. With this air comes dust and also pollen grains. The pollen grains are swept up sometimes more than 15 km (9 miles). Some of the grains return to the ground in raindrops, but the rest may drift for hundreds of kilometres.

LIVING CLOUD *The yellow cloud floating above a pine wood consists of billions of living pollen grains drifting slowly through the air.*

THE INCREDIBLE JOURNEY OF GRASS POLLEN

There are more than 9,000 species of grass in the world, and they all use wind to spread their pollen. A single grass plant can produce tens of millions of pollen grains, clouds of which are released into the air on breezy mornings in early summer.

The pollen drifts over fields and open spaces, swept along by the strengthening breeze. These grains are extremely small and have almost no fuel reserves. Unless they land on another grass flower within 24 hours, they die. This is exactly what happens to the vast majority of these tiny specks. More than 99.99 per cent travel only a few metres before they fall to the ground. Others are swept along by the wind, but are carried to a hostile destination, settling on water, concrete, or tarmac. And some end up in human noses and lungs, where they can trigger hay fever.

For a tiny number the journey ends in success. These grains settle on the female parts of other grass flowers, where they start to grow. They produce a microscopic thread that reaches through the flower towards its seed compartment, where immature seeds are waiting to be fertilized. The seeds can then develop, and the pollen grain's work is complete.

THE PERFECT MOMENT *Grasses do not release their pollen unless the weather is dry, so that the grains can be blown far away.*

SEAGRASSES PRODUCE 'SPAGHETTI' POLLEN

With the help of their very peculiar pollen, seagrasses manage to bloom and set seed under the waves. They rank among the very few plants that are able to flower under water.

Seagrasses have the biggest and strangest pollen grains in the world. Each 'grain' looks like a tiny strand of

POLLEN UNDER THE MICROSCOPE

Pollen grains are marvels of microscopic engineering, shaped to work in different ways. Windborne pollen grains from pine trees are usually flat and dry. Grains carried by animals are usually bigger, with a sticky surface that can 'hook' into fur. Another variant is orchid pollen, whose grains are carried in special packets, called pollinia. These are designed to fasten themselves to insect legs and feelers, or even to the beaks of hummingbirds for transportation.

Although they are extremely small, pollen grains are remarkably tough. Their outer casing is made of sporopollenin, a hard material with a complex microstructure, which can withstand heat, cold, and even strong acids. Each kind of plant produces pollen grains of a particular shape – such as the spiky grains of the sunflower – with characteristic surface patterns. Pollen grains fossilize very easily, and these fossil grains can be used to tell which plants grew where thousands of years ago.

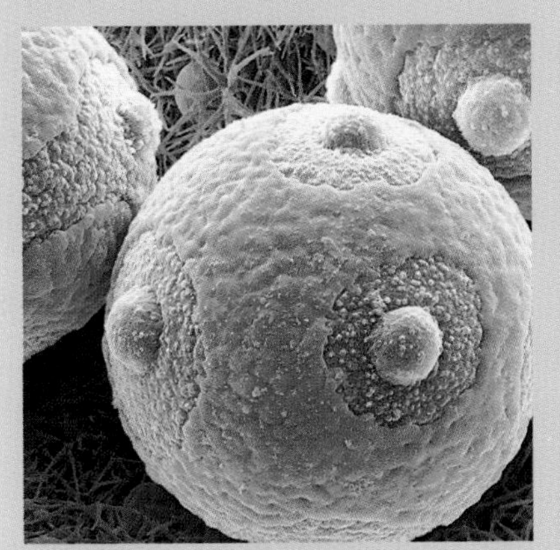

Flowering currant

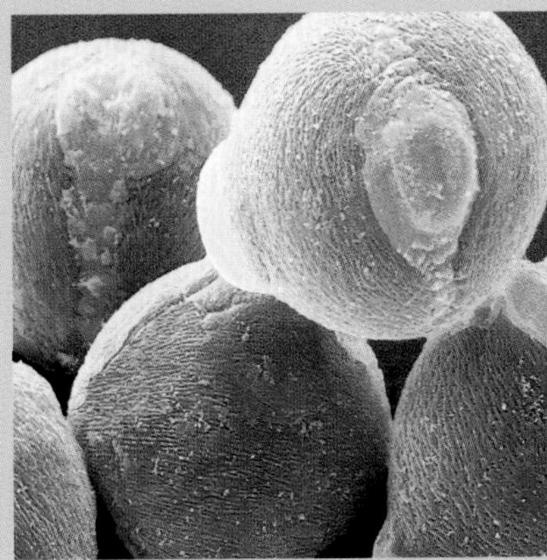

Garden rose

Hollyhock

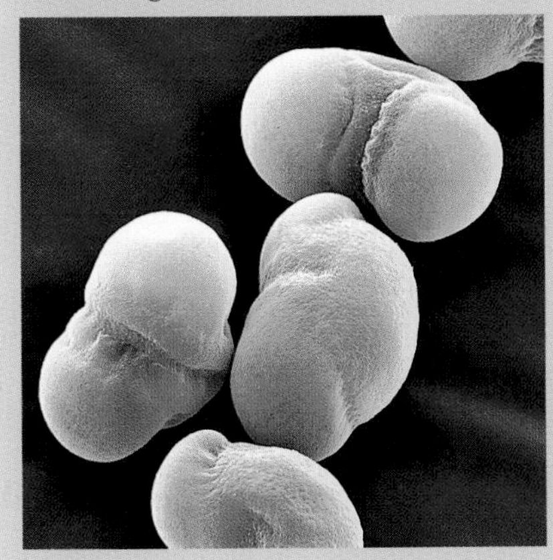

Silver fir

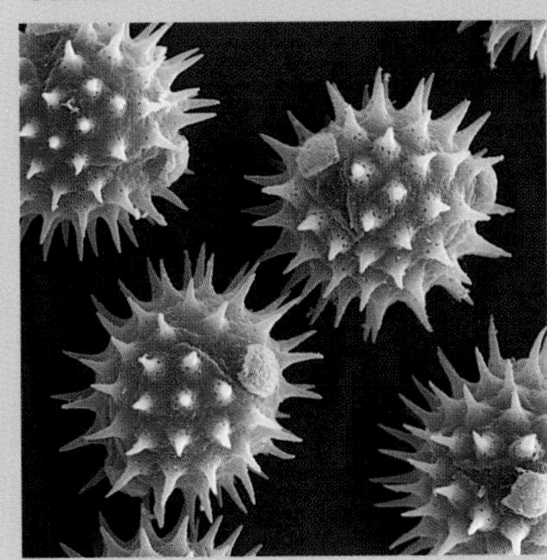

Sunflower

Tiger lily

spaghetti, and is up to 3 mm (1/8 in) long. The strands are exactly the same density as sea water, which means that they neither float nor sink. Ordinary pollen grains, however, tend to become waterlogged and are unable to produce seeds.

Once the pollen is released by the male flowers, it drifts with the current close to the shore, tangling up with anything it meets. Female seagrass flowers have fleshy spikes that stick into the water – a perfect shape for catching the pollen as it drifts past.

RIBBONWEED POLLEN SETS SAIL TO FERTILIZE

Ribbonweed lives in fresh water, and has a remarkable pollination system. It releases its pollen onto the surface in tiny 'boats'.

Like sea-grasses, ribbonweed has separate male and female flowers. Each female flower is attached to a long coiled stalk, which stays wound up until the flower is ready to bloom. The stalk then unwinds, letting the flower rise to the surface. Meanwhile, the male flowers develop near the base of the plant, inside fleshy cups. When they are mature, they break away from the cup, and float upwards laden with a cargo of water-repellent pollen.

Once the male flowers reach the surface, the most ingenious part of the system comes into operation. The water forms a dimple around the floating, waxy, female flower. If one of the pollen boats drifts into the dimple, it slides down towards the flower. The boat then transfers its pollen, and the flower can produce seeds.

ANIMAL POLLINATION

Flowers that are pollinated by animals are like living shop windows, designed to appeal to particular visitors, which spread a flower's pollen in return for a tasty reward. But a few plants play deceitful tricks on their animal helpers.

ODIOUS ODOUR DRAWS WINGED POLLEN CARRIERS

Carrion flowers on cactus-like plants in dry parts of Africa attract flies by smelling and looking like rotten meat. The star-shaped flowers open on the ground, and have fleshy, dull brown petals with hairy edges. The hairs tremble in the slightest wind, just like the fur on a decomposing corpse.

For blowflies, carrion flowers seem like the perfect place to lay eggs. They settle on their fleshy petals, tasting them as they move about. As a fly approaches the centre of the flower, small pollen packets clip themselves onto its legs. Having laid its eggs, the fly takes off to find another flower, where the pollen packets are removed by grooved stigmas. The flower sends off its pollen successfully, but for the fly's eggs, the story does not end well. The maggots find no meat when they hatch, and soon die of starvation.

ROUGH DEAL *Unlike most flowers, which produce nectar, the carrion flower offers the insects that pollinate it nothing in return.*

MOTH DRINKS DEEP TO HELP OUT FLOWER

In the forests of Madagascar, a remarkable orchid attracts a visitor with an extra-long tongue. The orchid has creamy white flowers, which produce sugary nectar at the end of a slender tube. The tube is up to 30 cm (12 in) long, but as narrow as a pencil-lead, which means that the nectar is almost impossible to reach.

Bees and butterflies cannot drink the orchid's nectar, but once darkness falls, another kind of pollinating insect arrives to feed. It is a Madagascan sphinx hawk moth, and is equipped with what is probably the longest insect tongue in the world at 32 cm (13 in). The moth flutters in front of the flowers and extends its tongue, pushing it into their tubes like a drinking straw. While it feeds, it becomes dusted with pollen, which it carries from flower to flower.

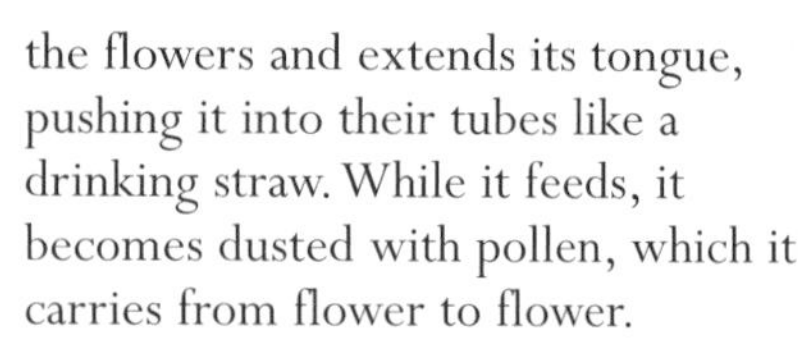

THE BEE AND THE BUCKET ORCHID

Bucket orchid flowers from Central America have an ingenious method of trapping bees and 'planting' pollen on them. The flowers look rather like roosting butterflies, with mottled yellow or orange wings, and a fluid-filled bucket beneath. They open at dawn and their

A BEE IN A BUCKET *Clambering out of a bucket orchid flower, this bee has two pollen packets firmly attached to its back.*

sweet odour quickly fills the tropical forest air. The bee homes in on the flowers from as much as 8 km (5 miles) away. It lands on a platform above the bucket from where it is drawn down by a waxy substance and then tumbles into the fluid.

Once immersed, there is only one way out – through a tunnel towards the front of the flower. As the bee clambers through the tunnel, it is held in place by a clamp. A packet of pollen is fastened to its back, and then the creature is released to fly away. When the bee visits the next bucket orchid, it repeats the process, but this time the pollen packet is deftly removed by a hook at the tunnel's exit.

DUTCHMAN'S PIPE IMPRISONS VISITORS

A plant called the Dutchman's pipe recruits animals to transfer its pollen. It lures visitors into its flowers with one-way doors, and then imprisons them inside.

This plant comes from Europe, and has small, yellow pipe-like flowers with a bulbous base topped with a flared 'chimney'. Their smell attracts tiny gnats, which eagerly fly inside. The gnats move down the tube into the base, where they find themselves trapped by downward-pointing hairs. Their prison chamber contains the reproductive parts of the flower, and a translucent window lures them farther inside in their attempts to escape.

The plant imprisons the gnats for several days by feeding them nectar. Meanwhile, the flower's pollen-bearing organs mature, dusting the captives. Finally, the hairs in the tube wither, releasing the prisoners to fly to the next flower.

ENTICING PERCH *The red and orange colours of the bird-of-paradise flower are particularly attractive to birds, which act as its pollinators.*

FEATHERED STAND-INS AID BIRD-OF-PARADISE PLANT

The bird-of-paradise plant is pollinated by birds' feet. As far as we know, it is the only plant in the world that enlists the help of birds in this way.

The plant, from Southern Africa, produces its flowers on waist-high stalks. Each flower has a set of upright orange flaps, which contain a tempting store of sugary nectar. A horizontal 'perch', positioned close to the flaps, contains the flower's pollen-producing organs. For most of the time these are hidden inside a special groove, where they are protected against the rain.

When a bird lands to feed on the flower's nectar, it settles on the perch. The perch bends under the bird's weight, and the groove suddenly springs open, exposing the pollen to the air. The bird collects some of the pollen on its feet as it feeds, then carries it to the next bird-of-paradise flower that it visits.

A CASE OF BREAKING AND ENTERING

The nectar-laden blooms of Central American pentstemons attract a bird that specializes in floral robbery. The glossy flowerpiercer has a short beak that is equipped with a sharp tip. Instead of reaching into flowers to feed, it pecks holes into their bases, so that it can reach the nectar from outside. The flowerpiercer does not collect any pollen when it feeds, so it raids the flowers' nectar stores without giving anything in return.

FURRY FLIERS GET TOGETHER ON THE CANNONBALL RUN

Nectar-eating bats find the cannonball tree hard to resist. This South American rain forest tree has saucer-sized, fleshy orange blooms that grow directly from its trunk. The flowers are designed to appeal to some of the world's biggest and least manoeuvrable airborne pollinators.

As the sun sets, the cannonball tree's flowers emit a sickly-sweet scent. This lures bats from several kilometres away. They have long tongues like hummingbirds, but instead of hovering, they usually land to feed. They use the claws on their wings to grip the flowers, then lap up the nectar. The tree gets its name from its round fruits, which are as big as a human's head.

THE DRAGON ARUM'S DEVILISH FLYTRAP

The flower head of the dragon arum from Mediterranean Europe has a devilish appearance. Growing up to waist-high, it has a tall narrow column, called a spadix, partly wrapped inside a furled leaf-like mantle. The outside of this structure is green, but the spadix and the lining of the mantle are a lurid reddish mauve.

Flies, attracted by the plant's putrid smell of rotting flesh, become trapped inside the mantle by downward-pointing hairs. Here they come into contact with the tiny female flowers, which are clustered around the bottom of the spadix waiting to collect any pollen that the flies bring in.

Meanwhile, higher up the spadix, the male flowers begin to mature. After a day or two the hairs wither, enabling the flies to clamber out. As they do so they brush pollen off the open male flowers and carry it away.

PRIMATES PICK ON TRAVELLER'S TREE

The traveller's tree from Madagascar has leaves arranged in the shape of a huge fan, up to 8 m (26 ft) across. For the tree's size, its flowers are tiny. They are hidden at the base of the leaves, inside cases made up of tough scales.

Brown lemurs have no difficulty prising the cases apart to get at the nectar within. These monkey-like creatures then push their pointed snouts into the flowers. They become covered with sticky pollen, which they carry to the next tree that they visit.

TWO-SIDED TREE *In a traveller's tree, young leaves start life at the top of the fan, then slowly fold back down the sides before falling away.*

Possum taps banksia's honey

The honey possum is one of very few mammals that relies on pollen and nectar for its food. The miniature marsupial, barely bigger than a mouse, is found only in the sandy heathlands of Western Australia.

Under the cover of darkness, it climbs up *Banksia* plants to reach their cone-shaped nectar-rich flower heads, and probes into the flowers with its slender pointed snout. Its tongue has a brush-shaped tip – a shape that is ideal for lapping up its food. As it feeds, the possum's fur becomes coated with pollen, and it carries this from flower to flower.

Some of Western Australia's banksias depend entirely on this furry visitor, and they are specially adapted to help it to survive. Instead of flowering in one burst, they have some flowers open all year round, ensuring that the honey possum has a steady supply of food.

SEED STORIES

Seeds are the plant equivalent of spaceships, packed with everything a young plant needs for survival on its journey from lift-off to landing. Inside each one is the embryo plant and enough food to sustain it until it can start to grow.

MICRO ORCHID SEEDS FIND SAFETY IN NUMBERS

High up in the treetops of tropical forests, orchids make the world's tiniest seeds, and scatter them in vast numbers. The orchids are epiphytes, or plants that live by perching on other plants. These treetop passengers use the wind to spread their seeds, so it helps that their seeds are light. But the wind is unreliable, and the chances of their seeds landing on suitable 'perches' are extremely small. As a result, epiphytic orchids concentrate on numbers, producing up to 10 million seeds in a single year.

Seeds this small cannot be seen clearly with the naked eye – they look like dark brown dust. A matchbox full of this dust would contain at least 7.5 billion seeds – enough for the world's entire human population to have one each, with some to spare.

TREE HOUSE *One-day orchids perch on trees in the Costa Rican rain forest and fill the canopy with their blooms. Their seeds are wind-blown.*

THE ANCIENT ART OF SEED PRODUCTION

Pine nuts are among the most ancient seeds on Earth. This is because they come from conifers – trees that 'invented' seeds, more than 300 million years ago.

By looking at pine nuts, it's easy to see why seeds were so successful. Most seeds, including freshly picked pine nuts, are enclosed in a tough outer coat. This enables them to survive heat, cold, and drought until conditions are right for germination. Each seed contains a prepacked store of food that gives the seedling a good start in life.

Compared to conifers, flowering plants are newcomers to the seed-production business. They evolved about 150 million years ago, and are now the most successful seed plants to be found in the world.

DEADLY POISONS GIVE PROTECTION

The seeds of the castor oil plant contain a substance called ricin, which is at least 5,000 times more poisonous than cyanide. An amount weighing as much as a grain of salt is enough to kill a fully-grown man.

This lethal poison is present in the seed's outer coat, not in its inner flesh which is used to produce oil. It helps to protect the seeds against hungry animals. Many plants, including the castor oil plant, use animals to spread their seeds, but they cannot afford to let animals use their seeds as food. By adding poisons, they make sure that most animals leave them alone.

Even some of the seeds that humans eat are toxic when raw. Red kidney beans, for example, have to be soaked in water and then boiled for at least 10 minutes to remove poisons.

ISLAND PALM PRODUCES WORLD SEED CHAMPION

The world's biggest seed is so enormous that it can fill a garden wheelbarrow. It weighs up to 20 kg (44 lb), and is produced by the coco de mer, a rare palm that grows in just one place, the Vallée de Mai on the island of Praslin in the Seychelles.

Coco de mer palms have gigantic fan-shaped leaves and, unlike many palms, each tree is either male or female. Once the female tree's flowers are pollinated, each one starts to develop a colossal two-lobed nut, which can take between seven and ten years to grow to maturity. The nut is enclosed by a massive green husk, and is shaped like a pair of buttocks.

There are only about 5,000 of these extraordinary trees on Praslin island, and they are now a major tourist attraction. But in days gone by, the hidden valley was unknown to people in the outside world, and traders sold the empty nuts for huge sums of money as fertility symbols. According to legend, they came from a tree that lived under the waves, which is how it became known as coco de mer, or 'coconut of the sea'.

GIANT PUZZLE *It is not known why coco de mer palms should have such huge seeds. Unlike coconuts, they are so dense that they cannot float.*

TOUGH NUTS TO CRACK

The egg-sized seeds of the African doum palm are so tough that it takes a hacksaw to cut them open. The interior of the seed is made of a pearly substance that resembles ivory, and this can be carved and polished just like ivory. Despite its toughness, this substance is the seed's food reserve. When the seed falls on moist ground and germinates, it softens and the seedling absorbs it.

The hardness helps to protect the seed from tropical animals, such as monkeys and parrots, which specialize in cracking open seeds and eating the flesh inside.

SEEDS ON THE MOVE

To give their seeds the best start in life, plants need to spread them far and wide. Some do this with the help of animals, water, or the wind. Others launch their seeds in miniature explosions, or catapult them through the air.

HEATED BROOM GOES OFF WITH A BANG

In warm weather the snapping of exploding broom seedpods can be heard many metres away from the plant. Brooms are particularly common in parts of the Northern Hemisphere that have dry summers.

Broom seedpods have two flat sides, which are joined together at their edges. When first formed, each pod is soft, but as its seeds start to grow, the sides dry out and turn black. As they dry the sides twist in opposite directions, but because they are joined the pod stays intact. Finally – often when the sun is shining – the twisting force becomes so great that the sides suddenly tear apart, and pop up like springs, flicking the seeds into the air.

TIGHTLY PACKED *Broom can spread quickly with each plant firing off 18,000 seeds every year into the surrounding area.*

PRIMED FOR ACTION *Stork's-bill 'beaks' can be as long as a finger. Each one has a cluster of seeds at its base, ready to be hurled into the air.*

CUPS FULL OF SEED ARE FIRED BY CATAPULT

Stork's-bills have seed heads which look like birds' beaks pointing upwards into the air. These curious structures are designed specially for spreading seeds.

A stork's-bill's 'beak' has five seeds around its base, each set in a papery cup. Every cup is attached to a long strand that reaches to the beak's tip. As the seeds mature, these strands dry out and start to curl. The beak finally gives way under the force. The strands snap free and hurl their seeds forwards like miniature catapults. The seeds rarely travel more than a metre, but this gets them away from the parent plant and onto suitable ground.

MISTLETOE'S ARTILLERY IN THE TREETOPS

Of all the world's explosive seed-scatterers, dwarf mistletoes hold the absolute distance record. Dwarf mistletoes live parasitically on conifers, particularly in North

THIEVING INTRUDER *A dwarf mistletoe growing on a pine tree. Mistletoes are parasites, drawing water and nutrients from the plants that they live on.*

America, and they fire sticky bullet-shaped seeds to help them to spread from tree to tree. Their seeds hurtle through the air at more than 100 km/h (60 mph), and can travel up to 15 m (50 ft) before hitting trees or dropping to the ground.

Each seed develops inside a leathery-skinned berry. Eventually, pressure from the build-up of fluid in the berry causes it to burst, shooting the seed into the air. A single mistletoe can fire 1,000–5,000 seeds every year.

EXPLOSIONS ALONG THE RIVERBANK

The Himalayan balsam, or policeman's helmet, has seed capsules that burst at the slightest touch. When one explodes, it can start a chain reaction, with others exploding in turn. A single head-high plant might have several dozen capsules primed to explode at any one time, so walking through a clump of them along a riverbank can trigger a fusillade of seeds. The Himalayan balsam is a relative of the garden busy lizzie. Its seeds were taken to Europe 150 years ago, where it now grows as a weed.

CUCUMBERS WAITING TO BURST

The fruit of the squirting cucumber plant from southern Europe explodes like a tiny land mine to spread its seeds. If anything touches the ripe, bristly, finger-sized cucumbers, they detonate, scattering seeds and juice up to 5 m (16 ft) through the air. The cucumbers hang from small upright stalks, at ankle-height above the ground. As they ripen juice builds up inside them. Although thick-skinned, the cucumbers have a weak point at the joint where the stalk is attached. If an animal knocks a ripe cucumber, the joint breaks, the fruit drops off, and its contents spill out.

UNEXPLODED BOMB *This squirting cucumber has split open, revealing the sticky seeds inside its fleshy skin, which burst out when the fruit falls.*

WILD OATS SOW THEMSELVES

Once they have landed on the ground, the seeds of wild oats literally bury themselves. This avoids the risk that something will find them and eat them. Each wild oat seed has a stiff bristle attached to it. The bristle coils up when the weather is dry and uncoils when it is damp. By coiling and uncoiling, it slowly works the seed 1–2 cm ($^1/_2$–$^3/_4$ in) into the ground.

THE DANDELION'S PARACHUTES

Of all nature's ways of making seeds fly, the dandelion 'parachute' is one of the most successful methods. The parachute's canopy is made up of a tuft of feathery hairs and, with the seed suspended beneath, the parachute is very stable as it drifts through the air. One dandelion seed in a hundred manages to travel 10 km (6 miles), or more.

Thanks to their parachutes and their tough taproots, dandelions are among the world's most efficacious weeds. When Europeans first crossed the Atlantic, they accidentally took the seeds with them aboard their ships, and the plants soon spread. Today, dandelions can be seen on roadsides and in lawns across North America.

SAILING AWAY *Caught by a summer breeze, dandelion seeds break away from their moorings and sail off on their feathery parachutes.*

HITCHING A LIFT BY RAIL

The Oxford ragwort is one of the few plants that has managed to spread its seeds by train. Originally from southern Europe, it was planted in Oxford University's botanic garden more than two centuries ago, but its wind-blown seeds helped it to escape. Soon after, Britain's railway network was built. The seeds were sucked aboard trains, and today mile after mile of railway track is lined with the ragwort's bright yellow flowers. Thistles are another hitchhiker whose seeds are spread in a similar way.

TUMBLEWEED SEEDS ARE ON A ROLL

A tumbleweed has a remarkable way of spreading its seeds after its death. When it dies, its roots shrivel up and snap off, and the rest of the plant is blown along by the wind, scattering its seeds as it goes.

There are several kinds of tumbleweed, all of which grow in dry, open parts of the world. Their seeds germinate in spring, when the soil is still moist, and they soon flower and set seed. When the soil turns dry, the leaves wither and their stems curl up to form a ball. The roots eventually snap and the spherical tangle is released to roll across the ground.

In some parts of the world, such as the American prairies and Russian steppes, tumbleweeds can be a

high in the treetops. If a sudden gust of wind catches them on their descent, they can be blown several hundred metres away.

COCONUTS SURF THE TROPICAL OCEAN WAVES

For coconuts, the sea is like a giant conveyor belt, carrying them away to distant shores. When a ripe coconut falls from a tree, it is surrounded by a thick husk that acts as a float. In the centre of the husk is the nut itself, protected by a very hard shell. The nut contains a supply of water, or coconut 'milk', and its own onboard store of food. It can survive at sea for a year or so and may drift more than 5,000 km (3,000 miles).

The journey is a hazardous one, and many coconuts float across the tropical oceans without ever coming to land.

But even if a nut is washed up on a beach, it is not necessarily home and dry. Robber crabs specialize in eating coconuts, tearing open the husks with their massive pincers. Storms can smash coconuts against rocks, or wash them back into the sea. If a nut escapes these perils, it grows its first leaves, and puts down roots. Within 20 years, the ocean wanderer is well on the way to maturity, and can be producing coconuts of its own.

SEA HEARTS FLOAT FROM DISTANT LANDS

Among the floating seeds, sea hearts, produced by a tropical vine, probably travel the farthest distance. Carried along by the Gulf Stream, these heart-shaped beans often journey from the Caribbean across the North Atlantic. Some even find their way to the beaches of northern Norway – a pointless journey of 12,000 km (7,500 miles), because sea hearts cannot grow in this cold climate.

Sea heart vines produce the world's biggest seedpods at 2 m (7 ft) long.

JOURNEY'S END *After a voyage across the seas, a coconut germinates on a tropical beach through a cavity which opens up at its 'sharp' end.*

problem for farmers. They pile up against fences in their thousands, and can even block roads.

DESCENT OF THE GIANT GOURDS

Imagine an object that looks like a miniature Stealth bomber, spiralling downwards from the treetops in large, unhurried curves. Its translucent wings measure nearly 15 cm (6 in) across, and slung between them is a precious payload: a huge seed.

These giant gliders can be seen in the jungles of South-east Asia. They are produced by a climbing gourd, and are the biggest airborne seeds in the world. They develop inside fruits that are about the size of a football, and are released when the fruits split open

ANIMALS THAT SPREAD SEEDS

Plants often enlist the unwitting help of animals, sometimes as small as ants or as big as elephants, to spread their seeds near and far. In return for their efforts, plants provide animals with a guaranteed supply of food.

THE COMPELLING LURE OF THE NUTMEG TREE

The slightest hint of scarlet attracts birds to sample the fruit of the nutmeg tree. But unlike many plants, which lure birds with juicy berries, the nutmeg tree has hard, inedible seeds. The attraction for birds is that each one is wrapped in a basket of scarlet flesh called an aril – an irresistible delicacy.

Because the arils are stuck to the nutmeg seeds, birds have to swallow both the seeds and their arils together. The arils are digested inside the birds' stomachs, but the nutmeg seeds are not. Birds cough them up, often dropping them onto the ground far from their parent tree.

Nutmeg arils are not only eaten by birds. In South-east Asia, they are collected and dried along with nutmegs themselves, and are crushed to make the fragrant spice called mace.

READY TO TRAVEL
A nutmeg's red aril is its 'fare' for travelling across the forest inside a bird's stomach. The bird eats the aril and spits out the nut.

THE PLANT THAT HANGS ON WITH GIANT HOOKS

Hooked seeds are designed to fasten themselves to animals, which carry them here and there. In places where large animals are common, the seeds reach a giant size.

Some of the largest are those of the grapple plant, or devil's claw, found in tropical Africa. Its seeds are housed in cases as big as a fist, armed with finger-length spikes that have triple-barbed tips. These fearsome-looking objects are dropped onto the ground where they lie in wait for unwary passers-by. When a large animal walks on a grapple plant seed case, the hooks fasten themselves to its hoofs or paws. Hours or days later, after maybe 5–10 km (3–6 miles) of limping, the animal manages to remove its unwelcome encumbrance and the seeds come to rest in their new home.

AN ATTRACTIVE PACKAGE FOR HELPFUL ANTS

Along with many other plants, wild cyclamens rely on ants to spread their seeds. To attract the ants, the seeds are coated with a tasty substance. As the seed capsules mature, the cyclamens coil up their flower stalks, bringing the seeds down to the ground where the ants can reach them. Ants are attracted by the seeds' smell and they carry them off to their nests, where they feed on the outer coating, but leave the rest of the seed intact. In this way the seeds not only get distributed but are planted out of reach of birds or rodents.

CATFISH SOW SEEDS IN THE FLOODWATERS

The buriti palm, from the Amazon rain forest, uses fish to spread its seeds. The palm lives in places where the forest floods for several months each year, and its oily fruits often drop into the water below. Catfish swallow them whole, and scatter the undigested seeds in the shallows. Once the floodwaters have subsided, the seeds germinate, and the young buriti palms take root.

There are probably more seed-spreading fish in South

America than anywhere else on Earth. Many of them have good hearing, and they find food by listening for the sound of fruits hitting the surface.

BURIED FRUITS FOR DRY TIMES

The desert melon has evolved a means of growing its seed-bearing fruits underground to protect them from the searing sun. This might seem like a difficult place for scattering seeds, but the desert melon of South Africa can count on an underground ally – the aardvark.

Aardvarks got their name from the Afrikaans meaning 'earth pig', since they are pig-like in shape, and also slightly pig-like in behaviour. They dig for their food, and normally eat termites, but in dry times they also unearth desert melons for their thirst-quenching juice. The melons' seeds pass through the aardvark's gut and are planted in the animal's dung. Aardvarks often leave their droppings near their burrows, which means that they grow their own doorstep 'melon patches'.

THE ELEPHANT AND THE ACACIA

Elephants destroy trees to satisfy their immense appetites. But they also help to plant them, by scattering the seeds of their favourite foodplants in their dung.

One common African tree that relies on the elephant is the ana tree, or

GIANT GARDENER *Elephants chew their food with enormous teeth, but some seeds pass through the animal's body without being crushed.*

apple-ring acacia. Elephants love the tree's twisted seedpods, and they use their trunks to pull them down from high branches. The succulent pods soon break down in the elephant's cavernous stomach, but the seeds are much tougher. They travel through the elephant's digestive system unharmed, and are passed out intact in its dung.

Elephant dung is the perfect growing medium for the seedlings, and within a few weeks they start to sprout. These elephant-sown trees sometimes grow in lines, marking the course of elephant paths.

STOWAWAYS IN SEEDS

With animals on the lookout for food, seeds often get eaten before they can develop. But plants make many more seeds than they need and sometimes have to sacrifice a number in order to win the help of animals in their dispersal.

WEEVILS' JAWS PERFORM DEADLY SURGERY

Weevils are among plants' worst enemies, damaging countless billions of seeds every year. The hazelnut weevil is only as long as a fingernail, but half of this length is made up by a curved 'snout', tipped with a pair of tiny jaws. The weevil uses this like a precision surgical instrument, cutting neat holes into hazelnuts so that it can lay its eggs inside. The weevil's grubs then feed on the kernels, before turning into adults and falling to the ground inside the nut in autumn. They enlarge the original hole and fly away.

THE SECRET OF THE JUMPING BEAN

The Mexican jumping bean plant is renowned for having lively seeds. In warm weather, the seeds skip across the ground in hops up to 5 cm (2 in) long, and they can keep up these botanical acrobatics for several months. This bizarre phenomenon is produced by a grub inside the seed.

The jumping bean moth lays its eggs on the plant's female flowers, and the caterpillars feed on the developing seeds. By the time the infested beans fall to the ground they contain plump, well-fed caterpillars and little else.

The caterpillar spins a cocoon inside the bean and jumps along by pulling and releasing the silken threads. The powerful movements of the tiny grub can propel the bean and its passenger over remarkable distances. The caterpillar may jump dozens of times an hour. It is thought that the caterpillar jumps to move out of the sunshine and into places where it is less likely to be spotted by predators. The grub eventually breaks free by eating its way out.

WHY YUCCA AND ITS MOTH ARE INSEPARABLE

Yuccas are desert plants from North America that rely entirely on a small white moth for survival. The yucca moth flutters from flower to flower in the cool night air, spreading pollen, so helping yuccas to form seeds. But there is more to this partnership than meets the eye. Unlike most pollinating insects, the yucca moth lays its eggs in the flowers that it visits and its caterpillars use the developing seeds as food.

Generally, this would be detrimental to a plant, but yuccas and yucca moths have a very close partnership. The caterpillars eat some of the seeds made by each yucca flower, but they always leave some to germinate in the normal way.

In return for providing the caterpillars with food, yuccas can rely on the moths to spread their pollen. Without each other, neither the moth nor the yucca would survive.

YUCCA AND MOTH *Compared with a mature yucca plant, which grows up to 5 m (16 ft) tall, the yucca moth is minute, yet its existence is vital to the plant's survival.*

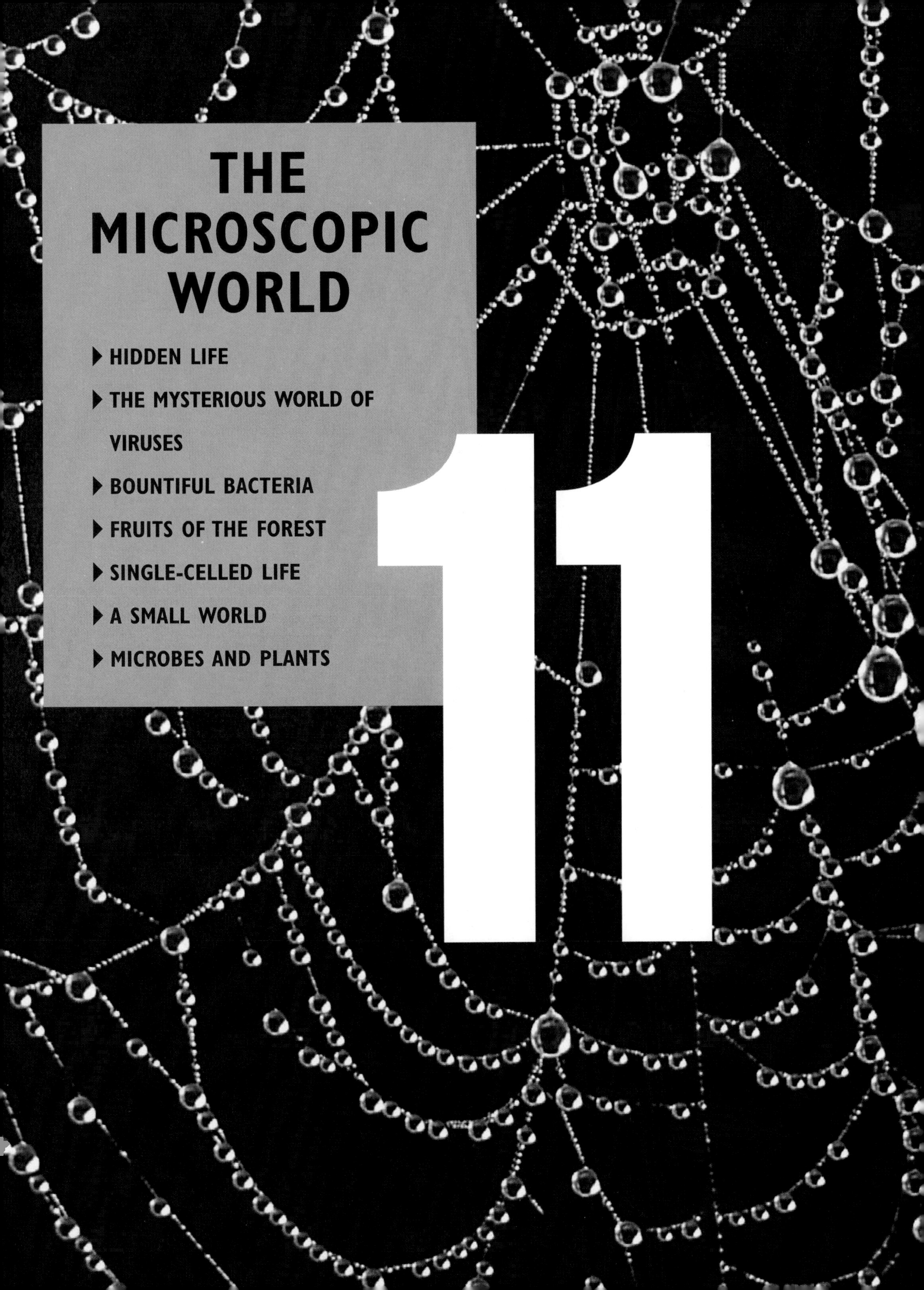
THE MICROSCOPIC WORLD
HIDDEN LIFE
THE MYSTERIOUS WORLD OF VIRUSES
BOUNTIFUL BACTERIA
FRUITS OF THE FOREST
SINGLE-CELLED LIFE
A SMALL WORLD
MICROBES AND PLANTS
11

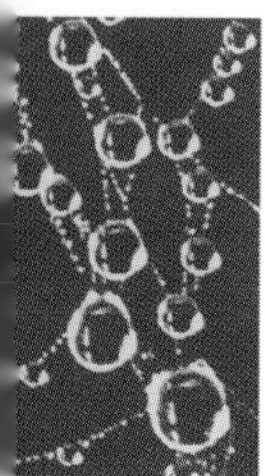

HIDDEN LIFE

Every habitat on Earth, from the sea to the ground beneath our feet, is teeming with microscopic life. Some microbes can be harmful, but most play a beneficial role, recycling raw materials and helping plants and animals to survive.

TARDIGRADES DRY OUT AND WAIT FOR RAIN

Tardigrades, microscopic animals that live in pools and puddles, survive drought by losing almost all the water in their bodies. When their pool or puddle begins to dry out they take emergency action, retracting their four stumpy legs and becoming barrel-like. Then they dry out with the pool. Tardigrades can remain in this dormant state for many years until the rain brings them back to life. Dormant tardigrades are able to survive in a vacuum and can even withstand temperatures as low as –272°C (–458°F) – far colder than anywhere on Earth.

TINY HUNTER *Clambering across a mat of algae, a tardigrade searches for food. Most tardigrades are less than 1 mm (1/32 in) long.*

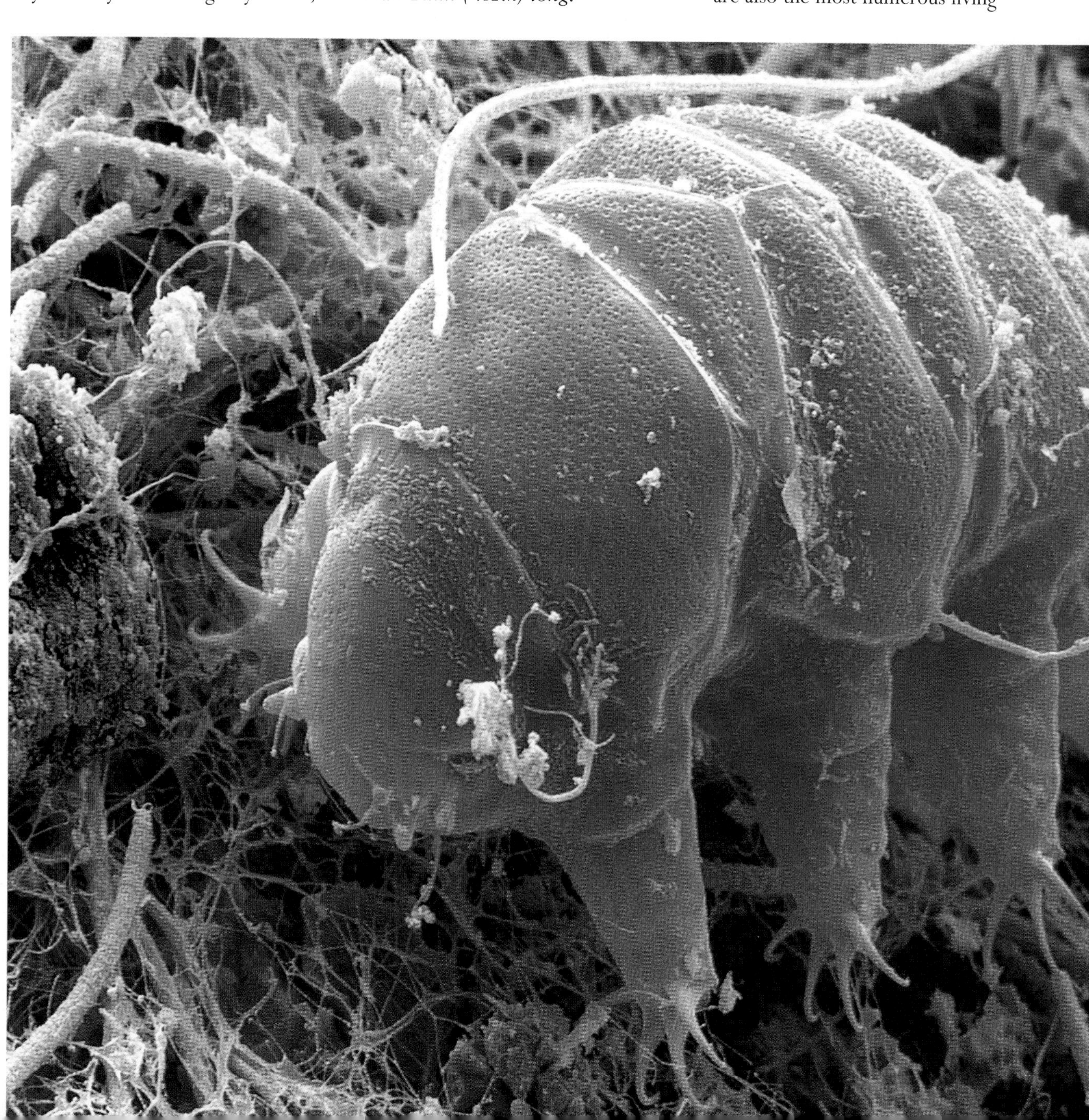

EARTH'S MOST NUMEROUS INHABITANTS

Bacteria are the smallest fully independent living things. If you were to place a single bacterium on the point of a needle, which was then enlarged to the size of a space rocket, the bacterium would only just be visible to the naked eye. Bacteria are also the most numerous living things on Earth. A single teaspoon of typical garden soil contains at least 5 billion of them, while the number of bacteria living on a healthy person's skin usually outnumbers their body's cells by about ten to one. Fortunately, most bacteria do no harm.

MICROSCOPIC ALGAE MAKE A LIVING DUST

The bright green dust sometimes seen on tree trunks is actually alive – and consists of millions of microscopic algae. These are simple plant-like organisms that live by soaking up the energy in sunshine. They do not have roots or leaves, but they do contain chlorophyll – the green chemical that plants use to collect and utilize solar power. The most common alga on tree trunks, called *Chlorella*, is so small that it would be lost inside a full stop.

There are many different kinds of algae, but *Chlorella* is unusual because it can live in a variety of ways. As well as growing on tree trunks, it can survive in water and on soil. It can also live as an internal part of other living things such as lichens (*see page 302*) and simple freshwater animals called hydras.

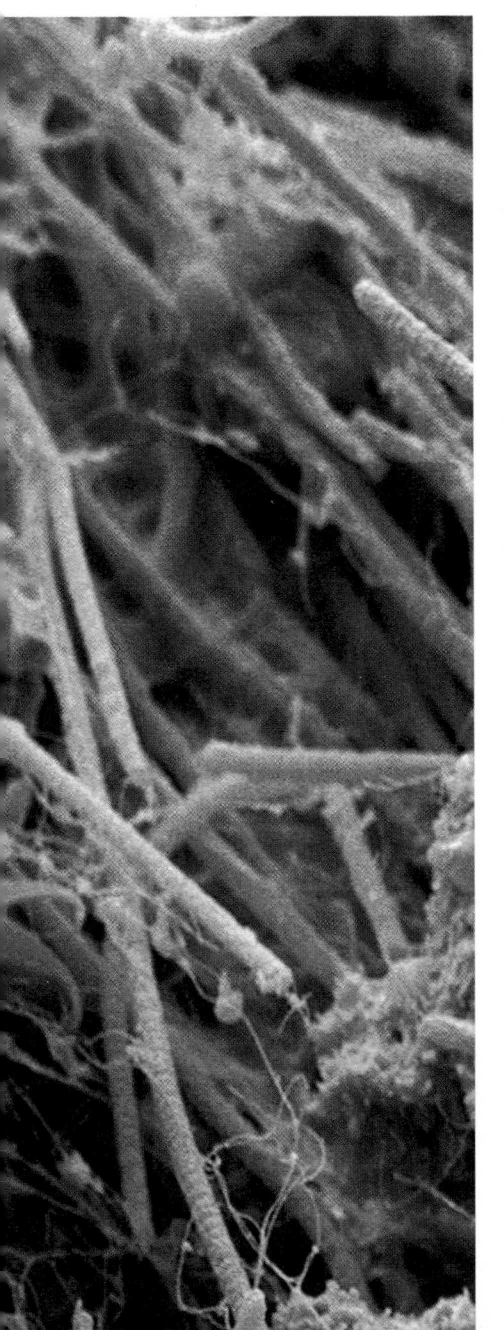

INVISIBLE CREATURES IN A WATERY WORLD

Rotifers are among the smallest animals on the planet, measuring less than 0.25 mm (1/1000 in) long. They live in wet or damp habitats and can even survive in the film of water around individual particles of soil. Despite their small size, most rotifers are big eaters, preying on bacteria and other forms of microscopic life. They do not have limbs and move by beating a crown of fine hairs on their heads.

In many species the females reproduce without having to mate.

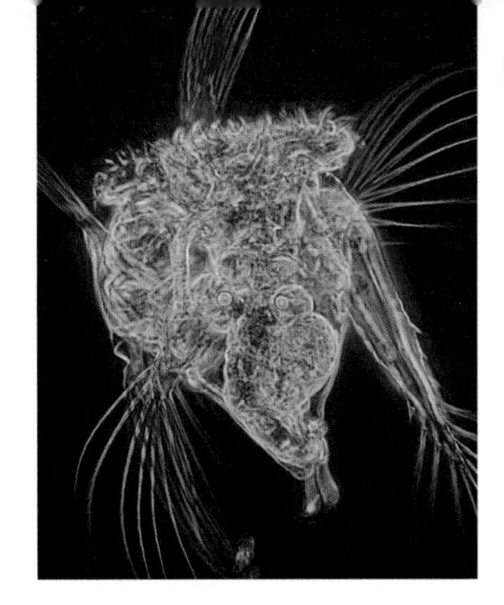

CLEAN SWEEP *Swimming through the water of a pond, a rotifer uses its microscopic hairs to sweep up even tinier forms of life.*

A QUESTION OF SCALE

The largest inhabitants of the microscopic world are microanimals and algae: on average, ten of them laid end to end would reach across the width of a hair. Microanimals are often transparent, while algae are usually bright green, but even if you have exceptional eyesight, a single line of them would be impossible to see without the use of a microscope.

Bacteria – the most common microbes by far – are much smaller, so many more would be needed to bridge the same gap. It would take about 200 average-sized bacteria to do the job, but with mycoplasmas – the smallest bacteria – the total needed rises to about 2,500.

Average bacteria are easily visible under a standard light microscope, but if you wanted to look at mycoplasmas you would need an electron microscope, which can magnify hundreds of thousands of times.

Viruses are smaller still and can be seen only with the highest magnification. Some are as big as mycoplasmas, but others are so small that 10,000 would fit across a hair. Viruses lie at the very edge of the living world: anything smaller than this is so small and simple that it cannot be alive.

LIFE'S LIMIT *Less than a thousandth of a millimetre across, mycoplasmas are the smallest things made of cells.*

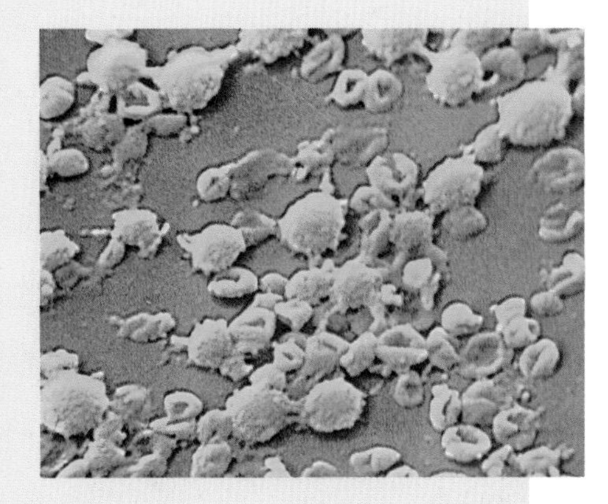

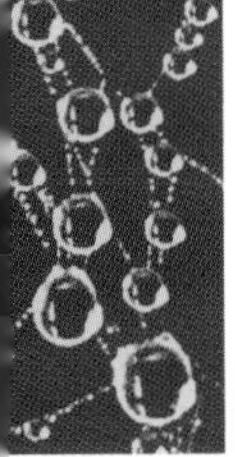

THE MYSTERIOUS WORLD OF VIRUSES

Viruses are microscopic packages of chemicals that have a reputation for causing disease. Unlike other microbes, they are not fully alive, but they are not dead: they inhabit the twilight world separating living things from lifeless matter.

PLANT VIRUSES HITCH A LIFT WITH AN APHID

Tiny insects called aphids make perfect virus carriers. As an aphid sucks the sap from a plant, having pierced its stem with its needle-like mouth-parts, it also picks up any viruses that the plant contains. The aphid then flies to another plant, carrying the viruses with it, and injects some of them when it next feeds.

More than 1,000 different viral diseases in plants have so far been discovered, including many that affect crops. Some of these diseases do little harm, but many make plants grow slowly – one reason why aphids are such unwelcome guests.

THE SECRET OF A LONG SHELF-LIFE

Viruses behave quite differently from things that are fully alive: they do not eat, cannot grow, and cannot reproduce on their own. More extraordinary still, they can be dried out and turned into crystals, just like nonliving substances such as salt. If crystallized viruses are kept dry, they can be stored indefinitely, in the same way that salt can be stored on a kitchen shelf. But if the viruses, such as the tobacco mosaic virus, are allowed to dissolve in water – even decades later – they are instantly ready to resume their parasitic way of life.

EVEN BACTERIA SUFFER FROM PARASITES

The T4 viruses that attack bacteria, land on them feet-first, like a spaceship settling on the Moon. These T4 viruses are a thousand times smaller than the smallest thing visible to the human eye. Once they have landed, they inject their genes into the bacterium and begin to take it over. The genes are chemical instructions that force the bacterium to assemble new T4 viruses.

It takes just 20 minutes for the new viruses to be built. When they are ready, the bacterium bursts open and the viruses drift away to find new victims of their own.

Many viruses are smaller and simpler than the T4, but they all work

MICROBE CARRIERS *For viruses, aphids are the ideal way of hitching a lift from one plant to another. When aphids breed, the viruses are handed on to their young.*

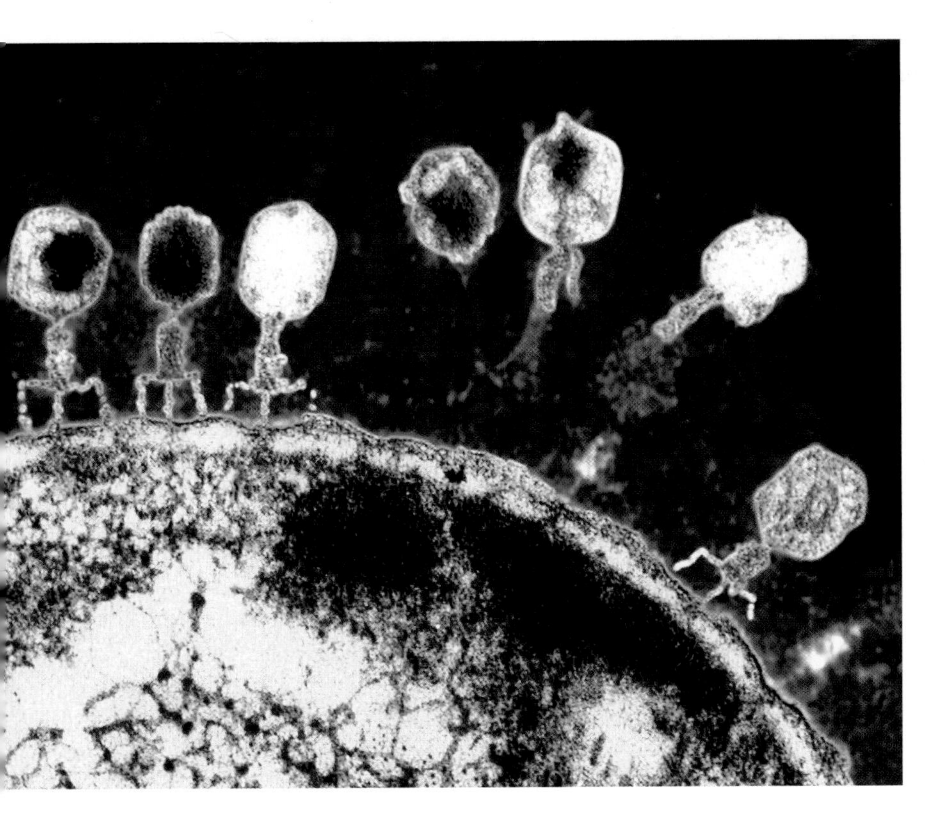

ON TARGET *T4 viruses on the surface of an* E. coli *bacterium – a common microbe that lives inside the human digestive system.*

in a similar way, by attacking the cells of living things. When they are outside living cells, they behave like packages of chemicals and show no signs at all of being alive.

THE VIRUS THAT GETS UP YOUR NOSE

It only takes a single rhinovirus to start a cold, but once the infection is under way the virus multiplies rapidly. Rhinoviruses attack the lining of the nose and throat (*rhino* comes from the Greek word for nose), making the cells produce copies of themselves. The infection causes the body's cells to produce more mucus, resulting in a runny nose, which triggers a sneeze and huge numbers of copied viruses are ejected into the air, giving them a chance to infect other people nearby.

Rhinoviruses are among the smallest viruses known, and like all other viruses they are very particular about the cells that they infect. They need to be warm, but they are not good at surviving at full body heat. That is why they attack the nose and throat, rather than the cells that line the lungs.

Rhinoviruses also stick to one particular host – another feature shared by viruses as a whole. The ones that attack humans rarely attack animals, which is why people do not give colds to their pets. More than 100 different rhinoviruses target people.

HOW A VIRUS CREATES BEAUTIFUL FLOWERS

Gardeners have found one plant virus they can positively encourage because of the striking blooms that result. The virus, known as the 'tulip-breaking virus', attacks tulips making them produce flowers with contrasting streaks. These magnificent blooms are much prized by horticulturists. Streaked flowers are also common in wild plants, such as brambles and honesty.

Mottled leaves are another common sign of virus attack. They are produced by mosaic viruses, which look like slender rods. Some of these viruses are 0.001 mm long and are among the 'giants' of the viral world.

WINNING STREAKS *The 'tulip-breaking virus' is the cause of these streaked blooms. The virus can be passed on in seeds and young bulbs.*

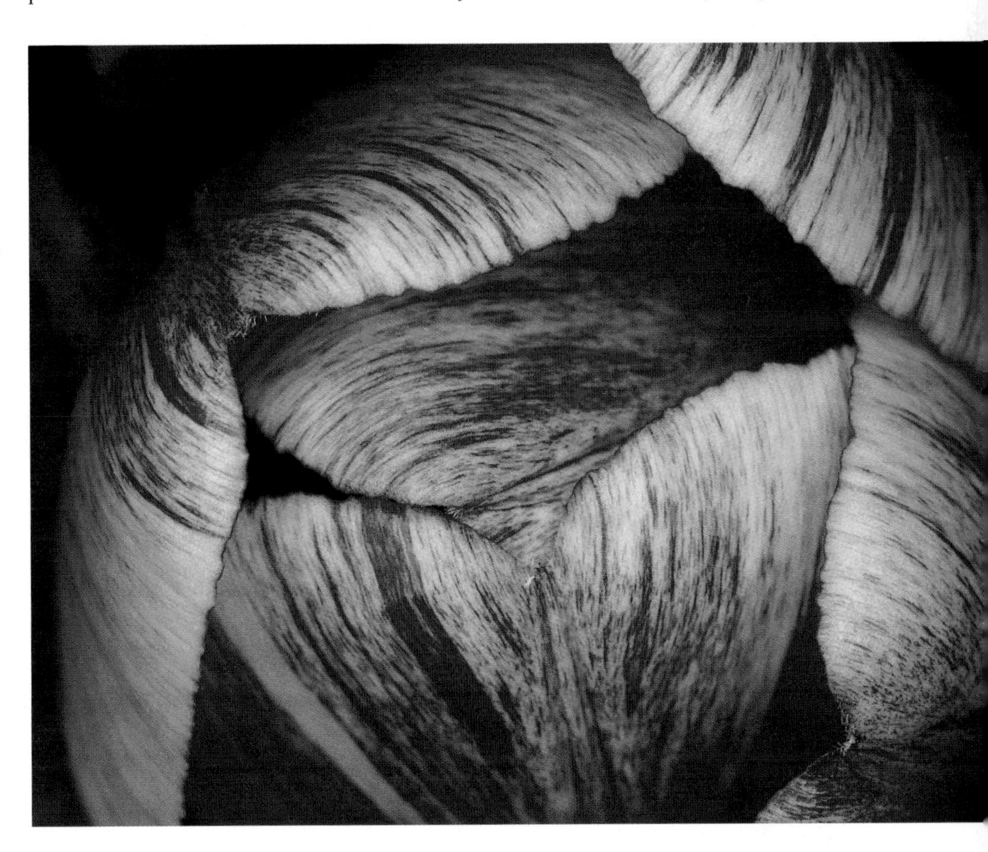

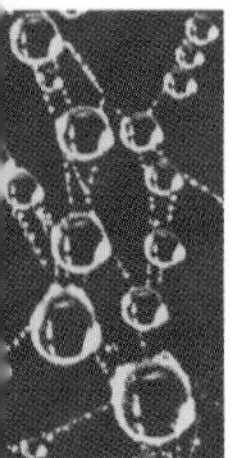

BOUNTIFUL BACTERIA

Bacteria were probably the first living things to appear on Earth. Relatively few bacteria are dangerous; the majority perform a vital role in maintaining life by recycling material and fending off infection. Some even produce light shows.

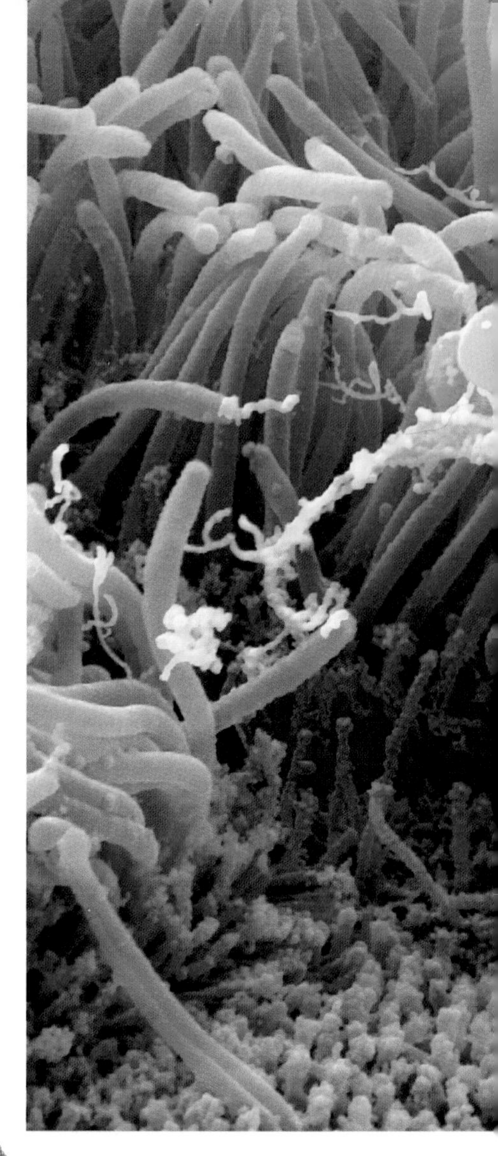

WHY BACTERIA ARE FUSSY ABOUT THEIR FOOD

A diet of iron or sulphur is a perfect recipe for survival for bacteria such as *Thiocystis.* Like tiny chemists, these bacteria take up minerals from their surroundings and make them react together, collecting the energy given out in the process. The bacteria use the energy to live and grow – just as we use the energy we get from our food.

This way of life dates back to a time when bacteria were the only living things on our planet. But mineral-eating bacteria are not the only ones with very particular diets. Many other bacteria feed on substances made by living things, or on their dead remains, and they can be extremely fussy about what they eat. Some feed only on substances in waterlogged mud, in rotting wood, or even in decomposing fur. By eating foods like these, bacteria break down waste and turn it into substances that plants and animals can use.

SULPHUR-EATERS *These tadpole-shaped* Thiocystis *bacteria live in lagoons and feed on sulphur in the water. Like plants, they also need light to survive.*

COULD THEY TAKE OVER THE WORLD?

Theoretically, bacteria can produce more than a thousand billion billion offspring in just 24 hours. Bacteria reproduce by dividing into two, which some can accomplish every 20 minutes. Continuing at this rate, the entire planet would be filled with bacteria – and nothing else – within a week. This does not happen, however, because as soon as their food runs short, bacteria stop dividing and the population explosion ceases.

Bacteria that infect animals, such as those that cause scepticaemia (blood poisoning) from wounds, face another hazard: their host's immune system. This defence system targets invaders with deadly efficiency. It often disables or destroys bacteria before they take hold and it 'memorizes' them in case they stage another attack. If the invaders reappear, they risk being instantly recognized and destroyed.

HOW GOOD BACTERIA KEEP OUT THE ROGUES

The human body hosts more than 100,000 billion bacteria, found on the skin, in the nose and throat, and in the digestive system. After a thorough wash billions of bacteria will

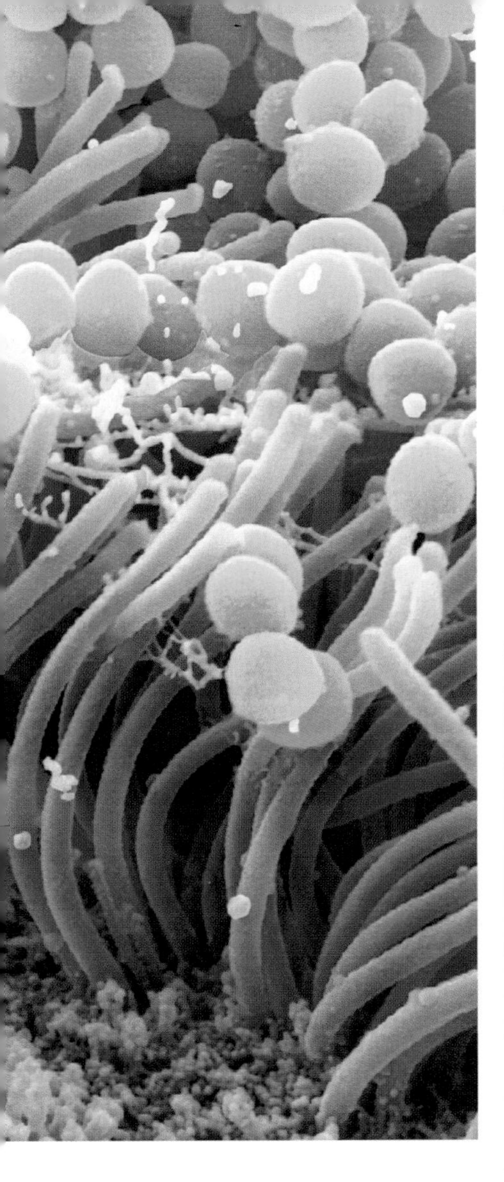

AIRBORNE ARRIVALS *The yellow spheres are* Staphylococcus *bacteria, trapped by microscopic hairs in the lining of the nose.*

be removed from the surface of your body, but even more will remain behind. The bacteria that live on our bodies are often very useful: they make it much harder for dangerous bacteria to find a place to live.

These helpful bacteria are only beneficial if they stay in their normal home. If they manage to get into other parts of the body, they can change from allies into enemies. For example, a bacterium called *Staphylococcus aureus* often lives harmlessly on the inner lining of the nose. But if the same bacterium gets into the body – through a wound, for example – it can be dangerous, causing infections of the ear, food poisoning, and inflammation of the heart.

MINI-TURBINES POWER FAST-MOVING MICROBES

For their size, bacteria include some of the fastest-moving living things on Earth. Corkscrew-shaped spirochetes, for example, that cause syphilis and tick fever, can swim 100 times their own length in a second. Scaled up, this is equivalent to a human swimmer racing through the water at 650 km/h (400 mph). These microscopic record-breakers are powered by slender hairs, which spin like propellers. Each of these hairs is fuelled by a chemical 'turbine', located at the point where the hair is attached to the bacterium's cell wall.

The turbine can run at different speeds and it can also spin in either direction, allowing the bacterium to accelerate, slow down, or change course. These rotary motors are unique; to date nothing else like them has been found in the living world.

SPEEDY SWIMMER *By spinning its turbine-powered hairs, or flagella, this* Salmonella *bacterium can move at speed. The hairs trail behind it.*

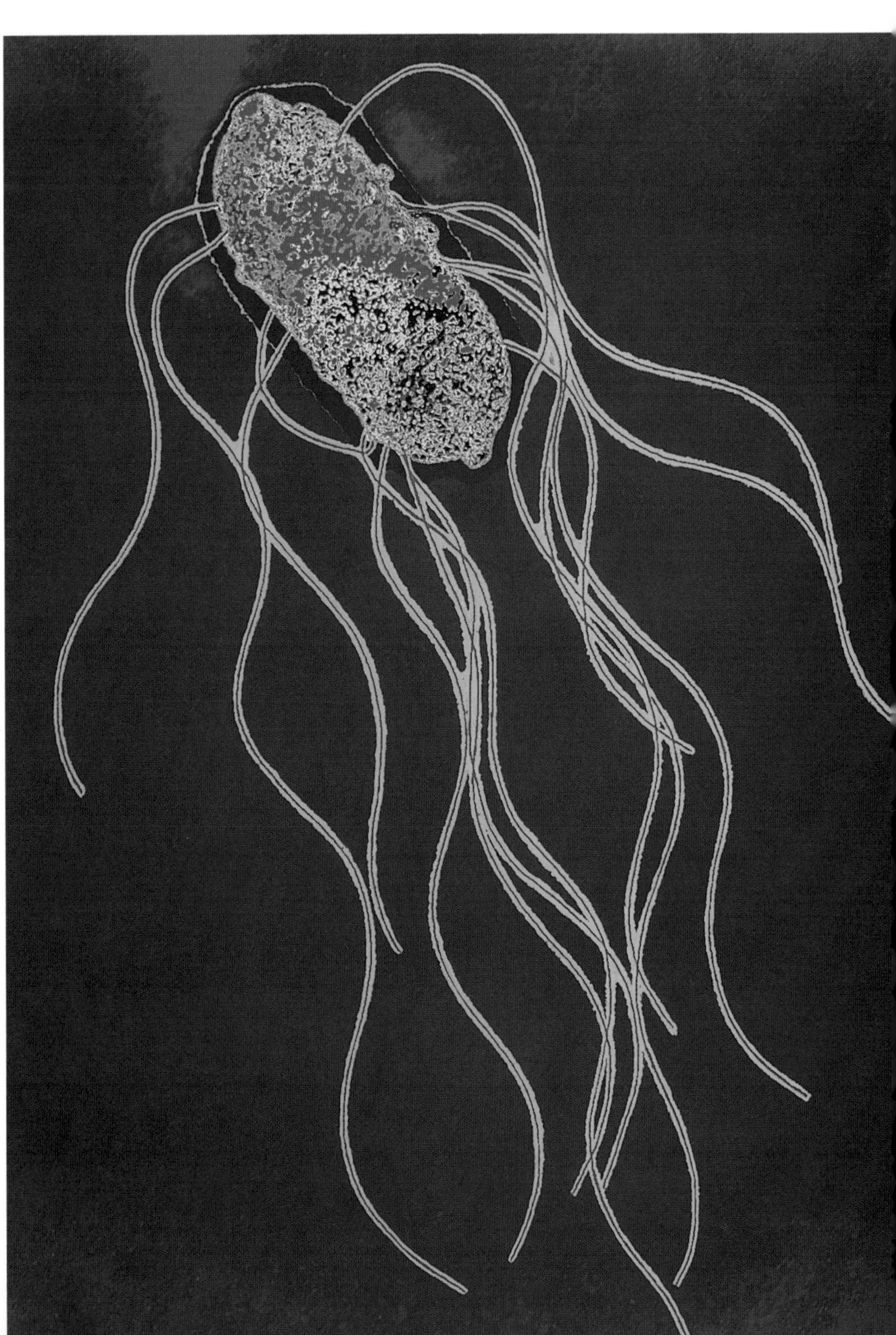

THE ANCIENT WORLD OF THE STROMATOLITES

A series of rocky mounds in the warm, shallow waters of Shark Bay, Western Australia, were formed by bacteria. These knee-high mounds, known as stromatolites, can be thousands of years old. Yet they are youngsters compared with fossilized specimens, which go back more than 3.5 billion years and are among the earliest signs of life on Earth.

Stromatolites are formed by cyanobacteria – bluish green microbes that grow in long, hair-like strands. Cyanobacteria live by harnessing the energy in sunlight and the ones that form stromatolites trap particles of sediment as they grow. The bacteria and the sediment build up in paper-thin layers, forming a rock-hard mound that may grow less than a hair's breadth every year.

BACTERIA BUILD-UP *Stromatolites in Shark Bay, Western Australia, lie exposed at low tide. The tallest mound is as high as an adult man.*

SHINING LIGHTS IN THE DEPTHS OF THE OCEAN

In the darkness of the deep sea many animals rely on luminescent bacteria to lure prey or distract would-be predators. Luminescent bacteria make light by combining oxygen with a substance called luciferin. When millions of bacteria are crowded together, the light can be seen from many metres away.

Sea animals, such as the flashlight fish, often house their bacteria in special light-producing organs and some use movable flaps of skin to make the light blink on and off. No one knows why these bacteria should glow, but by living aboard animals they have a safe, mobile home.

THE MICROBES THAT DWELL DEEP UNDERGROUND

Bacteria have been discovered in rocks 2.8 km (1$^{3}/_{4}$ miles) below the ground's surface. Subterranean bacteria live in extreme conditions: the pressure is intense, there is no light, and the temperature of the rock can be close to boiling point.

The bacteria live in microscopic pores inside the rock, where they feed on dissolved minerals. Compared to bacteria on the surface, these buried microbes lead long but extremely frugal lives. Their meagre food supply means that they get by on almost no energy and may reproduce as rarely as once every 50 years.

Heat is the greatest threat to their survival. On dry land some bacteria may be able to survive as far as 4 km (2$^{1}/_{2}$ miles) underground, but below this point the rock becomes too hot for living things to exist.

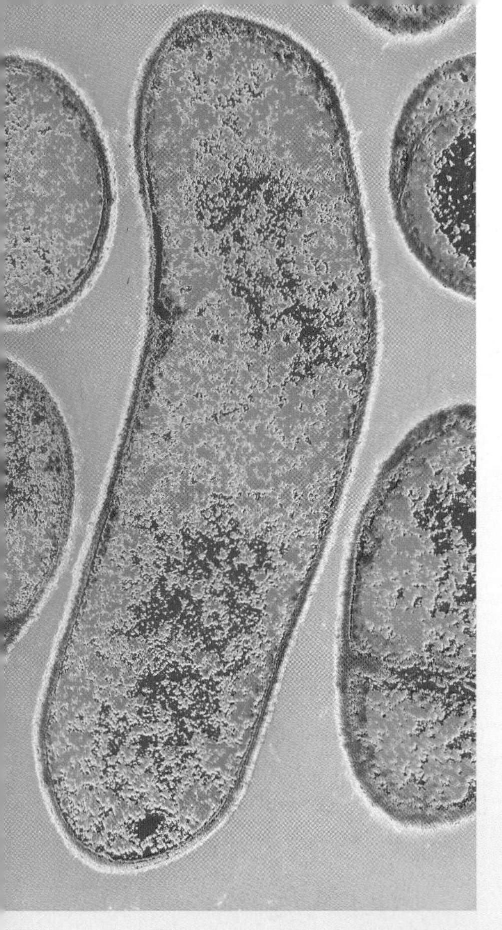

DEADLY BACTERIA SHUN OXYGEN

One of the most hazardous forms of bacterium, *Clostridium botulinum*, is responsible for causing botulism. It produces a powerful poison: a minute amount – about one-tenth of a millionth of a gram – paralyzes the body's muscles, leaving victims unable to breathe.

Bacteria such as this are known as oxygen-haters. They live in places where oxygen cannot reach, such as stagnant mud under ponds and lakes, or in wet soil. Here, they do no harm, but if they accidentally get into food, some of them can be deadly. The botulism-causing bacterium can survive in canned food, because cans contain little air. To kill the bacterium, the contents are heated to 129°C (264°F) after the cans are sealed.

HIDDEN HAZARD *The botulism bacterium causes a fatal form of food poisoning, but only if food has not been properly preserved.*

THE BACTERIAL ORIGINS OF WILL-O'-THE-WISP

The ghostly blue flames that sometimes dance over the surface of ponds and marshes on still, dark nights, are produced by bacteria. The bacteria, known as methanogens, live deep in underwater mud and produce bubbles of methane, an inflammable gas. When the bubbles surface, they can catch fire. These eerie flames give off the faintest light and are blown out by the slightest breeze, often disappearing from one place only to reappear in another.

In times gone by, the will-o'-the-wisp was thought to be a bad omen, but seeing this phenomenon is actually a lucky event, because it is one of the rarest spectacles in the natural world.

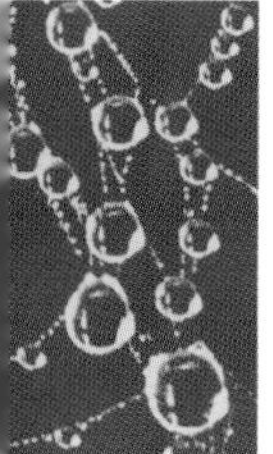

FRUITS OF THE FOREST

Fungi are all around us, but most are usually hidden from sight. Often we notice them only when they reproduce, because this is when they form the mushrooms and toadstools that are designed to scatter their spores.

THE LIVING INGREDIENTS IN BREAD AND BEER

Yeasts are a highly unusual type of fungi that each consist of a single cell. They use sugar as food, producing alcohol and carbon dioxide as waste as it is broken down. Yeasts are used for brewing and winemaking, and for making bread, because the carbon dioxide they release helps the bread to rise. Yeasts reproduce by growing tiny buds, which develop into new fungi. They can do this as often as every two hours.

Over the centuries, brewers and bakers have selected their own strains: *Saccharomyces cerevisiae* is used for baking; the one used for brewing lager is *Saccharomyces carlsbergensis*, named after Denmark's Carlsberg Brewery, where it was first isolated in the 1880s.

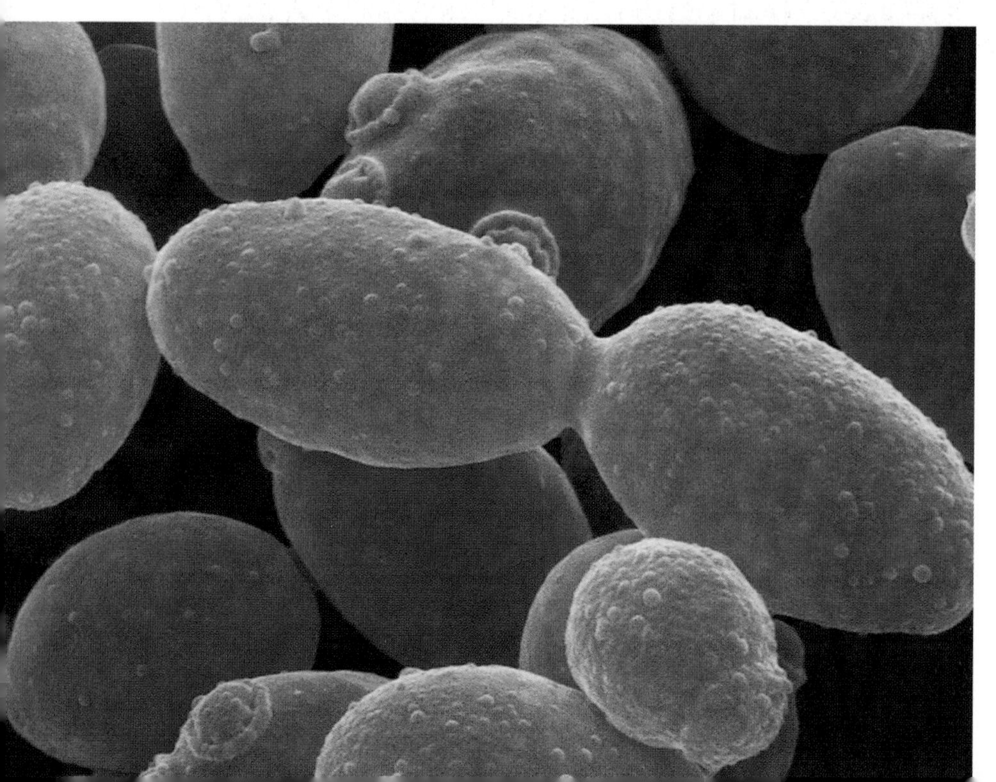

BUDDING FAMILY *Yeast cells feed on sugar and, in the wild, make up the 'waxy' bloom on the surface of grapes and other sugary fruits.*

ARE FUNGI MORE LIKE PLANTS OR ANIMALS?

Though they are often confused with plants, it has recently been discovered that fungi are genetically more closely related to animals. Fungi do not have roots or leaves, and they cannot use sunlight to stay alive. They live by sending out microscopic threads through soil or wood, feeding on anything edible that they find by pouring digestive juices over it.

To reproduce, a fungus grows a 'fruiting body' such as a mushroom, or a giant bracket big enough to fill a carrier bag. The fruiting body releases spores which drift away in the air to start life elsewhere.

MUSHROOM CAUSES ACCIDENTAL DEATH

The world's most lethal fungus is the death cap: just 5 mg (0.00017 oz) of its toxin is enough to kill a human. Found in woodland across the Northern Hemisphere, the death cap is unremarkable to look at, with a pale olive-brown cap and typical mushroom shape. If eaten, it has no ill effects initially. But once it is swallowed, the toxin starts to work. It attacks the victim's kidneys and liver, producing symptoms after 6 to 12 hours. Without treatment, most people die within ten days.

Many of the death cap's relatives, including the panther cap and the

DEADLY FUNGUS *A large ragged ring on its stem distinguishes a death cap from an edible mushroom. Death caps appear in woodland in early autumn.*

destroying angel, are also poisonous. The poisons are thought to be a natural 'accident', rather than a form of self-defence, because they take so long to have an effect.

DUNG-LOVING FUNGUS SHOOTS AT PLANTS

A tiny fungus that lives in animal dung ensures that its spores reach fresh food supplies by firing them on to nearby vegetation. The spores can then be eaten by a passing animal, thus ending up in the animal's dung. The fungus, known as *Pilobolus*, holds its spores in a transparent capsule on top of a short stalk. The capsule's sides act like lenses, focusing sunshine on to the top of the stalk, so that the capsule points upwards towards the light. Meanwhile, water pressure builds up inside the capsule, eventually causing the lid to burst off. The spore package can travel as far as 2 m (7 ft).

If the fungus is in luck, its spores land on a blade of grass, where they stick fast until they are eaten by a passing animal.

The spores are specially adapted to pass through the digestive system unharmed and they emerge in a handy pile of dung, ready to start life anew.

INVISIBLE GIANTS ATTACK TREES

In 1992 a honey fungus known as *Armillaria bulbosa*, that extended underground over 15 ha (37 acres), was discovered in Michigan, USA. The fungus was estimated to weigh 100 tonnes – the same as a blue whale – and to be 1,500 years old. A related species, *Armillaria ostoyae*, found in Washington State, USA, covers 600 ha (1,500 acres) of forest. Honey fungus consists of a network of feeding threads that spread up trees, attacking the wood. At intervals the threads produce clusters of yellow toadstools.

A RING REVEALED *Autumn rain encourages toadstools to sprout in a fairy ring formation. At other times, rings are visible as dark green circles.*

THE SECRET OF FAIRY RINGS

In fields and lawns, mushrooms often grow in 'fairy rings', as if they have been carefully planted. The ring is, in fact, a natural result of the way that fungi spread.

Each ring starts when a single spore lands in the grass and germinates. The spore produces a network of threads, which creeps outwards through the soil searching for food. In the centre, the oldest threads die away as their food is used up, but fruiting bodies (mushrooms) are produced around the edges, forming a ring. As the fungus grows, the circle of mushrooms gets bigger each year. The rings can reach more than 30 m (100 ft) across, by which time they may be centuries old.

THE MUSHROOM THAT DIGESTS ITSELF

The shaggy ink cap, or lawyer's wig, lets its spores flow away in a special 'ink'. The ink is produced by the mushroom dissolving itself – a process that starts as soon as the mushroom is mature. The cap's edges are the first to go, turning into a black liquid, containing millions of spores, that drips on to the ground. The spores may also hitch a lift on insects that land on ink caps looking for food. Eventually, the only thing left of the mushroom is a small 'button' perched on the stalk.

INKY SOLUTION *The shaggy ink caps at the front of this clump are new and intact; the one at the back has almost finished shedding its spores.*

THE DESTRUCTIVE APPETITE OF DRY ROT

Dry rot is one of the most feared fungi in the world. It can destroy the timber in buildings and cause collapse. The secret of its rapid spread lies in the way it grows. Unlike most fungi, dry rot can 'leapfrog' through indigestible substances, such as brick, tiles, and concrete, in its search for wood. The fungus's feeding threads keep probing until they find a way through.

The first signs of trouble are cracks and splits as woodwork is digested from within. Weeks or months later, bracket-shaped mushrooms appear. Each one produces millions of microscopic spores which germinate if they land on damp wood. By this stage, the only way to stop the fungus is to remove and burn the wood.

THE GIANT PUFFBALL'S SPORE EXPLOSION

The giant puffball is one of the biggest spore factories in the fungal world. It grows in temperate areas and can measure 80 cm (31 in) across and weigh as much as 20 kg (44 lb). As the spores ripen, the puffball dries out and the skin at the top breaks open. The puffball acts like a trembling powder-puff, blowing out spores when pressure is applied by wind, raindrops, or passing feet.

A 35 cm (14 in) diameter specimen probably produces about 5,000 billion spores, but only a minute number ever germinate successfully.

LIFE REKINDLED IN SCORCHED EARTH

After the devastation of a forest fire some of the first signs of life are fungi. The reason for their prompt arrival is that their spores were already present in the ground.

Temperatures of more than 40°C (104°F) make the spores of forest fire fungi, such as *Rhizina undulata*, germinate, giving the fungus a head start over other types of fungi.

Sometimes camp fires trigger the spores into life, and they have even been 'woken up' by hot tar being poured on to roads.

SINGLE-CELLED LIFE

Bigger than bacteria, but still too small to see with the naked eye, protists are the most complex kinds of single-celled life. Many of these creatures are housed in elaborate shells, while some have no hard parts and no fixed shape.

MYRIAD MICROSCOPIC SCULPTURES

The exquisite shells of single-celled algae known as diatoms have the appearance of the finest cut glass. They are covered in a complex pattern of struts and perforations, although most are far too small to be seen with the naked eye. Made of silica – the substance used in glass – the shells have two parts, which fit together like a case. There are more than 5,000 different kinds of diatom, each with its own unique style of shell.

They are found in fresh water and seas worldwide and provide a source of food for microscopic animals and larger species, such as fish. In an area the size of a small postage stamp there may be up to 100 million diatoms.

MINIATURE SKELETONS ON THE SEA FLOOR

Far out to sea, the ocean floor is covered with a layer of sticky ooze made from millions of microscopic shells and skeletons. Some of the shells come from diatoms (single-celled algae), but many belong to organisms called foraminiferans, which live in tiny cases made of chalk. These shells float when their owners are alive, but after death they begin their descent towards the seabed.

The downward journey can take a shell more than a year to complete. When it lands on the bottom, it adds its minuscule weight to a carpet of ooze that may be 500 m (1,640 ft) thick. Because diatoms and foraminiferans are so small, the ooze builds up slowly, often adding just 1 mm (1/32 in) every 100 years.

SINGLE-CELLED CREATURES THAT GO WITH THE FLOW

Single-celled animals called amoebas move by simply flowing forwards while the rest of their body slowly catches up. They do not have a fixed shape. At full speed, an amoeba can move at about 2 cm (4/5 in) an hour, which is impressive for something that does not have any real muscles.

Most amoebas live in damp or wet places and feed on microorganisms

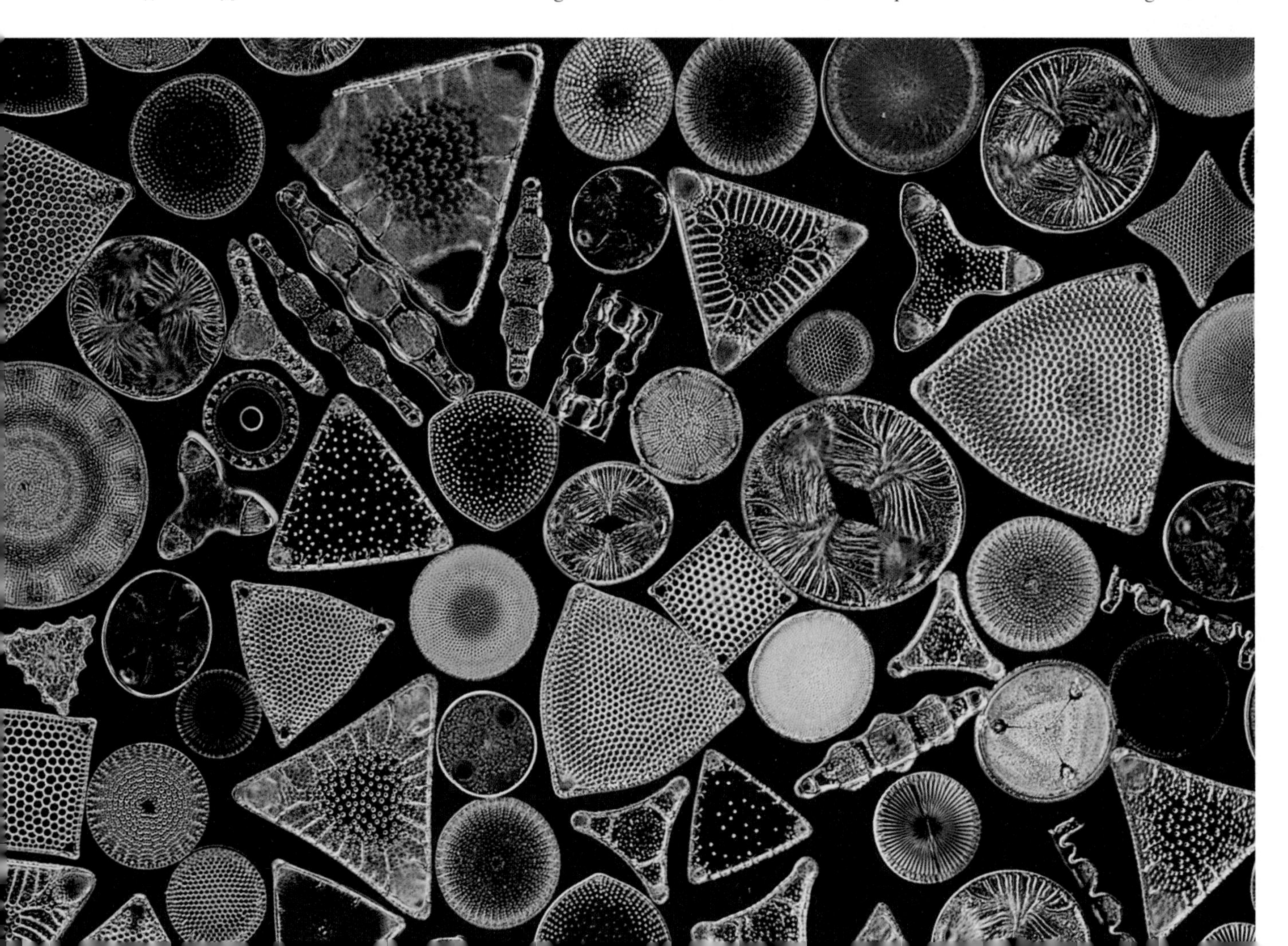

MINIATURE MARVELS *This collection of marine diatoms is just a tiny selection of the thousands of different types that live in the sea.*

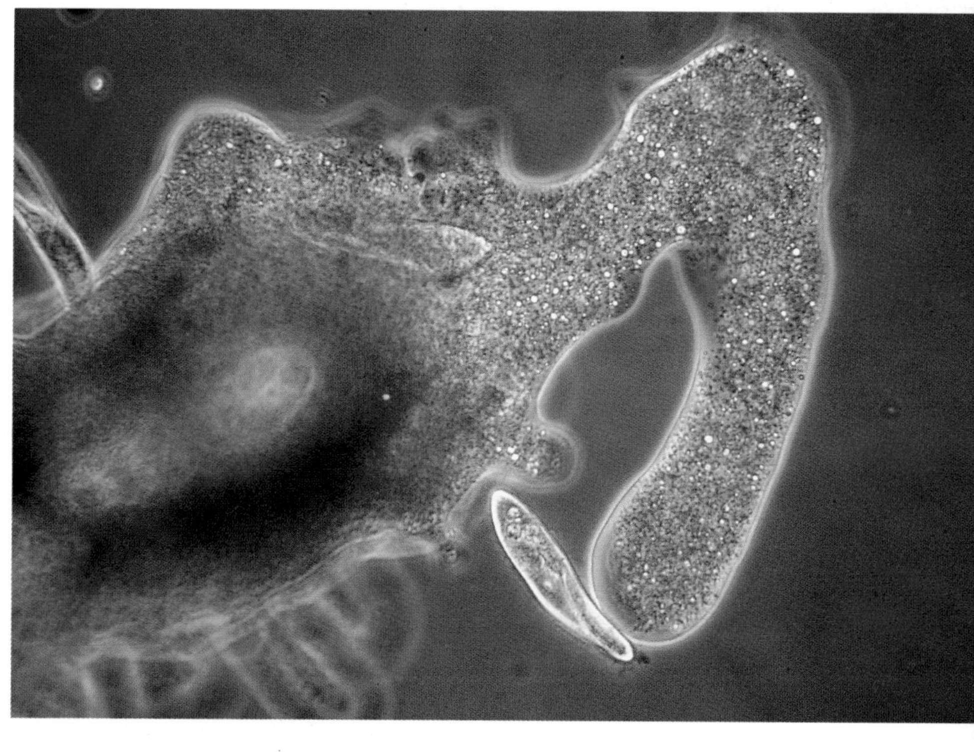

MICROSCOPIC MIGHT *Oozing out an arm-like mass of jelly, an amoeba engulfs its prey – another single-celled creature called* Paramecium.

such as rotifers that are even smaller than themselves. They track down their prey by feeling vibrations in the water and flowing around the prey so that it cannot escape. The amoeba then sucks the prey into its cell, storing it inside a bubble where it is digested.

To reproduce, amoebas simply divide in half, creating two new individuals that go their own ways. They can do this indefinitely, and as long as they avoid being eaten, they can theoretically live for ever.

DINOFLAGELLATE MULTITUDES TURN SEA INTO DEATHTRAP

In some tropical and subtropical regions summer brings the risk of toxic tides, when the sea turns a murky yellow or red. Thousands of dead fish may be washed up on the shore as a result. These 'red tides' are caused not by pollution, but by tiny organisms called dinoflagellates, which release poisons into the water.

Dinoflagellates behave partly like plants and partly like animals, using light to carry out photosynthesis, but also catching and eating other kinds of microscopic life. Most of the time they are thinly spread, so their poison does no harm. In warm, sunny weather, however, when the sea is calm, the dinoflagellate population can explode.

When this happens, poison builds up in the water, with catastrophic results for other forms of life in the region. After several days, the dinoflagellates disperse and the deadly tide breaks up.

MICROBE MARRIAGES OF SLIME MOULDS

When the time comes for social amoebas, also known as slime moulds, to reproduce, up to 200,000 individuals gather together. They form a brightly coloured 'slug', measuring up to 1 cm ($^{3}/_{8}$in) long, which can be seen with the naked eye.

The 'slug' of slime mould crawls across the ground towards the light, leaving a slimy trail. Eventually, after a day or more, it comes to a halt. Its 'back' bulges upwards and produces a ball containing spores which rises up on a stalk. Several hours later the ball bursts, releasing microscopic spores which drift away in the air.

When the spores land, they hatch into new amoebas, and the cycle begins again. Slime moulds are found throughout the world.

SLIMY MOVERS *A mass of slime mould travels across the ground (left), then raises its 'fruiting bodies' on stalks before releasing its spores.*

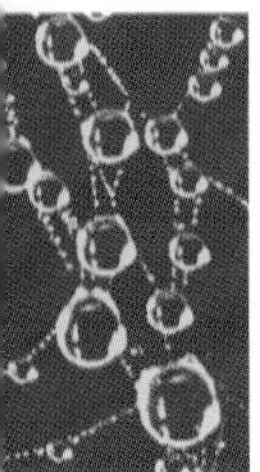

A SMALL WORLD

We often think of microbes simply as dangerous 'germs', but they can also be beneficial. In their microscopic environment, relationships can be just as varied and complex as the ones that are visible to the naked eye.

ESSENTIAL AID FOR PLANT-EATERS

Many animals feed on grass, but without microbes most of them would soon starve to death. This is because grass contains much cellulose – an extremely tough, but nutrient-rich substance that very few animals can digest. To get around this problem, grass-eating animals harbour microbes in their digestive systems which help them to break down food.

In hoofed mammals microbes are stored in a large stomach called a rumen. Here, billions of microbes live in a sea of chewed-up grass mixed with warm saliva. It is a perfect habitat for these tiny organisms and they multiply rapidly, using special chemicals to crack open the cellulose in the grass.

Hoofed mammals often 'chew the cud' (they regurgitate their food, and then chew it a second time), which turns the grass into even smaller pieces for the microbes to work on. By the time the microbes have finished, nearly all the cellulose is converted into useful substances that their host can digest.

LIVE-IN PARTNERS THAT SUPPLY FOOD

By day, reef-building corals rely on food made by microscopic algae living inside their bodies. These algae, called zooxanthellae, live very much like plants. They soak up the energy in the warm tropical sunshine and use it to make sugars and other nutritious substances.

Corals are made up of clusters of tiny animals called polyps. In return for giving the algae a safe place to live, the polyps take a share of their homemade food. Some coral polyps take more than 90 per cent of the food the algae make – which pays for choice accommodation close to the sea surface. At night, when there is less chance of having their tentacles attacked by fish, corals catch the drifting larvae of molluscs, crabs, and lobsters.

MICROBES WITHIN MICROBES

The microbes found in the guts of most species of termites comprise a tiny ecosystem of protozoa and bacteria. Termites need these microbes to help them to digest wood.

Some of the protozoa carry bacteria inside them, while others have them on their outer surface. These surface

REEF-BUILDING TEAM *Algae set up home in a lettuce coral off the Cayman Islands. They produce nutrients which the polyps feed on.*

bacteria often operate like tiny oars, speeding the protozoa through their miniature home, tucked away inside their termite host.

THE PUFFERFISH'S LITTLE HELPERS

Pufferfish contain a nerve poison that is 10,000 times more toxic than cyanide. It is produced by bacteria that live in the creature's body and is used to keep predators, including other fish, at bay. The pufferfish itself is immune to the poison, but anything that tries to eat it risks a very unpleasant death.

In Japan raw pufferfish is a great delicacy even though eating one can be potentially fatal. The poison, called tetrodotoxin, is found in the fish's skin, reproductive organs, and digestive system, all of which have to be removed before the fish is safe to eat.

FUNGI THAT CATCH THEIR OWN FOOD

Some carnivorous fungi grow special traps which they use to catch their prey. One genus of soilborne fungi (*Dactylaria*), found worldwide, grow loop-like snares at intervals along their feeding threads. The loops consist of three touch-sensitive cells. If a roundworm pushes its head into a loop, the cells swell up to three times their normal size in just a tenth of a second, gripping the worm so tightly that it dies. The fungus then spreads into its victim's body to digest it from within.

Some fungi even 'go fishing'. One species, *Zoophagus tentaculum*, lives in ponds, where it hunts for rotifers and other microscopic animals. The fungus grows tiny branches that are tipped with an edible-looking blob, but if an animal tries to feed on this, it becomes stuck fast. The blob then releases digestive juices and absorbs the catch.

COMPLEX LIFESTYLE OF A CLEVER WORM

The microscopic brainworm has a complicated life cycle, involving three different animal hosts, one of which it 'brainwashes'. Found in temperate regions, it starts off inside a plant-eating mammal, such as a deer, then moves on to a snail, via droppings that fall to the ground. Next it passes from the snail to an ant and when the ant is eaten by a deer, the entire cycle begins again.

Ants usually stay near the ground where they have little chance of being eaten by deer. Once the worm gets into the ant's brain it produces chemicals that induce the ant to laze in exposed places, like the top of a blade of grass. Here it is more likely to be eaten and the worm's life cycle can continue.

MALARIA PARASITES' PERFECT TIMING

Malaria is caused by microscopic parasites with an extremely specialized life cycle that relies on precision timing. When they get into a human body via a mosquito bite, they are carried to the liver, where they reproduce, releasing thousands of offspring. These circulate through the body in the red blood cells. Every few days, and always at night, the parasites simultaneously break out of the blood cells and infect new ones. The mosquitoes that suck up the parasites during this phase and carry them to a new host are nocturnal, hence the need for a night-time break out. The parasite tells the time by sensing its human host's body temperature, which rises and falls regularly every 24 hours.

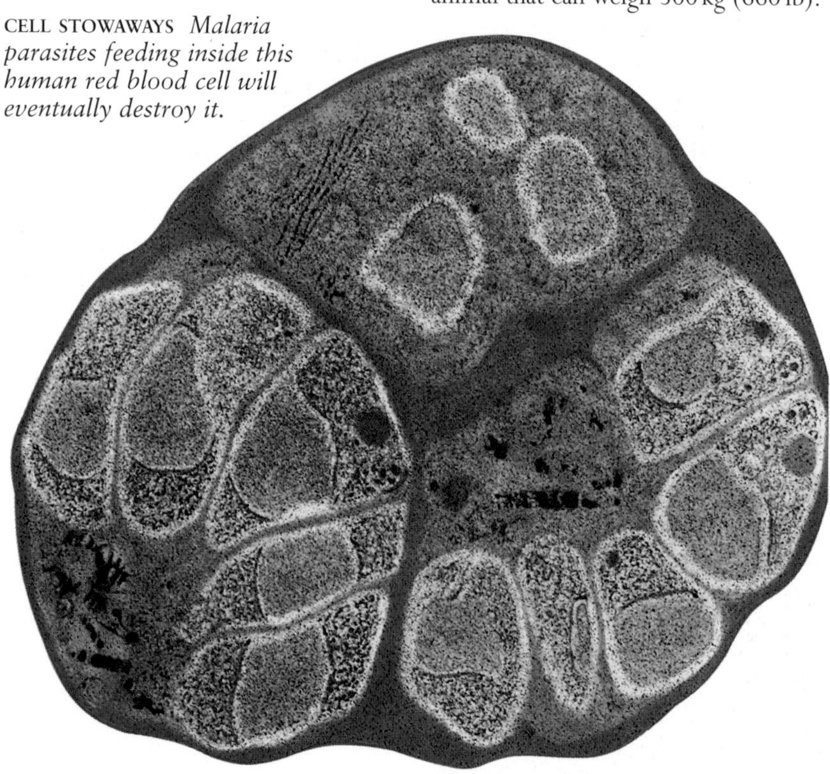

CELL STOWAWAYS *Malaria parasites feeding inside this human red blood cell will eventually destroy it.*

LIFE ABOARD A GIANT CLAM

With a shell up to 1 m (3 ft) across, the giant clam is one of the biggest and heaviest molluscs in the world. The fleshy mantle within the shell looks like a pair of bright blue-green lips due to millions of algae 'passengers'. The clam harbours the algae in return for some of the food they make.

At high tide, the clam opens its shell and spreads out its lips so that its algae can bask in the sun. But at low tide, it shuts its shell to protect its minute partners from prying animals and dry air. The clam also shuts if it is touched, reacting with surprising speed for an animal that can weigh 300 kg (660 lb).

GENTLE GIANT *Algae live within the giant clam's fleshy mantle providing it with carbohydrates made via photosynthesis. It gives them shelter.*

THE FLYKILLER THAT ATTACKS FROM WITHIN

The fly-killing fungus eats its victims from the inside. First a sticky spore attaches itself to a fly's body, germinating soon after it lands. The spore produces a thread that breaks into the fly, dissolving a tiny opening in the insect's hard body case. Once inside, the thread releases cells into the fly's bloodstream and the fungus quickly spreads.

When the fly dies, the fungus releases spores around its remains. The sticky spores then attach themselves to any other flies that come close and the destructive process turns full circle.

LAST RITES *The fluffy mass around this fly reveals that it has been killed by an insect-eating fungus – one of nature's effective pest controllers.*

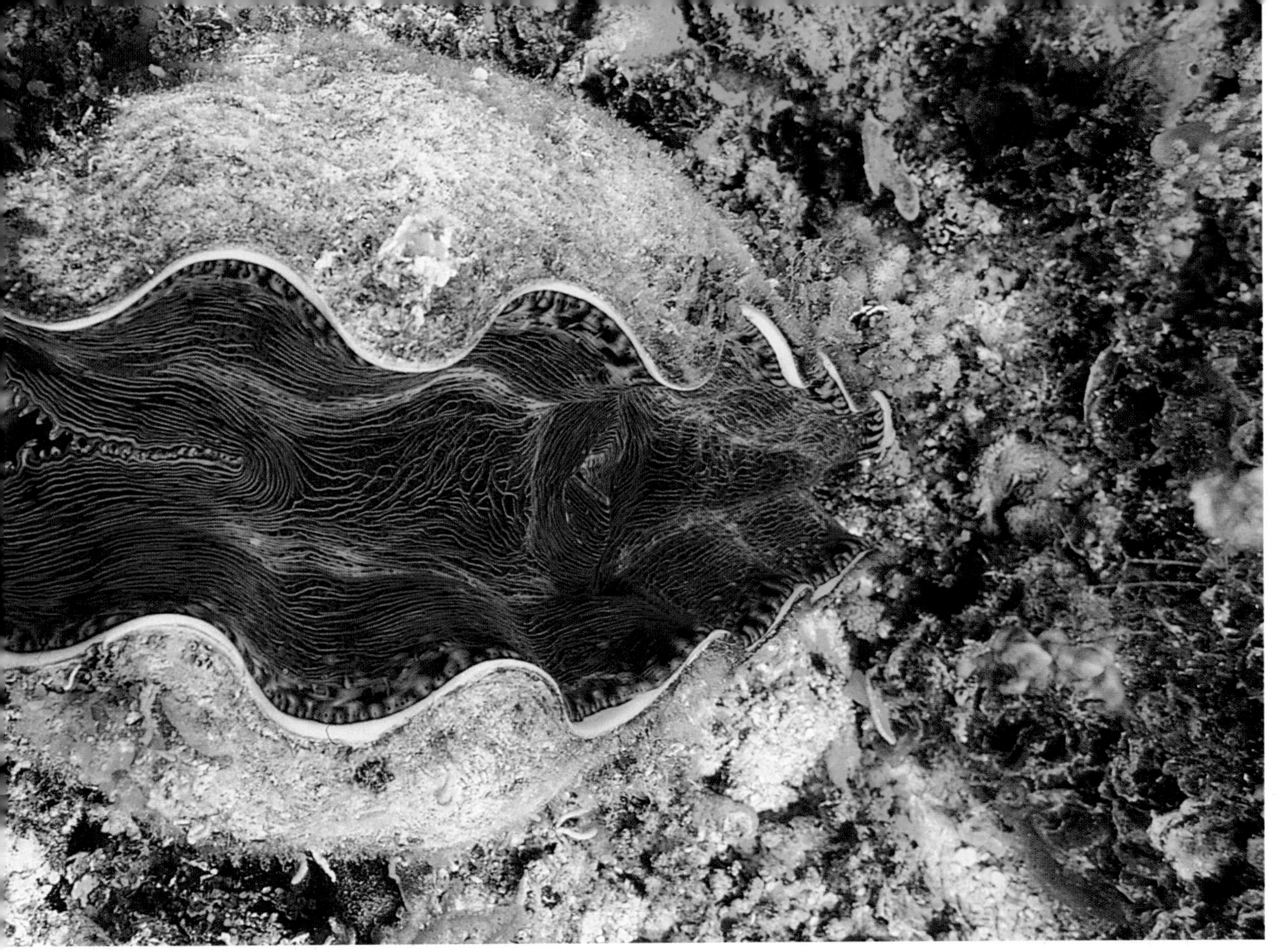

ANTIBIOTICS: NATURE'S CHEMICAL WEAPONS

Microbes have some highly effective weapons to keep their rivals at bay. The most important are antibiotics – chemicals that kill other microbes, or that make it difficult for them to grow. Soil is teeming with microbes and this is where most of the antibiotic-makers are found. Some are bacteria, but others are fungi that feed on dead matter.

Antibiotics were discovered by Alexander Fleming in 1928, when a fungus contaminated a dish of bacteria that were being grown in a laboratory, killing the bacteria where they grew. The fungus was identified as *Penicillium* and the result was penicillin, the first of many antibiotics to be used successfully in medicine.

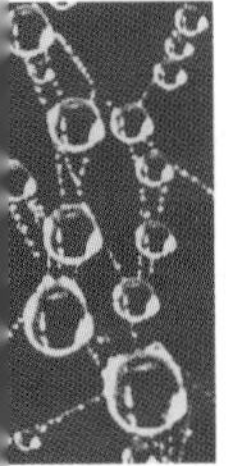

MICROBES AND PLANTS

For plants, just as for animals, microbes such as fungi can be both useful partners and deadly adversaries. Some kill trees, others search for vital nutrients to help them to grow. One particular fungus has become a gourmet's delight.

THE MOST EXPENSIVE FUNGUS IN THE WORLD

The black truffle from western Europe is the most sought-after fungus in the world. A gastronomic delicacy, it commands vast prices. Truffles can grow as big as a fist, but because of their underground habitat, humans have to rely on specially trained dogs or pigs to sniff them out.

The truffle's distinctive scent is designed to attract wild creatures, such as foxes and squirrels, which dig them up and carry them away, helping to spread their spores.

HIDDEN TREASURE *Pigs have a keen sense of smell – just what is needed to nose out nature's hidden store of highly prized black truffles.*

HOW BIRCH TREES GROW WITCHES' BROOMS

Birch trees often have several dozen strange growths that look like large birds' nests made out of tightly packed twigs. They are caused by fungi, such as *Taphrina turgida*, and bacteria in the tree which trigger a branch to produce hundreds of extra buds. The buds develop into twigs, which get more tangled every year. The growths can weigh branches down, but apart from this, they do the trees no real harm.

These growths have many traditional names, including 'witches' brooms'. But anyone trying to make a broom from them would face a difficult task. The twigs are gnarled and distorted, and are fastened together in a ball that is very difficult to pull apart.

FRIENDLY FUNGI HELP TREES TO GROW

Though many fungi are deadly enemies of trees, there are others that help trees to grow by procuring minerals. These helpful fungi spread their slender feeding threads through forest soil. When they find a tree's roots, they slowly wrap around them or even spread inside. Once the tree and the fungus are linked up, they help each other to survive.

The fungus acts as the tree's mineral location service. It tracks down essential substances such as copper and zinc, which trees need to stay healthy. The fungus is efficient at scouting for these soil minerals and it can bring them in along its threads from more than 100 m (110 yd) away. Meanwhile, the tree pumps sugary sap to its roots, and lets the fungus siphon off some of this high-energy fuel.

BIZARRE BROOMS *In winter, when birch trees lose their leaves, fungi-induced 'witches' brooms' stand out like spiky orbs against the sky.*

UNDERGROUND LIFE OF THE GHOST ORCHID

The ghost orchid spends most of its life underground, only becoming visible every ten years when it blooms. This rare orchid, from the temperate regions of the Northern Hemisphere, has no green parts and its pallid flowers, held on gaunt white stems, live up to its name.

The ghost orchid cannot photosynthesize, and lives on the nutrients supplied by a fungal partner as it breaks down decaying vegetation. After flowering, it continues its hidden life beneath the woodland floor.

This is not the only species of orchid that manages to live without light. There is an orchid from Western Australia (*Rhizanthella gardneri*) that even flowers underground and never comes to the surface.

THE NITROGEN CYCLE

Nitrogen makes up nearly four-fifths of the atmosphere and is one of the essential nutrients that plants need in order to grow.

Plants cannot use nitrogen directly from the air. Instead they rely on specialized bacteria that bind the nitrogen into the soil so that the plants can absorb it through their roots. Without these nitrogen-fixing bacteria, few of the world's plants would survive.

Farmers have known for centuries that some plants are good for improving the fertility of the soil – including peas and clover – and as a result they have planted them on a rotational basis. These plants have small nodules on their roots which contain nitrogen-capturing bacteria. The bacteria provide the plants with as much nitrogen as they need.

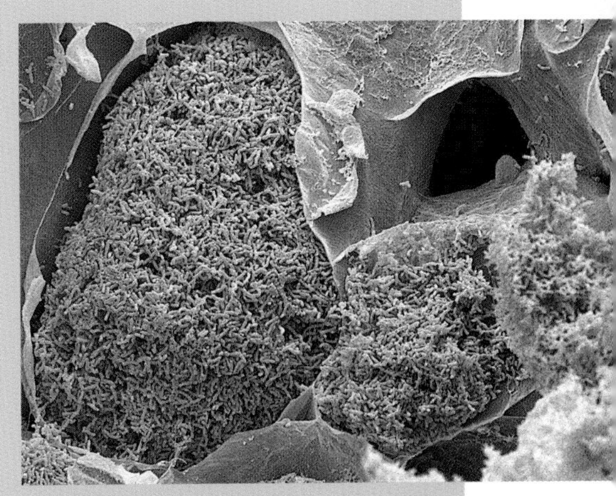

PERMANENT PARTNERS *Part of a root nodule from a clover plant seen through an electron microscope. It contains millions of nitrogen-fixing bacteria inside a protective 'skin'.*

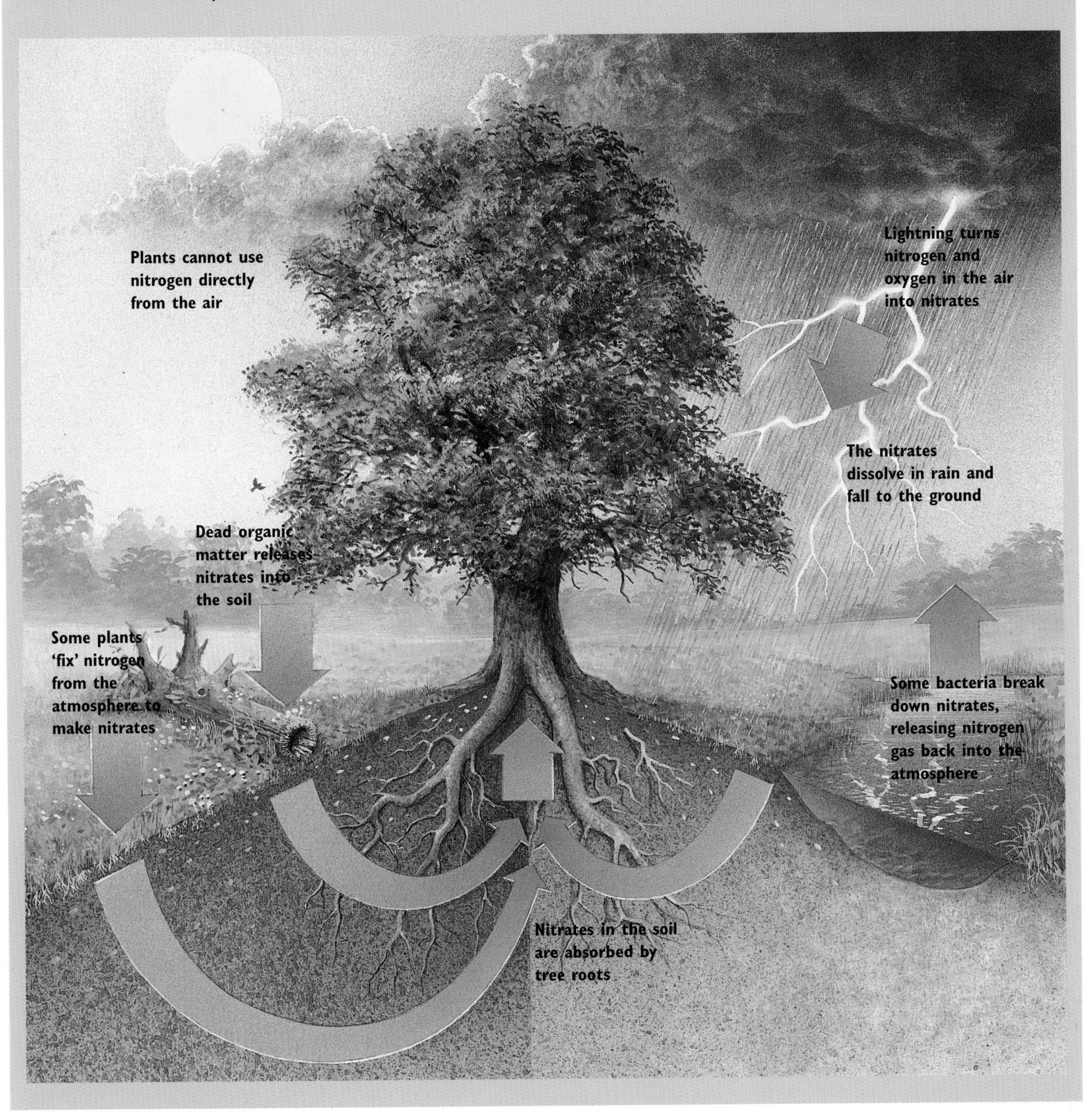

LEAFY HOME OF THE MINI-MULTITUDES

Billions of microscopic creatures live inside the leaves of trees and plants. Roundworms dine on the sugary sap, sucking it up with their sharply pointed mouths, while the minuscule caterpillars of micromoths – tiny insects whose adult wingspan is sometimes less than 1 cm (1/2in) – create wandering tunnels or 'mines' that become wider as the caterpillar grows. The mines eventually come to a halt at the point where the caterpillar climbs out of the leaf and lowers itself to the ground on a thread of silk. Micromoth mines are almost transparent, which makes them easy to see if you hold a leaf up to the light.

MINER'S MEAL *Home to the caterpillar of the bramble leaf miner moth is a winding, paper-thin tunnel, which it carves between the upper and lower surfaces of the leaf.*

WHEN SUCCESS IS A TEAM EFFORT

Lichens are a remarkable partnership between two types of microbe. There are more than 15,000 different kinds of lichen, all adapted to grow in places where it is hard for other plants to survive. Many spread over rocks or walls, growing very slowly, and sometimes living to be more than a thousand years old.

One of the partners is usually a fungus, which teams up with algae or bacteria. The fungus absorbs water from the air and provides a safe place to live, while its partner contributes by synthesizing food, using the energy in sunlight. Lichens are adept at coping with drought and cold, which is why they can exist on mountaintops and close to the poles.

A NEW FUNGUS CAN BE A KILLER

When a new fungus arrives from far away, it can have a devastating effect on its host. Early in the 20th century *Endothia parasitica* wiped out all the chestnut trees in North America, and a few decades later *Ceratocystis ulmi* – the cause of Dutch elm disease – ravaged elm trees on both sides of the Atlantic.

These epidemics are rare, because trees gradually develop a resistance to fungi. Although some die from fungal attack, others survive. But if trees have no time to build up any resistance, an epidemic can break out. This is exactly what happened with chestnut and elm trees, when deadly fungi arrived aboard shipments of wood.

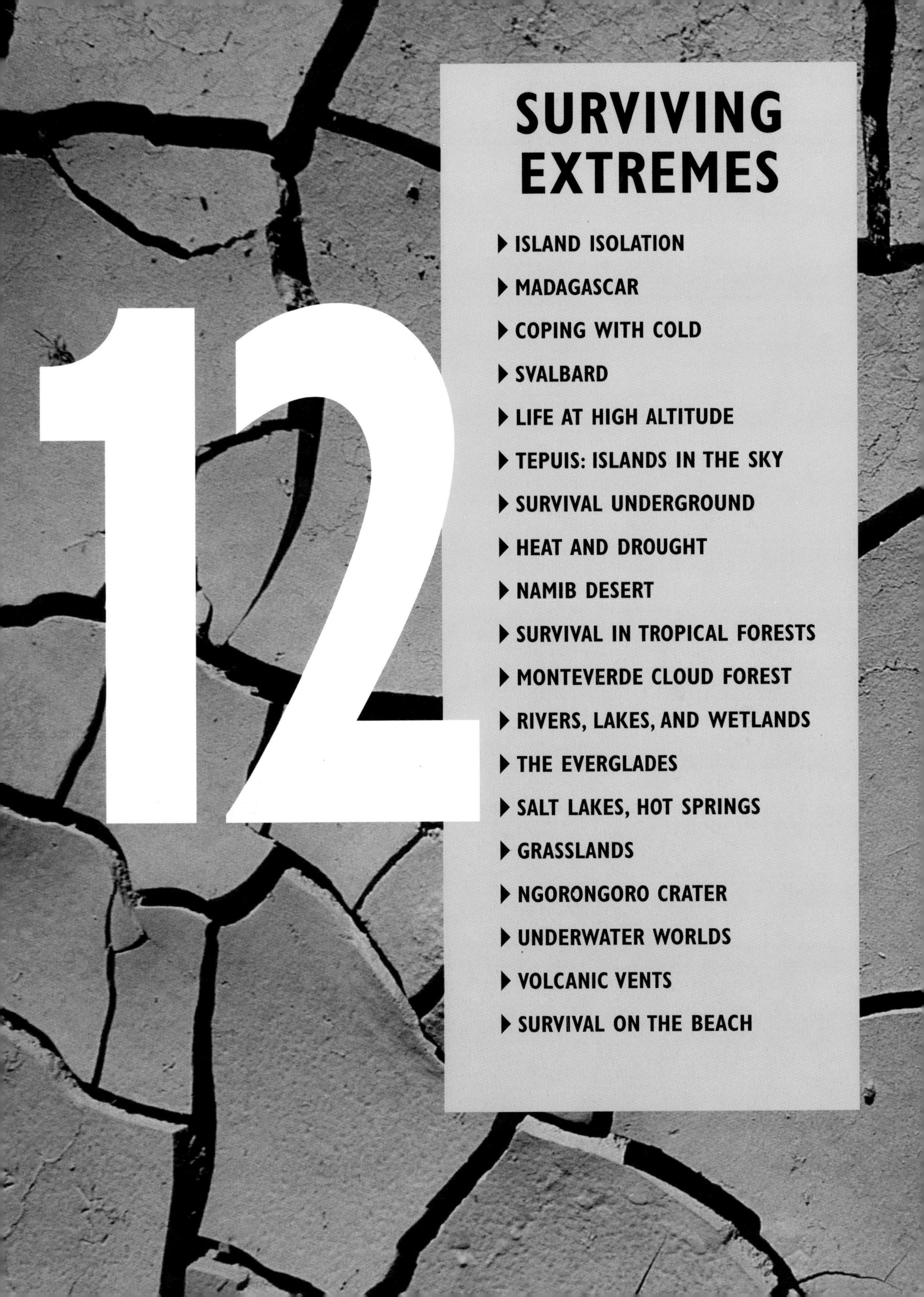

12 SURVIVING EXTREMES

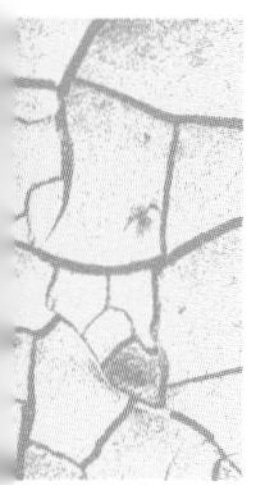

ISLAND ISOLATION

Remote islands are like life rafts that have been cast adrift for millions of years. They have evolved their own wildlife, with animals and plants found nowhere else, but time has caught up with them and some species are struggling to survive.

ANCIENT GIANTS OF THE GALAPAGOS ISLANDS

With a shell up to 1.2 m (4 ft) long, the Galapagos giant tortoise is the second-largest in the world. No one knows quite how these enormous animals, weighing more than 300 kg (660 lb), got to the Galapagos Islands, 1,000 km (620 miles) off the west coast of South America, but they were there long before humans first arrived – so long that each island has evolved its own local variety with distinctively shaped shells. In rainy weather, Galapagos giant tortoises often lounge in deep puddles, but no one has ever seen them venture into the sea.

Another remarkable Galapagos reptile, the marine iguana, gets its food from under the waves. It is the world's only sea-going lizard, swimming down to depths of 10 m (33 ft) or more, and holding its breath for 15 minutes. Although the Galapagos Islands are on the Equator, the water around them is surprisingly cold. Most lizards would come to a halt in these chilly conditions, but the marine iguana slows down its heartbeat when it dives, so that it does not lose too much warmth through its skin. Once back on land, its first priority is to sunbathe (*see page 14*).

TORSO-SHAPED TREE

One of the strangest plants to be found on a far-flung island is the sack-of-potatoes tree. It is unique to Suqutra, about 240 km (150 miles) off the Horn of Africa. Growing up to 5 m (16 ft) tall, the tree has short, stubby branches that protrude from a disproportionately fat trunk. In some specimens the trunk looks remarkably like a human torso. Suqutra is very dry, and the tree uses its trunk to store water during the long months of drought.

SEA-GOING LIZARD *Straddling an underwater boulder, a marine iguana in the Galapagos grazes on seaweed that grows close to the shore.*

EVOLVING APART

Why are island animals so different from ones in other parts of the world? One of the main reasons, as shown by the giant tortoises of the Galapagos Islands, is to do with food. On mainland South America, and on other continents, tortoises often have to compete with plant-eating mammals for food. When tortoises arrived on the Galapagos, there were hardly any land mammals, so they had most of the food for themselves.

Galapagos tortoises were not large to begin with, but without mammals the stage was set for this change. Bigger than average tortoises could get more food, so they produced the most young. The tortoises kept getting bigger, until they reached the size they are today. It is not known how or when the tortoises arrived. The most likely explanation is that they floated over the ocean from the mainland on trees washed out to sea by storms. Like other reptiles, they can go without food or water for many days, which would have helped them to survive. At one time, giant tortoises were also found on islands in the Indian Ocean, but today few are left.

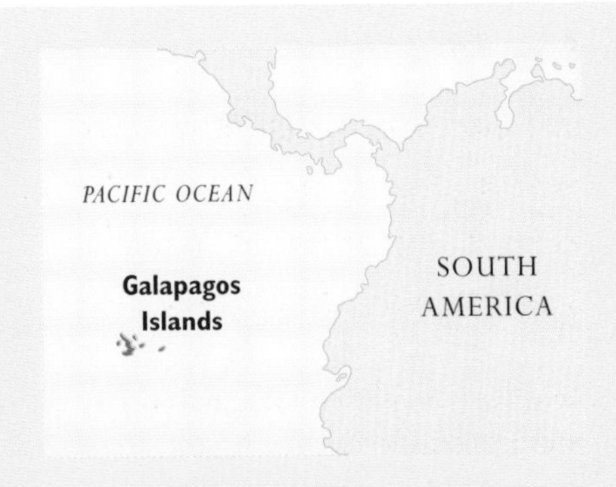

The Galapagos Islands are cut off from mainland South America by 966 km (600 miles) of ocean. Giant tortoises gather in the craters of old volcanoes to breed.

OCEAN WANDERERS THAT THRIVED IN PARADISE

A single flock of finches landed in the Hawaiian Islands at least 15 million years ago. They were one of the few creatures ever to breach the almost impenetrable barrier of the Pacific Ocean. Over the millennia, these pioneering birds have slowly evolved into more than 40 different species, each suited to a different way of life. When Polynesian seafarers arrived in Hawaii about 1,500 years ago, they became the first people to see the birds and give them names.

One of the birds that attracted their attention was a finch with an extraordinary bill. The bottom half is straight and short, while the top half is long and curved – the ideal shape for picking insects out of soft wood. It became known as the *akiapolaau*. Its relative, the *iiwi*, has a very unfinchlike beak: both halves are long and curved. It feeds on nectar, and its beak is ideally shaped for probing in flowers. Adult *iiwis* are a brilliant scarlet, and were favoured by Polynesian settlers to make ceremonial feather cloaks. The *akiapolaau* has suffered at the hands of hunters and is now endangered, but the *iiwi* is still thriving.

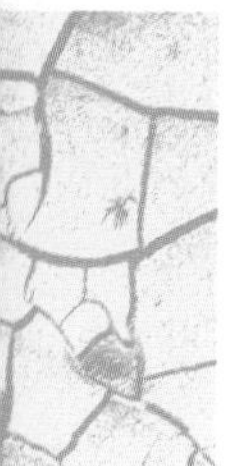

MADAGASCAR

Nearly 1,600 km (1,000 miles) long and 500 km (310 miles) wide, the island of Madagascar has been separated from Africa for millions of years. Its distinctive flora and fauna include spiky trees, minute lemurs, and pop-eyed chameleons.

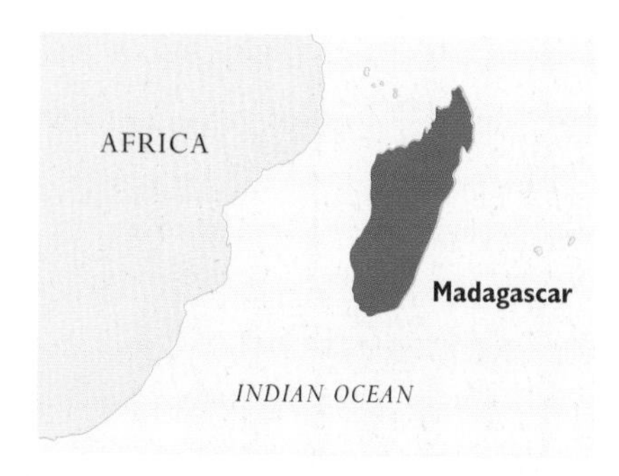

THE WEIRD AND WONDERFUL AYE-AYE

Imagine an animal the size of a cat, with large cupped ears, vicious teeth, and bulbous eyes. Add to this a bristly grey coat, a bushy tail, and slender fingers, one of which is exceptionally long. This is the aye-aye, Madagascar's strangest primate.

Its bizarre looks help to equip the aye-aye for its nocturnal tree-dwelling lifestyle. Large eyes give it good night vision. Big ears help it to locate insects. Sharp claws and incisors help it to climb and to chisel fruit, and its elongated middle finger is perfect for getting to those 'difficult to reach' places, such as inside a coconut.

CREATURE OF THE NIGHT *The aye-aye hunts at night, using its long, thin claws to climb. It has a good sense of smell, but it can pinpoint insect grubs by listening with its unusually large ears.*

THE PLANET'S SMALLEST PRIMATE

Once thought to be extinct, the world's smallest primate was rediscovered in Madagascar's forests in 1993. Looking like a mouse, and tiny enough to sit in a hand, the western rufous mouse lemur's body, excluding its fluffy tail, is just 11 cm (4¼ in) long.

Mouse lemurs are nocturnal tree-dwellers with forward-facing eyes like humans, and grasping fingers and toes for climbing. They enjoy a varied diet, including fruit, leaves, flowers, and sap, as well as insects and tree frogs.

CHAMELEON CAPITAL OF THE WORLD

A third of chameleon species live on Madagascar. Chameleons spend most of their lives in the trees, where their bodies are particularly well adapted for survival. They catch insects with their long sticky-ended tongues, which shoot out with astonishing speed to ensnare their prey in just a tenth of a second. Their swivelling eyes give good all-round visibility; the local people have a saying: 'Behave like a chameleon: look forward and observe behind.'

One of Madagascar's most unusual species is the panther chameleon, which uses a palette of vivid red, green, and turquoise skin pigments to change colour and attract potential mates.

FISH IS ANCIENT LINK TO HUMAN EVOLUTION

The coelacanth is a primeval deep-sea fish thought to have been extinct for 60 million years. Then in 1938 it was spotted by an incredulous museum curator in a South African fish market. It is now known that the seas around Madagascar are an important home for this slow-moving living fossil.

The coelacanth's remarkable feature is its fins. Most fish have fins directly attached to their bodies, but the coelacanth's fins have muscular bases like stubby legs. It is this limb-like arrangement that closely resembles that of the first land animals to crawl out of the sea millions of years ago.

THORNY BOUGHS POSE NO PROBLEM FOR PRIMATES

Madagascar's long isolation from mainland Africa has led to the evolution of unusual plants. The dry south-west corner is home to spiny forests, which contain specimens found nowhere else on Earth.

The octopus tree trails a mesh of spined branches up to 10 m (30 ft) into the sky. Its trunk is stubby, and its tiny leaves fall at the onset of the dry season. The Madagascan ocotillo is a close relative. It, too, has long wispy stems armed with ferocious thorns, which fail to deter lemurs from leaping among its branches.

SPINY FORESTS *Although covered in thorns sharp enough to rip skin, octopus trees are a favourite haunt of Madagascar's lemurs.*

Royal gathering in the Falklands

At St Andrew's Bay on the island of South Georgia, more than 75,000 king penguins come ashore every year to raise their young. South Georgia is surrounded by the Southern Ocean, one of the stormiest regions in the world and this bay is one of the few places that offers shelter and easy access to the sea. Female king penguins lay only a single egg each time they breed. The parents take it in turns to keep the egg warm by resting it on their feet and tucking it under a fold of feathery skin (*see page 32*). This picture shows a small section of the colony in December, which is midsummer in the Southern Hemisphere. Most of the birds are adults with eggs, but the brown ones are ten-month-old chicks that hatched the previous year. The chicks' fuzzy feathers are not waterproof and the young birds depend on their parents for food until they reach maturity at about 14 months. For penguins, and many other sea birds, breeding is safer in huge groups like this, because it helps to keep predators at bay. They value their space, however, remaining an exact, if short, distance apart.

COPING WITH COLD

In a polar environment where a naked human would soon die of exposure, warm-blooded animals thrive. Their secret is natural insulation: blubber, fur, and dens of snow. Plants, too, have adapted their lifestyle to bloom in adversity.

BLUBBERY IS BEAUTIFUL FOR POLAR MAMMALS

Beneath the icy seas of the Arctic and the Antarctic is a fertile world full of food. But for warm-blooded animals, even a brief dip is potentially deadly, because water soaks up body heat nearly 100 times faster than air. To survive in polar seas, warm-blooded animals have to stop their precious warmth draining away.

For whales, seals, and walruses, the answer lies in an oily fat called blubber. They make blubber from their food and store it under their skin. In whales, the layer of blubber builds up at a rate of 20 cm (8 in) a year, and can be as thick as 50 cm (20 in) in adults. This fatty jacket acts as a buoyancy aid and an emergency fuel supply, but its most important job is keeping heat where it is needed – in the animal's inner core. When spring arrives blubber cannot be taken off so the animals shed some of their excess warmth with the help of blood vessels, which can be opened or closed depending on whether heat needs to be lost or conserved. Even so, blubber is so efficient that in summer, when the temperature can rise to a balmy 5°C (41°F), basking seals and walruses can find themselves overheating and have to take a swim to cool off.

FULL-FAT JACKET *A thick layer of blubber keeps the walrus warm and comfortable even when it is stretched out on the ice.*

ULTIMATE FUR COAT KEEPS COLD AT BAY

Sea otters have the densest fur of any animal on the planet. They need it, because they have no blubber at all and spend their life at sea along the Pacific coast of North America from California to Alaska. Even in California the water is chilly and in Alaska the sea is close to freezing.

Up to 125,000 hairs are crammed into each square centimetre of the sea otter's coat. Some of the hairs are long and tough, but most are short and soft, and are so tightly packed that water never touches the otter's skin. These short hairs also trap a layer of air, which acts as an insulator.

A constant supply of fuel is essential to the otter's well-being. It dives for shellfish and other animals, and eats about a quarter of its own body weight each day – the equivalent of an average adult human eating 18 kg (40 lb) of steak.

FOOD SUPPLIES *Sea otters collect shellfish from the seabed. To eat, they bring the food back to the surface in their paws.*

LONG CLAWS FORM LEMMING WINTER TOOLKIT

Like sea otters, lemmings have thick fur coats, but they have evolved a different way of escaping the cold. During the long Arctic winter they use snow for insulation, and tunnel their way between it and the ground with special 'spades' on their front feet. These excavation tools are formed by claws that grow larger as the wintry weather begins.

The snow keeps the lemmings warm and safely out of view of predators. When it melts in spring, the lemmings' runs are a secret no more, and they tunnel underground instead. To a trained eye, the remains of their winter homes are easy to identify, because trails of leftover plant food from inside can be seen snaking over the ground.

FEATHERED LEGGINGS KEEP ARCTIC BIRDS SNUG

In cold conditions, the most vulnerable parts of an animal's body are those that stick out, such as legs and feet. To survive, polar residents have evolved ways of keeping their extremities warm. Snowy owls and rough-legged buzzards have feathered legs, while rock ptarmigans, being mainly ground-dwellers, also have feathered feet. These 'snow shoes' allow ptarmigans to spend hours foraging for food over frozen ground, and they also help to spread their weight as they walk over soft snow heaped up by the wind.

If things get really cold, ptarmigans dig hollows in the snow banks, and crouch down with their feet tucked up close to their chests. These birds are so well insulated that they can survive winters in places like northern Greenland – farther north than any other land bird.

WINTER WARMERS *Showing off its feather-covered legs, a snowy owl lands beside its exposed nest in the Arctic tundra.*

SECRET OF SURVIVAL *Antarctic ice fish look pale because they have colourless blood. They are slow swimmers and tend to live mostly along the seabed.*

WHY POLAR FISH HAVE NATURAL ANTIFREEZE

Ordinary fish would run the risk of being frozen solid in the Antarctic's salt-laden sea water. Yet more than 260 species of fish manage to survive temperatures as low as –1.7°C (29°F), thanks to two different ways of defeating the cold.

For Antarctic ice fish, the secret of survival lies in a special blood protein, which works like a natural antifreeze. This keeps an ice fish's blood liquid at temperatures of –2.5°C (27.5°F), which is low enough to keep it out of trouble. Many other Antarctic fish are 'supercooled', which means that although their blood is below freezing point most of the time it does not freeze because glycopeptides in it ensure that crystals never actually form.

FLOWER OASES IN THE ANTARCTIC PENINSULA

In Antarctica just two species of flowering plant have been identified. Compare this to the rest of the world, where more than a quarter of a million flowering plants have been singled out and it is likely that there are still tens of thousands waiting to be discovered.

Antarctica is so cold that very little land-based life can survive. Lichens are scattered all over the continent, but flowering plants are found only on the Antarctic Peninsula, where summer temperatures creep above freezing for several weeks each year. One of the two species, Antarctic hair grass, forms dense tussocks in wet gullies; the other, Antarctic pearlwort, grows in cushion-like clumps near the shore. In their very short season they reproduce just like other plants by forming seeds.

Neither of these species would win any prizes for their beauty, but for Antarctica's land-based wildlife they are a precious resource.

Tucked away in the shelter of the tussocks and clumps, tiny animals feed on each other, and on plant remains. Many of these creatures are only just large enough to be visible with the naked eye; the largest of them all, a wingless fly, measures a mere 1.2 cm (1/2 in) long.

PRECIOUS SHELTER *Flourishing in the brief summer, clumps of Antarctic hair grass provide a vital mini-habitat for some of the continent's tiniest animals.*

SUPREME SURVIVORS OF THE SOUTHERN HEMISPHERE

Lichens are among the toughest living things on our planet. They grow on the islands around Antarctica, and they also survive in parts of the southern continent itself. In 1965 an expedition to Antarctica's remote Horlick Mountains discovered lichens growing on rocks just 400 km (250 miles) from the South Pole. Between the Horlick Mountains and the pole itself there is nothing but ice, so these lichens are almost certainly the most southerly 'plants' on Earth.

SURVIVING THE COLD *Lichens and mosses carpet the rocky landscape of the South Orkney Islands, lying 700 km (435 miles) north from Antarctica's frozen coast.*

Unlike true plants, lichens are a living partnership between an alga and a fungus (*see Microbes and Plants, pages 300-2*). They do not have roots and so they are able to scramble over bare rock, keeping a low profile. They can make do with a tiny amount of water, which is important on the driest continent in the world and enables them to withstand temperatures below –45°C (–49°F).

In laboratories, lichens have survived in liquid nitrogen at a temperature of –195°C (–319°F).

TINY TREES THAT LIVE ON THE EDGE

To survive, Dahurian larches shed their needles every autumn. The world's most northerly conifer trees, they are found in Russia about 1,600 km (1,000 miles) from the North Pole. Many are misshapen by the icy wind, and some achieve only head-height despite being more than a century old.

Conifer leaves are exceptionally tough in order to cope with the climate. In the Arctic, conifer forests sweep north like a dark green tide, until the cold begins to bite. Then the trees become scattered and stunted, until they finally cease growing altogether.

RAINBOWS IN THE SNOW

In high mountains patches of old snow sometimes change colour, turning bright red, orange, or green. Looked at under a microscope, millions of microscopic snow algae can be seen. These minute single-celled plants are related to seaweeds, and live in the same way as all plants, by harnessing energy from sunlight.

Green and orange algae often live around trees; red algae thrive in the open. In spring, as the days start to lengthen, the algae multiply and the colour becomes stronger. When the snow melts, the algae live on in water.

THE ARCTIC POPPY'S SOLAR-POWERED INCUBATOR

When the snow melts at the beginning of the Arctic summer, the tundra briefly comes alive with ankle-high flowers. In the few weeks that follow, the flowers have to work fast to attract pollinators and make seeds.

The Arctic poppy uses its bright yellow flowers, shaped like miniature satellite dishes, to track the sun as it moves across the sky. Petals in each flower reflect sunlight into the middle, which is the part where seeds are made and where energy is needed.

After a few minutes of sunshine, the temperature in the centre of the flower can be several degrees warmer than the surrounding air. This floral warm-up has another advantage. The heat attracts a number of visiting insects, such as flies and bumblebees. After they have basked in the warmth of the poppy and taken a drink of nectar, they help to carry pollen from one flower to the next.

SUN TRAPS *Arctic poppies flower and set seed in the space of just a few weeks. Their hardy nature has made them popular with gardeners, who know them better as Iceland poppies.*

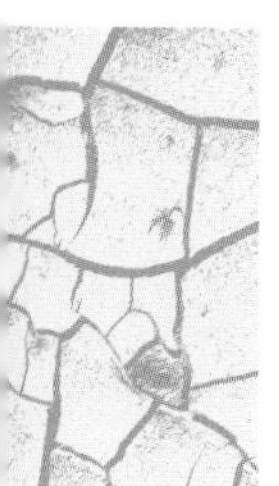

SVALBARD

Named after the Norwegian for 'cold coast', the remote Svalbard islands are set on the fringes of the Arctic Ocean. They form a forbidding and icy landscape in which only a few specially adapted animals are capable of surviving.

GREAT WHITE BEAR OF THE ARCTIC ICE

At the top of the Arctic food chain stands the polar bear, the biggest and most ferocious bear in the world today. It hunts by stealth, using its acute senses of smell and hearing to detect seals several metres beneath the ice. Padding silently above its quarry, it tracks down a seal's breathing hole, then patiently waits for the inevitable moment when the creature surfaces for air.

Even before the seal's nose breaks through the surface, the bear launches its attack. Adult ringed seals can weigh up to 95 kg (210 lb), but a polar bear weighing four times that amount and measuring up to 2.5 m (8 ft) tall at the shoulder, makes short work of dragging the animal out of the water. Within minutes the seal is dead, and the bear starts its calorie-rich banquet, sometimes consuming half the carcass in a single meal.

Every autumn when the sea freezes over, polar bears set off from the Svalbard islands on their long winter hunt. After the lean times of summer, they step onto the ocean ice and stay aboard for months, having good access to ringed seals and other prey.

EASY MEAL *A polar bear approaches its prey: a ringed seal pup lying on the sea ice. The pup has little chance of escape from its skilled predator.*

CHANGING COATS WITH THE SEASONS

Arctic foxes are renowned for being bold and inquisitive animals, unlike their more timid counterparts elsewhere. On Svalbard they sometimes trot right up to people in the hope of finding food.

Living year-round in the far north, these foxes are tireless runners and good swimmers. Their winter coat is usually white and is dense and luxuriant – an essential for survival in places where the air temperature can drop to –40°C (–40°F). In summer it turns brown, helping the foxes to dart unseen across open moorland. At this time of year they feed mainly on

AGILITY ON ICE *Wearing its lightweight summer coat, the Arctic fox leaps between patches of melting sea ice on the look-out for food.*

young birds and lemmings. Unlike polar bears, summer is their prime time for feeding, and for raising their young. If they catch more food than they can eat they often bury it in thawed ground and return to eat it days or even weeks later.

Arctic foxes breed in underground dens, which are passed on from one generation to the next. Some dens can be more than a century old and with as many as 24 entrances. They can be 25 m (80 ft) across and are easy to spot, because they are often surrounded by bright green plants, which have been manured by the remains of the foxes' feasts.

A VIGOROUS WARM-UP ACT

The Arctic bumblebee uses its furry coat like a survival blanket to keep warm. Before it takes off, it twitches the muscles that power its wings and the warmth they generate heats its body to 25°C (77°F). Only when it has warmed up can the bee take to the air.

Compared with other bumblebees, the Arctic bumblebee has an exceptionally tough life, despite its adaptations to the cold, but it does have one thing in its favour: when summer finally arrives, daylight is nonstop so it can gather pollen around the clock.

LICHEN FOOD FOR ARCTIC ANIMALS

The reindeer moss found growing in the Arctic is not a moss at all, but a bushy lichen with feathery fronds. Like all lichens, reindeer moss is extremely tough, and has no difficulty surviving winters that can last for more than six months at a time. This is because it is more dependent on the amount of hours of sunlight than on the air temperature.

In a region where cold air and waterlogged ground make it difficult for grass to grow, reindeer moss is vital for the survival of herbivorous Arctic animals such as reindeer and musk oxen.

A WARM START FOR YOUNG EIDER DUCKS

Young eider ducks enjoy the warmest and cosiest nests of any birds. This is because the nests are lined with a deep layer of fluffy down feathers, which female eiders pluck from their breasts. The feathers provide superb insulation, keeping the eggs and ducklings warm in temperatures far below freezing.

Eiders arrive on Svalbard every spring. They feed on crabs and other seashore animals. In the past, their nest feathers were collected and used for stuffing pillows and eiderdowns, but today artificial fibres have largely replaced feathers, so the ducks are left to nest undisturbed.

COSY QUARTERS *Insulated by the mother's down feathers and their own fluffy plumage, eider ducklings are well protected from the cold.*

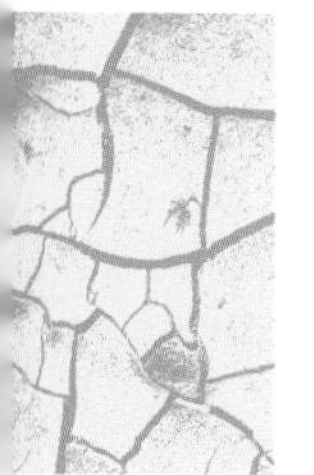

LIFE AT HIGH ALTITUDE

High-flyers and mountain dwellers face two serious problems. The first is cold – temperatures drop 1°C (1.8°F) for every 200 m (650 ft) you climb. The second is the thin air, which can leave terrestrial animals gasping for breath.

HIGH-FLYER MAKES USE OF PANORAMIC VIEW

Soaring above the open plains of Africa, Rüppell's griffon flies higher than any other bird. In 1973, a griffon collided with an aircraft flying at more than 11,000 m (36,000 ft) – which is high enough to fly over Mt Everest. At these heights, the bird has a vast panoramic view, and it can easily cover more than 150 km (95 miles) a day in search of food. But at altitude even the long-sighted griffon has difficulty picking out tiny details on the ground, so it watches other vultures flying beneath and is quick to follow them if they spot a meal.

To survive at such altitudes the griffon vulture needs some of nature's finest engineering. Its ultra-efficient lungs allow it to get enough oxygen to breathe, and its giant wings provide it with lift in the thin air.

GIANT WINGS *A griffon vulture's wings can be 2.8 m (9 ft) from tip to tip. This huge wingspan is supported by a lightweight bone structure.*

MARMOTS SETTLE IN FOR THE BIG SLEEP

In America, northern Europe, and central Asia, mountain-dwelling marmots spend most of their time in hibernation. They sleep for up to nine months each year. By hibernating in winter, animals avoid futile foraging trips when nature's larder is bare. For most animals, this dormant period lasts for three to five months.

Where marmots live, however, winters can be hard and very long. Rather than wake up prematurely, they sleep on until May or June, tucked away in long networks of underground burrows. They survive on fat reserves, which make up a fifth of their body weight. When summer finally arrives, marmots have to collect food, raise a family, and fatten up for their next hibernation, all within three months.

SUMMER HARVEST *Marmots need to put on weight in summer because food is so scarce during their record-breaking hibernation.*

VICUÑAS BREATHE EASY IN THE MOUNTAINS

Vicuñas live high in the Andes mountains of South America yet are never short of breath. At about 4,000 m (13,000 ft) above sea level humans would suffer breathlessness, but vicuñas can sprint effortlessly up rocky slopes, and on level ground they can run at nearly 50 km/h (30 mph).

These graceful cinnamon-coloured relatives of camels have normal mammalian lungs, but they have about three times as many red blood cells a litre of blood as humans. Their red cells last twice as long and because these cells contain an unusual form of the red blood pigment haemoglobin, they are better at collecting oxygen. Altogether, it adds up to an ideal high-altitude breathing system.

HUMMINGBIRD'S NIGHTLY SLOW-DOWN

Hillstar hummingbirds hibernate each night. Hummingbirds lead frenetic lives, and need a constant supply of nectar to fuel their activity. For species that live in mountains, such as the Andean hillstar, night is a crucial time: they must pay for heat loss with fuel consumption.

The hillstar's response is to treat each night as a mini-winter. As the air cools at sunset, its heart rate plummets and its body temperature drops, leaving it ticking over on the minimum amount of fuel. When the sun rises and the air warms up, its body returns to normal, and the hillstar sets off to feed. Thanks to this system, it can survive cold nights at an altitude of 4,000 m (13,000 ft).

SURE-FOOTED ROCK CLIMBERS

Ibex goats are born with nature's equivalent of climbing boots. These mountain goats from Europe and Asia live far above the tree line and have an unshakable head for heights. When alarmed, they can run uphill or down, taking 6 m (20 ft) leaps in places where one misjudged step could spell instant death.

Their hoofs are small and have hard edges with a recessed central pad. The pad works like an elastic cup to grip uneven or slippery rock, and all four feet can fit onto a ledge no bigger than a human hand. Climbing comes naturally to young ibex. When their parents move off to feed, the youngsters follow them up steep slopes and along precarious ledges with fearless ease.

DIZZYING DUEL *During the breeding season, confrontations between male ibex are common, but they rarely lose their footing.*

OUT OF HARM'S WAY *By growing just finger-high, moss campion avoids the wind, an essential adaptation for surviving on windswept mountains.*

HIGH-LIVING PLANTS KEEP A LOW PROFILE

To avoid wind damage, mountain plants are often cushion-shaped, hugging the ground for protection. Cushion plants tend to be small, but many are so tough that they can take a person's weight without damage.

Moss campion is a prime example of the cushion-shaped way of life. This red-flowered plant lives across the Northern Hemisphere, on exposed mountain rocks, and has many relatives that inhabit lower ground. While other campions are often more than 1 m (3 ft 3 in) tall, moss campion is only a few centimetres high.

UPWARDLY MOBILE SURVIVORS

During the last Ice Age, many of today's mountain plants lived nearer to sea level. When the climate became warmer and drier, they were able to survive only by growing at higher altitudes. For them, high ground became a refuge.

The most famous of these mountain refugees is the giant sequoia, the world's most massive living tree. At one time, giant sequoias existed across much of North America, creating forests that would have been the most majestic on Earth. Sequoias need moisture, and when rainfall began to decline, only those on moist mountain slopes managed to survive. Today, giant sequoias are found on the slopes of California's Sierra Nevada, where they live in about 75 isolated groves.

FLOWERS KEEP THEMSELVES GOING AT ALTITUDE

Mountain buttercups are the highest known flowering plants in the world. A 1955 expedition to the Himalayas found a group growing

at 6,400 m (21,000 ft) above sea level. At this dizzying altitude, the climate is far too cold for most animals, and the air much too thin. There are no butterflies, hardly any bees, and few flying insects of any kind. Yet the buttercup still puts on a display each year, trying to attract the rare insect pollinators found at this height.

So how does the mountain buttercup produce any seeds if its flowers go unvisited? Like many mountain plants it has a back-up system. Instead of using pollen from other buttercup plants, it can make seeds using pollen it has produced itself. Without this trick, the world's highest plants would soon disappear.

SNOW AND THE GREENHOUSE EFFECT

A covering of snow helps plants to pull through the winter. It works like an insulating blanket to keep out the much colder air above.

Throughout winter, the snowbound plants lie dormant, protected from frost and icy gales. As winter draws to an end, the snowy blanket starts to melt. Deep below, the pitch blackness of winter gives way to a gradual dawn, as the ever-strengthening daylight filters through. This steady increase in light tells the plants that spring is not far off, and they start to grow.

As plants grow they produce a small amount of heat, and this melts the snow from below. Soon, each plant is in its own miniature greenhouse, glazed over by a snowy roof. By the time the snow clears, the plants are well prepared for the new season ahead. Some buttercups flower less than a week after they first appear.

NATURAL PROTECTION *Dense 'fur' on giant lobelia leaves reflects sunshine, protecting the leaves from sunburn. At night, it keeps them warm.*

FUR COAT PROTECTS FROM HEAT AS WELL AS COLD

Giant lobelias survive where most plants would be grilled or frozen to death. On East Africa's Mount Kenya giant lobelias grow in rocky valleys at 5,000 m (16,000 ft) above sea level, almost on the Equator. During the day the sunshine is fierce, but after dark temperatures fall below freezing in the thin mountain air.

Giant lobelias cope with these conditions by having furry leaves. The leaves grow in a large cluster on top of the plant's sturdy trunk, and they open out into a rosette after sunrise. During the day, their fur acts like a sunscreen, reflecting light and preventing them from being burned. As the sun sinks, the leaves close up again, protecting the plant's tender buds from frost.

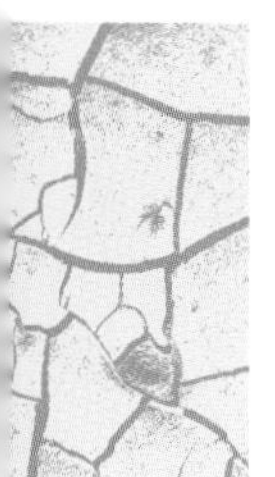

TEPUIS: ISLANDS IN THE SKY

Like flat-topped islands floating in a sea of clouds, Venezuela's tepuis tower up to 1,500 m (4,900 ft) above the surrounding forest. They are home to animals and plants that live in lofty seclusion, cut off from the world below.

THE ELUSIVE MOUSE OF MOUNT RORAIMA

In 1929, an expedition to Mount Roraima discovered a new species of mouse, which is exclusive to the tepui and extremely rare. Roraima's pocket-sized rodent is the only mammal native to the mountaintop that has been discovered to date. After the expedition, it was not seen again for another 60 years and remains one of the most elusive mice in the world. Most tepuis are so inaccessible that even today, with the dangerous ascent being carried out by helicopter, investigating their wildlife is far from easy.

THE TOAD THAT HAS NO TADPOLES

There is a toad on the tepuis that hatches out from its egg as a tiny fully formed toad, rather than beginning life as a tadpole. This particular toad has evolved a lifestyle to suit its habitat where food can be hard to find. On low-lying ground, freshwater pools often contain lots of food that tadpoles can eat, but rainwater pools high up on tepuis have very little food, which makes them a poor nursery for any animal's young.

Fully formed toads are more likely than tadpoles to find food, so in bypassing the tadpole stage the tepui toad has a greater chance of survival.

CARNIVOROUS PLANTS IN THE CLOUDS

Many of the plants that grow on tepuis are carnivorous, taking their nutrients from insects rather than the poor soil. Tepui sundews specialize in catching tiny flies and other airborne insects, trapping them with the glue-tipped hairs on their leaves. The insects are attracted by the plant's bright red colour which indicates a possible source of food.

Tepui bromeliads have a different technique: their upright leaves are clustered together, forming slippery fluid-filled funnels. Insects clambering around the rim fall inside, where they are drowned and slowly digested.

TINAMOUS – THE WORLD'S WORST AVIATORS

The tepui tinamou is one of the rarest members of a strange family of birds that is hopeless at flying any distance. They are renowned for crashing into obstacles while on the wing and landing clumsily on the ground.

Tinamous are found only in South America, in forests and grasslands as far south as Patagonia. They look like small, plump chickens, with short tails and strong scaly feet. If threatened by danger, they run rather than take to the air. Of the 50 different kinds of tinamou, the tepui tinamou is one of the least studied. It is restricted to just two tepuis and the last reported sighting was 20 years ago.

Tinamous nest on the ground and lay the most remarkable eggs with a brilliant sheen, resembling highly polished porcelain.

HOW THE TEPUIS WERE FORMED

More than 170 million years ago the tepuis we see today formed part of a vast sandstone plateau, 1,500 km (930 miles) across. The plateau sloped towards the Atlantic, an ocean that was still forming as Africa and South America separated.

Movements within the Earth fractured the rock, creating faults and cracks. Over millions of years these were eroded by rain, reducing the plateau to a series of flat-topped 'islands'.

Mount Roraima is one of the largest tepuis, with a flat summit 14 km (9 miles) across. Auyan Tepui or 'Devil's Mountain' is the site of the Angel Falls where rainwater plunges 1,000 m (3,300 ft), making it the tallest waterfall in the world.

Some plants and animals have evolved in isolation on the tepuis, including 40 endemic birds, hundreds of flowering plants, and one mammal.

ISLANDS IN THE SKY *The mighty Angel Falls, left, cascade over the Venezuelan tepui known as Devil's Mountain; below, Mount Roraima, a world within a world.*

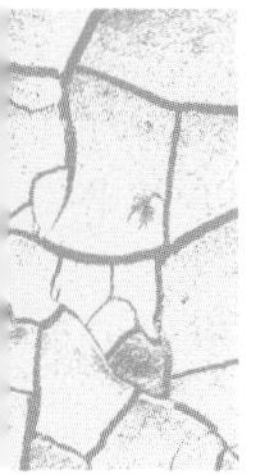

SURVIVAL UNDERGROUND

For some animals, caves are an ideal habitat. Although food may be scarce and there is little or no daylight, cave-dwelling has the twin advantages of a constant climate and good protection from hostile predators.

CRICKETS GET A FEEL FOR LIFE IN THE DARK

Cave crickets, unlike the majority of their outdoor relatives, have minute eyes and cannot fly. But they do have antennae or 'feelers', up to four times as long as their bodies, which help them to navigate and track down food.

There are 200 species of cave cricket worldwide. Unfussy about what they eat, they feed on bat droppings, hibernating butterflies, and fungi, as well as any smaller cave crickets that have the misfortune to cross their path.

These insects spend their entire lives underground, but their distant ancestors would have lived on the surface. It is thought that some cave crickets were driven underground during the last Ice Age, because caves provided a refuge from the cold. They adapted to a subterranean existence, and have remained there ever since.

BLIND FISH NAVIGATES BY SMELL AND TOUCH

Most fish have sensors to aid navigation, but those of the cave characin are especially acute. They need to be, as this particular fish is blind and inhabits a subterranean world of rivers and lakes.

Living in the watery underworld of Central America, the cave characin is a ghostly pinkish white, with very thin scales, and it has a featureless forehead without any eyes.

Cave characins are omnivorous, and feed on bat droppings, dead insects, and anything else they can find. They locate most of their food by its smell, using extremely responsive 'taste buds' on their lips and in their mouths. But they still have to navigate their way around obstacles, which they accomplish with a line of pressure sensors along their sides. By picking up tiny pulses of pressure as the fish moves about, they enable it to 'see' its surroundings.

TOUCHY SUBJECT *Using its giant antennae, a spotted cave cricket searches for food. The antennae can sense air currents caused by insects on the move nearby.*

THE OLM'S SUBTERRANEAN WAY OF LIFE

Deep under limestone mountains around the Adriatic Sea, olms forage for food in lakes or streams. These highly specialized salamanders have an unusual way of life. With tiny eyes, they rely on their good sense of smell to track down their insect prey. To breathe under water they have feathery gills – which most salamanders lose when they become adult and take up life on land.

Not much thicker than a pencil, the olm can grow to 30 cm (12 in) long. Its pale pink body has four miniature legs with tiny toes, but it never leaves the water, and moves mainly by beating its paddle-like tail.

Olms are peculiar in that they can reproduce as gilled larvae before they reach adulthood. The olm's olfactory courtship is played out in the dark: the male wafts his scent in front of the female's snout to attract her. Once the female's eggs are fertilized, she lays them among underwater rocks.

LIGHT THAT LURES IN A WORLD OF DARKNESS

At New Zealand's Waitomo caves, tiny points of light spread over the rocky ceiling like a starlit sky. These are living lures, the grubs of cave-dwelling fungus gnats.

Fungus gnat grubs are predators, and they use light to catch their prey. Each one spins a line of silk across the cave roof, and then lowers between 30 and 50 sticky silken threads from it. The grub then switches on its light-producing organs, which emit a blue-green phosphorescent glow. Tiny cave midges are attracted towards the light, and become entangled in the threads.

DEATH IN THE DARK *Hanging from a cave roof, fungus gnat 'fishing lines' are deadly snares. Flying insects are trapped in pearl-like blobs of glue.*

Once a midge has been ensnared, the fungus gnat grub hauls it up like a fisherman bringing in his catch.

The more successful their traps, the brighter the grubs glow, and the more quickly they develop into adults. The adult fungus gnats mate and lay their eggs in caves, producing the next generation of eerie fishers in the dark.

CAVE-DWELLING BIRDS SEE WITH SOUND

In the caves of Venezuela and Trinidad, dusk is the signal for an extraordinary event. Speeding through the air on narrow pointed wings, thousands of oilbirds (guacharos) emerge from inside the caves and set off in search of the oil-rich fruits of the fig, laurel, and palm. Just before dawn breaks the flock returns, and the oilbirds vanish deep underground, returning to their own nests and chicks, laden with supplies. These are some of the few birds in the world that can use sound to navigate when they are underground. Unlike bats, whose shrill squeaks are generally beyond human hearing, the oilbirds' echolocation clicks are clearly audible. So, too, are their calls – a weird array of wails, croaks, and screams, which fill the night air as they fly overhead.

Oilbirds, which measure 33 cm (13 in) from beak to tail, build nests on cave walls out of their own droppings. The adults' daily delivery of fruit turns the chicks into well-rounded individuals. The birds were once harvested by local Indians, who gave them their name.

SAFELY HOME *Oilbirds nest in caves 800 m (1/2 mile) underground. They make their nest from their droppings, mixed with regurgitated seeds.*

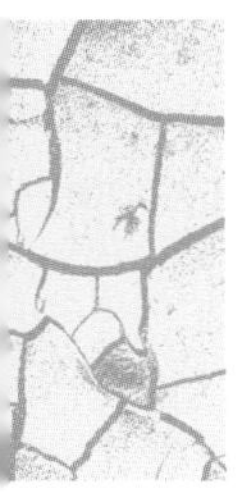

HEAT AND DROUGHT

No animal can survive without water. Yet, amazingly, some manage to live in the hottest and driest places on Earth. To do this, they have to be able to keep themselves cool, and know how to eke out every drop of water they can find.

A PORTABLE STORE OF FOOD AND WATER

The 'ship of the desert', as the camel is known, may sail on sand, but it needs water to drink. When a dehydrated camel does find water, it can consume up to 57 litres (14 gallons) at a time.

Water alone will not keep a camel alive. To cope with desert life, it needs to be able to store food as well. This is where the camel's hump comes in. It contains a store of fat, and works like a water tank and food hamper in one.

By breaking down some of this fat, the camel releases enough water and energy to keep going. It can last for several weeks without food and, in hot conditions, for seven to ten days without water. As the fat is used up, its body weight may drop by a quarter, but it soon recovers when food is available.

A RAT THAT DOESN'T DRINK

If there is no water to drink, one way to get by is to stop drinking. North American kangaroo rats have adopted this extreme solution to the water problem, allowing them to thrive in places where summer temperatures can reach 50°C (122°F) or more. These long-legged desert rodents need water, just like other animals, but they can survive indefinitely without a drop ever touching their lips.

The kangaroo rat's secret lies in the careful control of its water balance. On the debit side, it loses water through its highly concentrated urine and extremely dry droppings. It also loses water vapour in its breath, but it keeps this to a minimum by staying underground during the heat of the day, and by having a nose that condenses most of the vapour before it is lost. However, making savings alone is not enough: the kangaroo rat has to get water from somewhere in order to stop itself dying of dehydration.

Some of its water comes from moisture in the seeds that form its diet, but 90 per cent comes from water formed inside its body. Called metabolic water, this is released when food is digested. As far as humans are concerned, dry seeds would be little help in a desert, but they are all the kangaroo rat needs.

WATER LINE *Arabian camels drink in an orderly manner and can utilize fresh, brackish, or even salty water.*

SURVIVAL RATIONS *The food and water stored in a Gila monster's tail lasts for three or four months, and helps it to survive through the hot, dry summer.*

COOL SHADES FOR A DESERT SNAKE

The horned viper of the Sahara Desert has short projections above its eyes. These 'horns' act like sunshades when the viper ventures out into the fierce sun from its home in burrows or from under rocks.

Deserts are ideal places for reptiles, because they need warmth from the sun to make them active, but it takes a special kind of reptile to survive in extreme heat and shifting sand.

The horned viper is at home in this kind of habitat. Its skin has unusually rough scales, which help it to shuffle its way down into the sand to keep out of the heat if no cover is available. Its scales also make a rasping sound when rubbed together, warning would-be attackers to keep their distance. This is a warning worth heeding, as the snake is deadly poisonous.

LIZARDS WITH FAT TAILS FOR LEAN TIMES

In North America's Sonoran Desert the Gila monster can live off the fat of its tail. This brightly coloured lizard, one of only two poisonous species in the world, lives in a region where winters are cool and summers are hot and dry. Its fat tail is a combined food and water supply, helping it through times when both are hard to find.

Australian marsupials called fat-tailed dunnarts use the same system. The dunnart's tail is as long as the rest of its body. In times of plenty its tail becomes thick and carrot-shaped, but when times are hard it slims down as the fat stores are used up.

EAR SIZE COUNTS

How do you tell where a fox or a hare lives? One way is to look at its ears. The Arctic fox and Arctic hare both have small ears, but their desert-dwelling relatives, the fennec fox and black-tailed jackrabbit, have ears that seem far too big for their bodies.

The reason for this is largely to do with temperature control. Small ears are essential in the Arctic, because anything bigger brings the risk of frostbite. In deserts, outsize ears are a good way of getting rid of unwanted body heat. This is because they have lots of small blood vessels near the skin surface, so the blood's warmth escapes into the air. Large ears are also more sensitive, picking up the sounds of prey or predators after dark, when most desert animals are active.

SHADY CHARACTER *When life gets too hot, the horned viper wriggles its way into the sand, so that only its head is exposed to the burning sun.*

SOFT-HEARTED GIANTS *Saguaro cacti have tough skins but soft, water-filled flesh. For elf owls (right), old woodpecker holes make a safe haven from predators.*

SOFT-CENTRED CACTUS WITH A TOUGH EXTERIOR

The saguaros of northern Mexico and the American south-west are kings among cacti. These stately desert plants can live for a century or more, and grow to 20 m (65 ft) tall. With branches like surreal candelabra, they can weigh more than a tonne.

Like many cacti, the saguaro's fleshy, water-retentive waxy stems are ribbed, and covered in clusters of vicious spines. The spines deter animals from pilfering its valuable resources of water, and its thick skin helps to prevent evaporation.

The depth of the ribs shows how much water is in store. When water is abundant the saguaro swells like a stretched accordion. During droughts, it shrinks again, and the ribs stand out.

In a habitat with no real trees, saguaros appeal to many kinds of animals. Nectar-feeding bats visit the saguaro's large creamy-yellow flowers in spring, and later in the year fruit-eating animals help to spread its seeds. Gila woodpeckers peck holes in the saguaro's stems, and when they move out, their homes are often taken over by elf owls. As the world's smallest owls, they find the world's biggest cactus a perfect home.

DESERT PLANTS BREATHE MORE DEEPLY AT NIGHT

In the cool of the night, desert plants carry out a special trick to help to conserve their water reserves. All plants 'breathe' through microscopic pores, which take in the carbon dioxide the plants need for photosynthesis – the chain of chemical reactions essential for life and growth. At the same time, they release oxygen as waste, but along with the oxygen, precious water vapour also escapes. For desert plants, losing water can have disastrous results. So instead of opening their pores during the day, they open them at night, when the cool air keeps water loss to a minimum. They still carry out photosynthesis in daylight, like all other plants, but they do it with their pores tightly closed, like an animal holding its breath all day long.

STEMS THAT SERVE AS LEAVES

Cacti and other fleshy desert plants cope with lack of water by having no leaves. For them, leaves are a luxury they can do without, because leaves give off water, and the more leaves a plant has, the worse the problem becomes.

Instead, the green cactus stem works like a collection of leaves, trapping the energy from sunlight so that the plant can grow. Cactus stems have tough waxy skins, and a relatively small surface area – another means of reducing the amount of moisture that disappears into the air.

DIGGING DEEP AND SPREADING WIDE

Desert plants go to great lengths in their search for water. In North America, the mesquite, a small desert tree, has roots 10 to 15 times the length of its trunk that have been found at depths of 50 m (165 ft). A fig tree in South Africa set the record when its roots were found in a cave 100 m (330 ft) below the surface.

Other desert plants have a network of thin fibrous roots, which spread out widely near the surface. Like an absorbent mat, they collect the smallest amounts of rain, or droplets of moisture from mist and dew.

BIZARRE BOOJUM'S DESERT ADAPTATIONS

Of the few trees in the desert, the contorted boojum with its brush-like top is the most bizarre. Found in two regions of northern Mexico, boojum trees are well adapted to their desert world. They sprout tiny leaves soon after it rains, and drop them when it turns dry once more.

The boojum is often unbranched, with a tapering trunk up to 20 m (65 ft) tall that bristles with small spindly twigs. Because boojum wood is relatively soft, the trunk often starts to lean over, then twist and sag as it grows taller. Some boojums bend so much that they end up with their tips almost touching the ground.

THE BENDY BOOJUM *Like living poles, boojum trees grow out of cracks between rocks. Some grow ramrod straight, but many sag.*

NAMIB DESERT

The Namib is a strip of sandy desert with one of the driest climates on Earth. Months or even years can go by without rain. For many animals the key to survival is the dense fog that rolls inland after dark from the Atlantic Ocean.

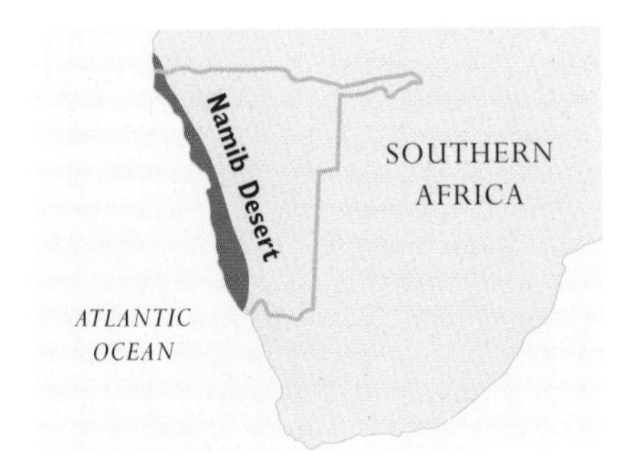

WELWITSCHIA'S SECRET OF A LONG LIFE

Scattered across the Namib are groups of welwitschias, strange plants, some over 2,000 years old. Welwitschias have just two leaves growing from a corky stem anchored in the desert floor. The leaves are grey-green, tough and woody, and are arranged like a centre-parted hairstyle flowing in ragged tresses on the ground. Although they win no prizes for beauty, leaves like this are very efficient at conserving any available water, which is why welwitschias live in places where few other plants are able to survive.

By the time a welwitschia reaches its 100th birthday, and is ready to reproduce, its two leaves look like a pile of dried-out compost heaped up by the desert wind. Yet century-old plants are mere youngsters. When the plants are 2,000 years old, their growth rate has slowed almost to a standstill.

SANDGROUSE HAS A BUILT-IN SPONGE

Most birds need a daily drink, so they live in places where water is close by. But the Namaqualand sandgrouse of north-western South Africa takes a different approach. This partridge-like bird travels up to 300 km (185 miles) a day to quench its own thirst and has a special way of collecting water to take to its nest. In the early morning and evening, chattering flocks of sandgrouse set off to the nearest waterhole, skimming over the desert on rapidly beating wings. They land some distance from the water's edge, and spend several minutes watching for danger before they walk down to the water to drink.

Once the birds feel safe they wade in chest-deep, drinking as much as they can. Most of their plumage is water-repellent, but their breast feathers are as absorbent as a sponge. Each adult soaks up three or four teaspoons of water, and then flies back to deliver it to its chicks, which suck the water from the feathers. The adults keep up this daily delivery for nearly two months, until their chicks are able to fly to the waterhole.

MOVING ON SHIFTING SANDS

The web-footed gecko and Namib sidewinder move across sand dunes with consummate ease. The web-footed gecko has slender toes connected by webs of skin, a perfect adaptation for running about where other animals are reluctant to tread. It spends its day in burrows and emerges after dark to stalk insects.

Unfortunately for the gecko, another reptile, the Namib sidewinder adder, also moves freely in these shifting surroundings. Like other sidewinders, it throws its body diagonally across the sand, leaving a

LIVING FOSSIL *Welwitschias are plants with no close living relatives. They have no flowers: to reproduce, they grow egg-shaped cones.*

PRECIOUS DROPS *With its abdomen tipped up in the air, a Namib darkling beetle collects droplets of moisture from clouds of fog.*

series of J-shaped marks on the surface. It sounds awkward, but sidewinders can be surprisingly agile and rapid, managing about 10 km/h (6 mph).

The Namib sidewinder adder has eyes in the top of its head. During the day it lies just under the sand, eyes peeping above the surface, and waits for animals to come within striking range. At night it sets off in search of prey, particularly geckos.

ATTACK FROM BENEATH *The desert golden mole feeds on a web-footed gecko, which it has ambushed by bursting out of the sand.*

HOW BEETLES DRINK IN THE NIGHT FOG

In the Namib, the darkling beetle has devised an ingenious way of getting a daily drink. The beetle lives on sand dunes, and uses its body as a water-collection device. At dusk, it goes to the crest of a dune, takes up position with its head down, and waits for the fog to spill in from the sea. During the night the desert cools, and so does the beetle's body. As the fog sweeps over the dunes, tiny droplets of moisture condense on the beetle's back, and run down towards its mouth. The beetle laps them up, slowly collecting the liquid it needs for another day in the desert. Before the sun is high and the fog burns off, the beetle sets off on its search for food.

TUNNELLING THROUGH THE DUNES

One inhabitant of the Namib has a novel way of negotiating the dunes. The desert golden mole 'swims' through the sand, like a fish hunting near the surface of the sea. With its pale yellow silky fur, blunt nose, and short, strongly clawed feet, the mole measures just 9 cm ($3\frac{1}{2}$ in) long. It spends almost all its life beneath the surface of the sand.

The mole is sightless – its eyelids are permanently fused – and it relies on its nose and keen hearing to track down termites, beetles, and lizards walking over the sand above. When something edible is within range, the mole launches a surprise attack, bursting out of the sand and grabbing its prey with its teeth.

During the breeding season male and female moles dig deep shafts in search of firm sand, and it is here that the young are born.

SURVIVAL IN TROPICAL FORESTS

With their year-round warm and wet climate, tropical forests have the richest variety of animal and plant life in the world. But with so many species competing for space, light, and food, the struggle for supremacy is remorseless.

PRIMATES' SAFE ROUTES IN THE TREETOPS

In tropical forests, many wild animals live high above the ground. But for larger animals, such as monkeys in the Americas, Africa, and Asia, and gibbons in South-east Asia, moving about among the treetops is potentially dangerous. A single mistake can send them plunging to the ground, resulting in broken bones or even death. To reduce the risk of falling, monkeys and gibbons tend to follow well-worn pathways through the canopy, just as we use steppingstones in a river. The branches along these routes are strong and safe, and they are kept free of debris by the animals' hands and feet.

Because monkeys and gibbons use these paths regularly, they know every twist and turn, and exactly where they need to jump. Whenever danger threatens, they make straight for one of these tried-and-trusted routes, ensuring a speedy getaway.

DEADLY HELICOPTERS ON PATROL

Whirring lazily through the forest on translucent wings, a helicopter damselfly searches for its prey of hapless spiders. Despite its delicate appearance, it is one of the most ruthless hunters in the insect

UNDERHAND ATHLETE *A white-handed gibbon swings through the forest in northern Thailand. Its arms are longer and stronger than its legs.*

world, patrolling Central America's forests. Its wingspan, of up to 19 cm (7½in), is the largest of any damselfly, and it is capable of rapid aerial manoeuvres, making it hard to catch.

The helicopter damselfly hovers just in front of a spider's web before suddenly darting in to make a kill. Remarkably, the spiders do not seem to notice the beat of its giant wings, or its large eyes looming towards them in midair. By the time they spot the damselfly, their fate is already sealed. The damselfly grabs a spider with its legs, eats its succulent abdomen, and drops its victim's severed limbs and head on to the forest floor.

MIMIC PASSIONFLOWERS DETER BUTTERFLIES

Heliconia butterflies in Central and South America's tropical forests feed and lay their eggs on passionflower plants. The adults drink the nectar of passionflowers, and their caterpillars have a taste for the leaves, so the adult females have to search out passionflower plants before they can lay their eggs.

When a female heliconia finds a passionflower, she checks the leaves for any other heliconiid eggs. If she finds none, she lays a small number of her own. When the caterpillars hatch out they have all the leaves for themselves. But if another heliconia has been there before her, her caterpillars may face a struggle for food, so she will leave the plant alone.

Some species of passionflower have evolved an ingenious defence that exploits this part of butterfly behaviour. They produce small yellow bumps on their leaves and tendrils that look exactly like heliconia eggs. Visiting butterflies are often deceived by these bumps, and they fly off to look for other plants, leaving the passionflower to grow unharmed.

IMITATION EGGS *These nodules on a passionflower leaf look exactly like butterfly eggs. They trick butterflies into laying their eggs elsewhere.*

FOOD FROM ABOVE *The nocturnal paca supplements its diet of roots, shoots, and stems with the fallen leftovers of animals in the canopy.*

THE PACA'S FEAST OF FALLEN FOOD

For forest floor animals such as the paca, a shower of leftover food creates an opportunity not to be missed. The size of a small dog, these South American rodents feast upon a variety of items discarded by wasteful monkeys and parrots.

Large groups of monkeys foraging in the rain-forest canopy during the day often drop almost whole fruits after taking only a few bites. Parrots can be just as untidy, scattering fleshy rind, nuts, and seeds. At night the paca forages for the substantial scraps left by these and other animals.

FEARSOME ANT ARMIES AND THEIR CAMP FOLLOWERS

Although each is no longer than a pin, South American army ants are dangerous predators. They live in nomadic bands 50,000-strong, and swarm over the Amazon Basin forest floor, attacking and eating anything they can overpower. Animals up to the size of small lizards rush to avoid a grisly fate. But escaping one set of predators often drives them straight into the mouths of others, for army ants attract camp followers, known as antbirds, which grab and devour the would-be escapees.

Up to 25 antbirds accompany the army on its deadly raid, with the largest in front of the squadron and others relegated to the periphery. Even if an animal escapes both ants and birds, clouds of parasitic flies may lay eggs on it, condemning it to a slow death.

THE KAPOK TREE'S GIGANTIC BUTTRESSES

The kapok tree has some of the biggest buttress roots on Earth to strengthen it against the wind. Found in the American tropics, Africa and South-east Asia, it starts life looking like any other tree. At 50–75 years of age, its shape begins to change. Massive branches spread out from the tree's crown, and the base flares outwards in sinuous ridges.

These ridges, each up to 5 m (16 ft) tall, are the tree's buttress roots. To complete this spectacular piece of natural engineering, in which each buttress meets the ground, it merges with a snake-like root that can run 50 m (55 yd) across the forest floor. Because the root is near the surface, it can intercept and absorb nutrients from decaying leaves before rain washes them away.

Strengthened by its buttresses, the kapok tree towers over other trees, reaching up to 60 m (200 ft) tall. When the centre of the trunk rots away, the buttresses continue to prop it up, just as man-made buttresses support ancient buildings.

PRIMEVAL PROPS *The buttressed kapok tree produces seeds containing downy kapok, once used to make a cotton-like material.*

INTRICATE JIGSAW OF RAIN-FOREST NEIGHBOURS

From above, tropical forests look so tightly packed that it is impossible to see the ground. Yet looking upwards from the forest floor, neighbouring trees interlock like pieces in a jigsaw puzzle, but the pieces never touch. Instead, each tree is surrounded by a gap, a wandering ribbon of 'no man's land', 1 m (3 ft 3 in) wide, that stands out against the sky.

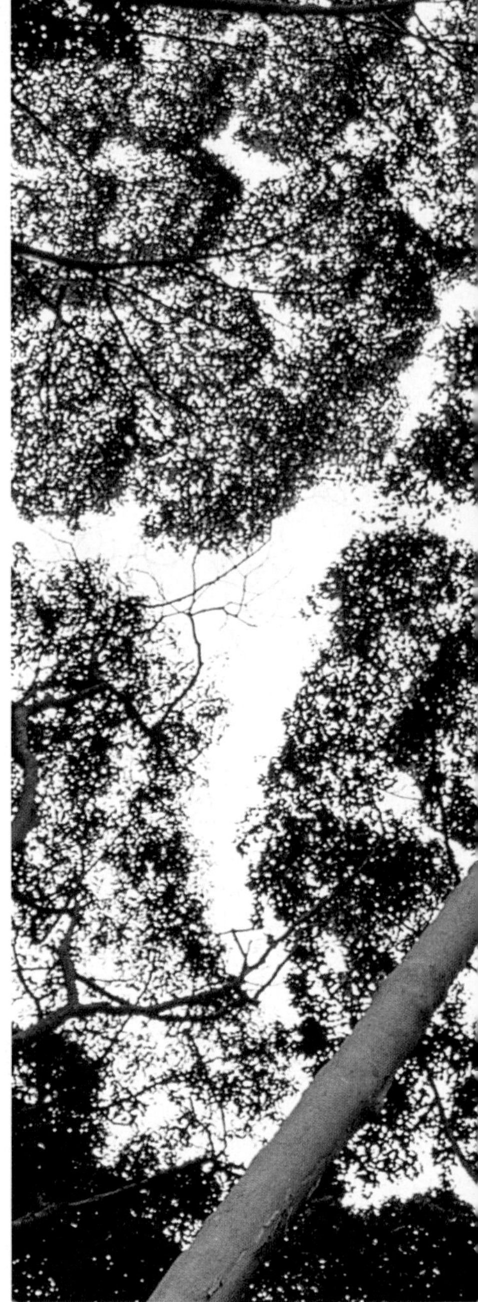

PERSONAL SPACE *By keeping their distance from their neighbours, tropical forest trees make it harder for fungal diseases to spread.*

Why do trees grow like this, and why do the gaps not fill in? In almost every other habitat, plants grow as quickly as they can, and often swamp each other in the constant battle for space. High up in the canopy of tropical forests, trees seem to disobey the rules. In fact, the trees keep their distance because it stops them clashing branches in high winds. It also deters leaf-eating insects from tree-hopping.

TREES THAT WAIT FOR THEIR MOMENT OF GLORY

Tropical forests are dominated by huge trees that are among the biggest on Earth. But beneath these giants, in the deep shade cast by the canopy, are quite different trees that are often overlooked. These are the forest's 'trees-in-waiting'. Like plants growing in the depths of the sea, they live on a meagre ration of light that filters down from above. They can be decades old, yet their trunks are often no thicker than a broomstick, and in some years they hardly grow at all.

For these trees, life takes off when a giant tree dies and falls, flooding their gloomy world with light and offering them a chance to grow up through the canopy and replace a fallen giant.

AFRICA'S SUPER-SHREWS

For small animals, being stepped on and crushed is one of the hazards of life on the forest floor. But although it weighs little more than a hen's egg, the hero shrew from Central Africa can survive being stood upon by a fully grown man.

The hero shrew's resilience lies in its spine. The vertebrae have extra-long interlocking bony struts, making the spine extremely strong, yet allowing it to bend in the normal way.

RECYCLING IN ACTION ON THE RAIN-FOREST FLOOR

Every year, several millions of tonnes of dead leaves fall from forest trees. In cool parts of the world, they build up in deep drifts, and help to create some of the richest soil on Earth. But in tropical forests, they tend to disappear quickly.

As soon as a leaf hits the ground in the tropics, bacteria and fungi get to work. Within a few days the leaf starts to break up, and tree roots absorb the nutrients that are released. Most leaves vanish within a month thanks to the remarkable recycling system that helps to keep the forest alive.

MONTEVERDE CLOUD FOREST

On the spine of mountains that runs through Costa Rica is a tropical forest where the constant drip of moisture creates a lush world teeming with life. Monteverde is home to more than 100 species of mammal and 400 species of bird.

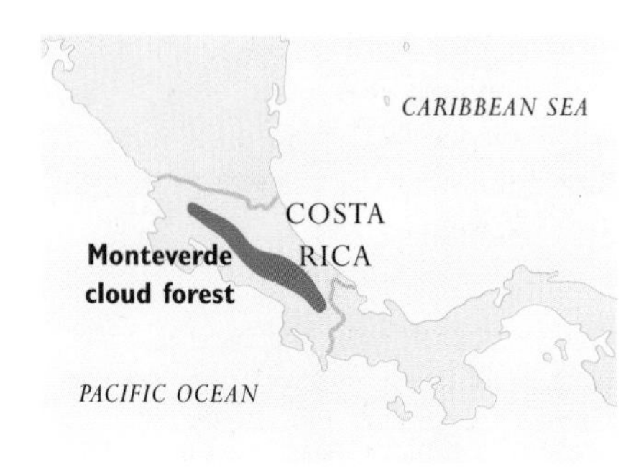

BUTTERFLIES WITH WINGS OF GLASS

Seen under a magnifying glass, most butterfly wings reveal a mosaic of overlapping coloured scales, arranged like tiles on a roof. But at Monteverde, some butterflies have very few of these scales and, as a result, their wings are almost transparent. Known as clearwings, the butterflies often cruise low over the forest floor, to feed on a diet of dead insects, bird droppings, and nectar from flowers.

The most likely reason for their glass-like wings is that these creatures are very vulnerable to attack by birds when feeding close to the ground. Having see-through wings makes them harder for birds to spot, giving them a better chance of survival.

THE BIRD-GOD'S FABULOUS PLUMAGE

In the time of the ancient Aztecs, more than 500 years ago, one cloud-forest bird, the quetzal, was revered as a god. The male quetzal's fabulously long, iridescent green tail plumes were used in special ceremonies, but instead of being killed for these feathery treasures, the captured birds would be released. Female quetzals are showy birds, but they pale by comparison with their mates. The 35 cm (14 in) males have silky red undersides, green breasts, and a tuft of green feathers, but their tail plumes are what truly catches the eye. Hanging like long tresses over a quetzal's true tail feathers, they trail as much as 60 cm (2 ft) in the air.

As if conscious of their finery, the male birds spend much of the day motionless, perching in the canopy. From time to time they flutter into the air to find oil-rich fruit, usually returning to the same perch to feed.

SEE-THROUGH BUTTERFLY *A clearwing butterfly probes a flower for nectar. Curiously, other related species are colourful and brightly patterned.*

FERN FORESTS FROM THE AGE OF THE DINOSAURS

Many millions of years ago, when dinosaurs roamed the planet, tree-like ferns were among the Earth's most abundant plants. Most of today's ferns are smaller, ground-hugging plants, but in Monteverde and humid parts of the tropics, tree-sized ferns have been able to survive.

At first glance, tree ferns look like a cross between 'normal' ferns and palm trees. They have dense fans of luxuriant bright green fronds, borne on trunks up to 12 m (40 ft) tall. Like palm trees, they grow only at the tops of their trunks: if the growing point is cut off, the tree fern dies. Yet these survivors from the past differ from other trees in several ways. They lack a branching root system, and their trunks are fibrous, without a protective covering of bark. In the moist atmosphere, these trunks soak up water like sponges, and are often covered with mosses and other plants.

THE CLEVER LURE OF THE LOBSTER PLANT

In the American tropics, hummingbirds pollinate many flowers, and over millions of years, flowers have adapted to their feathered visitors. At Monteverde, the most spectacular of these flowers are lobster claws, which are distantly related to the banana. Because hummingbirds have a poor sense of smell, lobster claws rely on their shape and colour to attract attention. Their flowers are relatively small, but they are enclosed by giant 'claws', up to 30 cm (12 in) long, which are vividly streaked with orange or red. The claws zigzag their way up the stem, creating a bizarre multicoloured flower head. Hummingbirds visit a number of widely dispersed flowers, and a colourful display ensures that the lobster plant's flowers are not overlooked.

FULL OR EMPTY? *Hummingbirds have to discover which of the lobster claw flower heads are empty and which are loaded with nectar.*

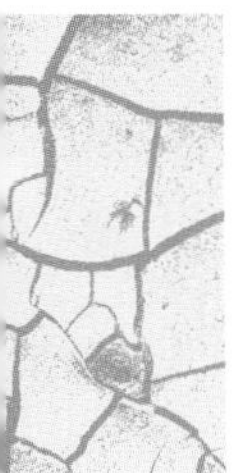

RIVERS, LAKES, AND WETLANDS

Freshwater animals, found all over the world, face many challenges in their struggle for survival. For some, the main problems are getting air or escaping predators; for others, home is a precarious habitat that can all too easily dry out.

RUSSIA'S LANDLOCKED SEAL

The world's smallest seal lives in one of the most landlocked places on Earth. Its home is Lake Baikal, a huge freshwater lake in Siberia, some 1,600 km (1,000 miles) from the sea.

How did the seals ever get to such a remote part of the world? The most likely explanation is that they came from the Arctic Ocean at the height of the last Ice Age, 300,000 years ago. As the glaciers spread south, the seals' ancestors swum before them, eventually reaching the lake. When the glaciers retreated, the seals stayed put, leaving them where they are today.

The Baikal seal is only 1.2 m (4 ft) long. It feeds on fish, and behaves just like a seagoing seal. During the winter, when the lake freezes over, it survives by making breathing holes in the ice. In early spring, females haul themselves onto the ice and give birth in lairs under the wind-driven snow. The newly born pups have long, silky white fur, which protects them from the intense cold. By the time they are ready to take their first swim, about eight weeks later, their original coat has been shed and they are steel-grey like the adults.

RUSSIAN RETREAT *After the long, harsh Siberian winter, Baikal seals make the most of the summer sunshine. These small seals weigh about 80–90 kg (176–198 lb) each.*

SPRING OF LIFE FOR DESERT PUPFISH

A tiny oasis of water, far out in the Nevada Desert, is the only home of the Devil's Hole pupfish. One of the world's rarest and most isolated freshwater animals, the pupfish lives in a pool 20 m (66 ft) square – the size of a swimming pool – and its population of 500 can be taken in with a sweep of the eye.

Pupfish have probably lived in the Devil's Hole for more than 20,000 years. Each about the size of a pen cap, they feed on tiny creatures and algae growing on rocks in the pool. The pool is fed by an underground spring, and the fish rely on the spring for their survival. If the water level falls, their home could shrink, quickly driving them to extinction. The fish's home is now protected as a wildlife reserve.

DETERMINED DIVER *An American dipper, one of five species worldwide, fights a stiff underwater current to reach the streambed.*

WALKING UNDER WATER

Fast-flowing streams and rivers in the Northern Hemisphere, South America, and Africa are home to an unusual waterbird. Looking like an extra-large wren, the dipper is the only songbird that has adapted to a watery life. It does not have webbed feet, but its thick feathers are thoroughly waterproofed, and it can close its nostrils when it dives. It also has a transparent third eyelid that can flick across each of its eyes, helping the dipper to see as it hops over water-splashed rocks.

Instead of feeding at the surface, as many other waterbirds do, the dipper walks and flutters along the riverbed, picking up insect grubs among the stones. With its light body, it faces a constant battle to stop itself bobbing up again like a cork, so it always dives against the current, then walks along with its head bowed and its tail pointing upwards. The current presses down on its back and wings, keeping the dipper submerged for as long as it can hold its breath.

Young dippers are able to swim and dive before they can fly. They plunge into the current as soon as they leave the nest, following their mother as she disappears beneath the surface.

HOW LUNGFISH SURVIVE THE SUMMER DROUGHT

Lungfish breathe in a different way from other fish, and can even survive buried in mud. If a river or lake dries out, most fish suffocate to death, because their gills can only collect oxygen from water, not from air. But for lungfish, drought is not much of a problem.

Found in South America, Africa, and Australia, these eel-like fish, 1–2 m (3–7 ft) long, live in swampy lakes, where the oxygen supply is often low. They survive by coming to the surface to breathe, and can do this because they have lung-like organs connected to their throats. The Australian lungfish has one lung, while others have two. When a lungfish takes a gulp of air, the air travels via its throat into its lungs, allowing oxygen to reach its blood.

If its home starts to dry up, the fish seals itself in a burrow 50–60 cm (20–24 in) deep, leaving a tiny air passage through which it can breathe.

WAITING FOR THE RAIN *Sealed up inside their muddy burrows, African lungfish have been known to survive for up to four years.*

TEMPORARY VISITORS *Water soldiers appear at the surface for just a few months. Yet the plant is classed as a noxious weed in some US states.*

LIVING SUBMARINES

Winter is a difficult time for freshwater plants, because ice can cut into their leaves and stems. Most floating plants die away at this time of the year, but the water soldier has a remarkable self-preservation technique: after it flowers, the plant sinks to the bottom like a submarine, and then comes back to the surface the following spring.

Submarines need ballast to make them sink, and so does the water soldier. It gets this by absorbing calcium carbonate from the water, the same substance that makes electric kettles 'fur up'. Once the plant has taken in enough of this ballast, it sinks out of sight. In spring, some of the ballast dissolves, increasing the plant's buoyancy enough to make it float.

HOW MANGROVES THRIVE AMID THE SALT

Mangroves thrive in salty water, unlike other trees for which salt is a deadly poison. They grow in river estuaries and muddy coasts throughout the tropics.

Mangroves cope with the salt by getting rid of it efficiently. They pump it out of their roots, and store it in leaves that they are about to shed. As a result, they can 'drink' seawater with no apparent ill effects. Since waterlogged ground is a poor source of oxygen, mangroves have another survival trick: 'breathing roots' that take oxygen from the air.

VITAL PROPS *Stabilized by arching roots, mangrove trees keep a firm foothold in seashore mud. At low tide, their trunks perch in midair.*

UNSINKABLE WATER FERNS

Water ferns spend their whole lives afloat, on the surface of lakes and ponds. These plants are small enough to fit in a jar full of water, and if you put on the cap and shake the jar as hard as you can, they always bob back to the surface, right-side up, with their leaves perfectly dry.

Water ferns live in rainy places, mainly in the tropics, and they need water-repellent leaves to stop themselves sinking during sudden downpours. Their leaves are covered with microscopic hairs, which trap air and allow water to roll off.

Armed with this flotation system, water ferns can clog up lakes and reservoirs, making it hard for other plants to survive.

A LIFE IN THE DAY OF A MAYFLY

Just one day of adulthood is all a mayfly can expect to survive. Mayflies start life as six-legged larvae with no wings. For up to three years, they crawl over the bottom of lakes, rivers, and streams, feeding on tiny plants. As summer approaches, the oldest larvae stop feeding, and climb up plant stems until clear of the surface. Their skins split open to reveal two pairs of filmy wings, and once these have dried, they take to the air. A short time later, they shed their outer skin a second time, and become fully mature and ready to mate.

A mayfly's life now becomes a race against the clock. Its digestive system no longer works, so it cannot feed. Instead, it concentrates on finding a partner and, if it is a female, on producing and laying eggs. Thousands of swarming males are joined by females in a mating ritual over the water. By sunset their work is over and most of the mayflies are dead.

FINAL FLOURISH *A mayfly sheds its skin for the final time. It will live for only a single evening, just enough time to find a mate and reproduce.*

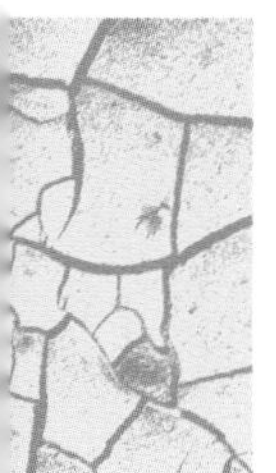

THE EVERGLADES

A sea of grass dotted with tree-covered 'islands', Florida's Everglades are drenched by subtropical storms in summer. The boggy terrain and humid climate make this a paradise for animals and plants that need abundant fresh water.

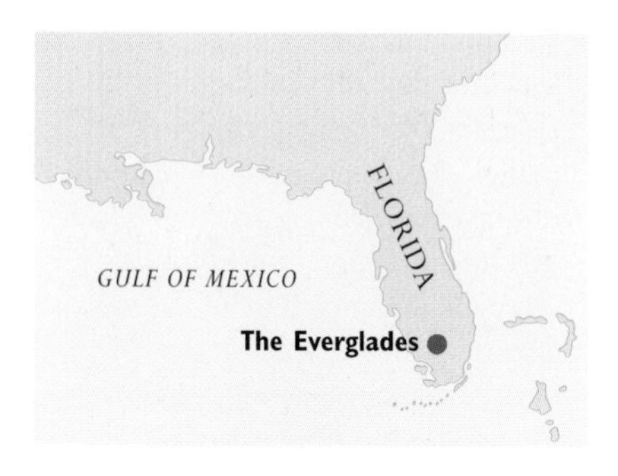

FUSSY EATER PICKS SLOW-MOVING PREY

Unusually for birds of prey, snail kites restrict their diet to a single species of water snail. The snail kite is a tropical species that, in the Everglades, is on the very edge of its range. It has a particular penchant for apple snails, which feed on algae and debris in the water. Every few minutes, apple snails have to visit the surface for a breath of air, and it is then that the snail kite strikes. Dropping towards the water, it snatches a snail with one of its feet, then flies back to a perch to feed. Its long hooked beak is specially designed to prise the snail from its shell, while it holds onto it with its claws.

GOURMET MEAL *Standing on a convenient post, a snail kite settles down to dine on its favourite food – a freshwater apple snail.*

TREES THAT BREATHE THROUGH THEIR KNEES

Most trees die of suffocation in waterlogged ground since it stops oxygen from reaching their roots. In the Everglades, the swamp cypress overcomes this problem with roots that have knobbly 'knees'. These protrude upwards through the water's surface, to collect oxygen from the air.

Cypress 'knees' help to trap silt and rotting vegetation, which after many years become small islands or cypress domes. Although these are often less than 1 m (3 ft 3 in) high, the trees growing on them can be as tall as a 15-storey building. They are perfect nesting sites for the Everglades' water birds, which use them as high-rise breeding grounds.

CONSERVATIONISTS OF THE EVERGLADES

American alligators are surprising but effective conservationists of the Everglades. Without their vital work, many swamp-dwellers in the region would die during the winter dry season. The alligators clear out muddy ponds, known as 'gator holes, using their feet and noses to push mud and plants aside. They do this to create reservoirs, which they use during the winter drought. These water holes provide a valuable habitat for a host of other animals, such as snails, turtles, fish, and amphibians, all of which provide food for each other, and for passing birds.

The alligators' excavations have a major impact on the Everglades landscape. Their ponds pockmark the scenery, while the excavated spoil acts like a natural compost heap. In time, these heaps become covered with grasses and young trees, creating more habitat for the Everglades' wildlife such as insects and birds.

KNOBBLY KNEES *Growing out of waterlogged mud, the aerial roots of a swamp cypress gather oxygen – essential for the tree's survival.*

HOW THE EVERGLADES WERE FORMED

Six million years ago, the area that is now the Everglades lay beneath a shallow sea. The seabed sediment turned into layers of limestone, which were drowned and exposed repeatedly as the sea level changed.

When the glaciers melted at the end of the last Ice Age, the sea rose and blocked the outlets from Lake Okeechobee, creating the silty swamps that exist today. The highest point in the Everglades is only 10 m (33 ft) above sea level, ensuring that water trickles through the swamps very slowly on its journey to the sea.

TROPICAL INTRUDER IS FISH OUT OF WATER

The balance of the Everglades' ecosystem is often compromised by foreign invaders. One such outsider is the walking catfish, a predatory freshwater species from South-east Asia that escaped from tropical fish breeders in the 1960s. True to its name, it really can walk out of water, using strong spines on its underside. On rainy nights, it drags itself from pool to pool, breathing air through its specially modified gills. Its impact on the complex ecology of the Everglades is hard to assess: though it is a formidable predator of other fish, it is also prey for herons and egrets.

EVERGLADES PIGEON'S POISONOUS PROTECTORS

The white-crowned pigeons in the Everglades like the berries of the poisonwood (hog gum) trees. In the fruiting season, they fly inland from their nesting colonies in coastal mangrove swamps to feed on the poisonwood berries. The pigeons have a special defence against predators: the trees' caustic sap keeps their enemies at bay. The pigeons are immune to the poison in the berries, and in return for food and protection, they help to spread the trees' seeds.

SALT LAKES, HOT SPRINGS

Life has existed on Earth for more than 3 billion years, and during that time living things have adapted to some extraordinary habitats. Even steaming springs, salt-laden lakes, and pools of oil have their own special kinds of life.

BACTERIA WITH A TASTE FOR SALT

Though a salt-rich environment is generally hostile to life, certain types of bacteria, known as halophiles and halobionts, thrive in such conditions. Species that live in the Dead Sea, one of the saltiest lakes on Earth, are so well adapted to a salty life that fresh water is poisonous to them. If they are put in pure water, even for a few seconds, they die.

Sea water is about 3.5 per cent salt – equivalent to a tablespoon dissolved in every litre (1 3/4 pints). In warm parts of the world, some lakes contain up to ten times as much and salt crystals can be seen on the shore where the water has evaporated in the sun.

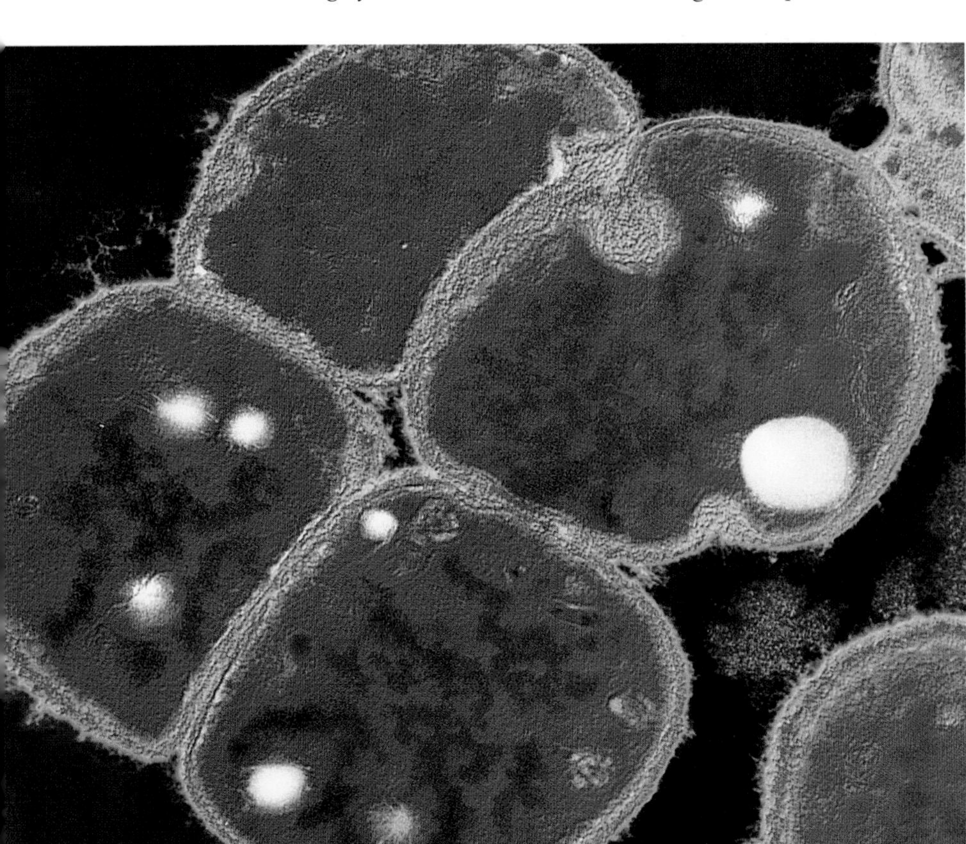

LIFE IN BRINE *Salt-loving bacteria thrive in habitats that would kill most forms of life. As a result, they have salt lakes largely to themselves.*

THE MARCH OF THE LESSER FLAMINGOS

For lesser flamingos, intense heat, salty water, and sticky mud are the ingredients of an ideal home. These birds live in Africa's salt pans and soda lakes.

Lesser flamingos feed almost entirely on cyanobacteria, or blue-green algae, which flourish in warm soda-rich water. To feed, the birds lower their necks until their heads are almost upside-down, with the tips of their beaks just beneath the surface. They walk forwards with their heads swinging from side to side, and filter out food by pumping water through special fibres in their beaks, which work like a sieve.

Their microscopic food is abundant in some lakes, supporting flamingo flocks that can be more than 1 million birds strong. Unfortunately for young lesser flamingos, safe places to nest and good places to feed are often far apart. Flamingos cannot fly until they are more than two months old, so they have to walk to their feeding grounds.

In some African lakes, hundreds of thousands of chicks make their way across the mud through the blazing heat to water some 50 km (30 miles) away. Up to half the young may fail to complete this arduous journey, but the survivors may live into their thirties.

LIFE NEAR BOILING POINT

Imagine diving into a pool heated to 85°C (185°F), filled with water that could eat into metal. A plunge like this would mean instant death for

ON THE EDGE *The hot springs fly is one of the few insects that can survive around the steaming springs at Yellowstone National Park.*

most forms of life, but not for bacteria that live in hot springs. One hot-spring bacterium, *Sulfolobus acidocaldarius*, thrives in temperatures close to boiling point, in water that is even more acidic than vinegar.

Compared with these organisms, plants and animals are much less resilient, yet some still manage to live in water that is too hot for humans.

In Yellowstone National Park, USA, home of some of the world's biggest hot springs, strands of an alga called *Zygogonium* form large mats that wave to and fro in the steaming current. The miniature jungles are too warm for most animals, but the grubs of the hot springs fly graze on the algal strands. The flies do not have this piping hot world to themselves. A tiny mite (*Partnuniella thermalis*) preys on their eggs, while another fly (*Bezzia setulosa)* produces carnivorous grubs that roam over the algae, attacking any other grubs they find. Even in hot springs, predators are never far away.

DRIED-OUT EGGS TAKE TO THE AIR

One problem of living in salt lakes is that they often dry up, leaving their inhabitants homeless. For brine shrimps, this is not a disaster but an ingredient of success.

These finger-sized crustaceans thrive in salty lakes. They collect food with their feathery legs, and are unusual in laying two kinds of egg. Some eggs have thin shells, and hatch soon after they have been laid. Others have much thicker shells, and sink into the lake-bed mud.

If the lake dries out, all the adult brine shrimps die, but the thick-shelled eggs can stay alive for up to five years. When strong winds blow over the lake-bed, they are often swept up into the air with the dust, travelling far and wide, and sometimes into water. If these dormant eggs get thoroughly wet, the embryos inside start to develop. The eggs hatch, and the young shrimps emerge.

OIL IS FLY'S LIFEBLOOD

The tiny petroleum fly of California lives in natural seepages of crude oil. It feeds on animals that get trapped in its sticky home.

Oil is a highly dangerous substance for animals: it clogs up fur and feathers, and sticks to legs and wings, with a hold that is as strong as glue. But for the petroleum fly it is the substance of life.

Petroleum fly grubs live inside the oil, and breathe through microscopic tubes that connect to the surface. They feed on anything that lands on the oil, including insects bigger than themselves. The grubs often swallow oil when they eat, but instead of poisoning them, as it would most other animals, the oil passes through their bodies without doing any harm.

When the grubs turn into adult flies, they are equipped with feet segments that attract moisture which repels oil. Using their oil-repellent feet they can walk over the oil's surface with no fear of getting stuck.

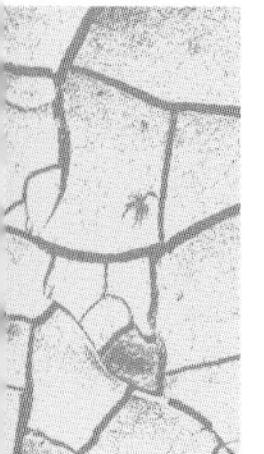

GRASSLANDS

Grasses, which cover 20 per cent of our planet, owe their success to being adaptable and tough, able to survive fire, drought, grazing, heat, and cold. They feed a huge array of animals, including the largest mammal herds in the world.

SCAVENGER STORKS FOLLOW FIRE TRAIL

Most animals fear fire, and run for cover at the first whiff of smoke or slightest flicker of flames. But in grasslands, some species profit from others' misfortunes. In Africa, foremost among them are large ground-feeding birds, such as storks.

Storks fly high in the sky to scour the plains for possible prey. If they spot smoke in the distance, they glide towards it, and land close to the flames. In this position they are ideally placed to intercept small animals trying to escape the blaze. Once the fire is out, the birds inspect the blackened ground, picking up any charred remains of animals that failed to get away.

BURNT OFFERINGS *Crowded together at a grass fire in Kenya, white storks watch for insects and animals flushed out by the flames.*

OWLS THAT GET TOGETHER UNDERGROUND

Grassland regions have few trees and lack safe places to nest so, to survive, burrowing owls set up home underground. These owls live in colonies of 12 or so pairs in the North American prairies and South American pampas, and are not choosy about the kind of grassland they inhabit. Sometimes they dig their own burrows, but usually they take over those abandoned by other animals.

Their nests are lined with grass and are about 1 m (3 ft 3 in) below ground, approached by a tunnel up to 3 m (10 ft) long. Once the nest is ready, the parents take turns to stand at the entrance, looking like feathery sentries on spindly legs. If an owl or its young are approached in or near their home, the sentries make a noise which sounds like a rattlesnake. This deters would-be intruders, leaving the young to grow up unharmed.

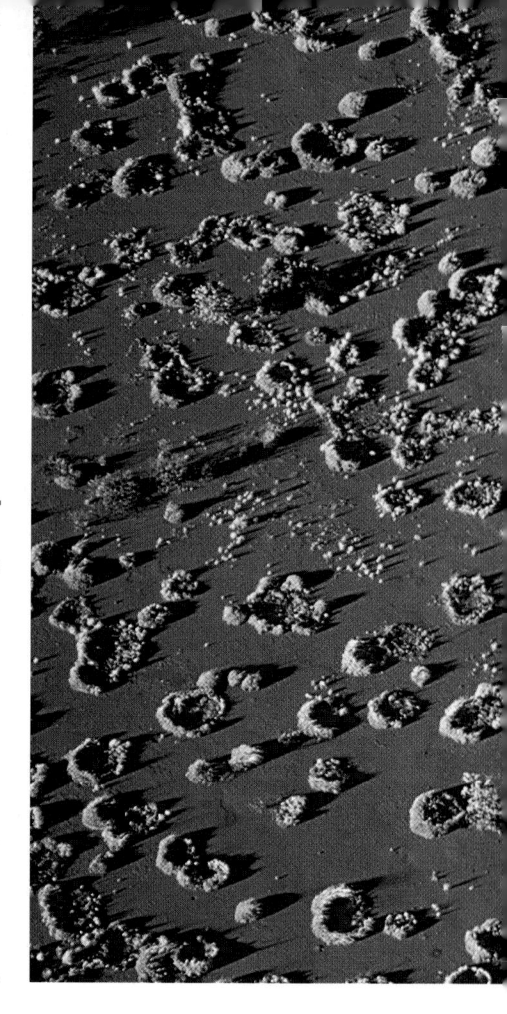

YOUNG WILDEBEEST ARE BORN TO RUN

Many grazing antelopes in Africa spend their lives on the move. Wildebeest, found mainly in Kenya and Tanzania, migrate in their hundreds of thousands, following an erratic path hundreds of kilometres long. They move in step with the seasons, their direction often governed by the sight or smell of distant rain.

Female wildebeest give birth to their calves while the herd is on the move to fresh grass. They do so in the open, surrounded by predators such as lions, leopards, and cheetahs. But young wildebeest can run with the herd when just a few hours old, and because they are all born within days of each other, each one has less of a risk of being singled out and attacked.

HOW GRASSES SURVIVE FIRE

Grasslands are often engulfed by fast-moving fires at dry times of the year. Yet within days of rain, the burnt grass is well on the way to

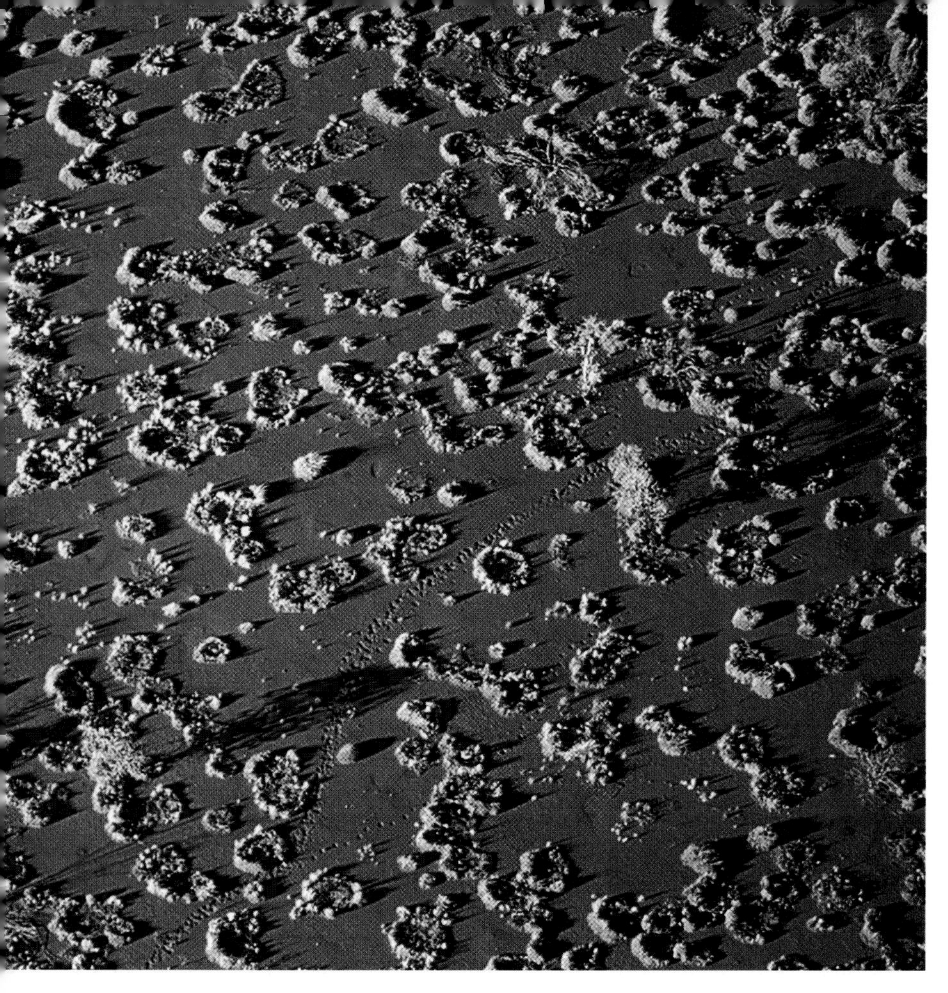

RINGS OF GRASS *Clumps of spinifex grass – each several metres across – are vital refuges for many small animals in the Australian outback.*

recovery, and nutritious green shoots are sprouting all around. Unlike other plants, grasses have most of their resources concentrated in their roots, safely out of harm's way. Also, the leaves grow from their bases instead of their tips, so if the tip is chewed off or burnt, the leaf continues to grow unharmed.

DEVASTATING NOMADS TURN GRASS TO DESERT

Locusts have been infamous since Biblical times for stripping vast areas of anything green. A single gathering of African locusts can contain 50 billion insects. Needing as much food as 30 million people, their appetite can have devastating results.

African locusts live in dry grassy places, and they are normally widely scattered. But if the weather turns unusually wet, as it does every few years, the way the locusts behave suddenly changes.

Nourished by a good supply of food, the females lay batches of eggs in quick succession. Soon the ground is teeming with tiny 'hoppers', which grow into adults within days. Unlike their parents, these locusts are highly gregarious, and gather together in swarms. Once the food runs out, the swarming locusts take off, following the rain, and eating anything green that lies in their path.

READY TO ROAM *Young locusts have prodigious appetites. Once their wings have developed, they will fly hundreds of kilometres to find food.*

FINDING SANCTUARY IN THE SPINIFEX

Viewed from the air, much of Australia's red desert seems to be covered in ring-shaped growths. These are actually clumps of spinifex, a plant unique to the continent, and one of the toughest grasses on Earth. Spinifex grows where few other plants can survive, providing desert animals with food, and also a home.

Spinifex leaves are narrow and hard, but this does not stop termites and ants turning them into a meal. They, in turn, are prey for spiders and geckos, which hunt among spinifex clumps during the cool of the night.

Birds also benefit: the spinifex pigeon, a small partridge-like bird with a jaunty crest, feeds mainly on spinifex seed, which it finds around the tussocks months after they have dropped from the parent plant. Unlike other pigeons, it rarely flies far: when confronted by danger, it disappears among the spinifex clumps.

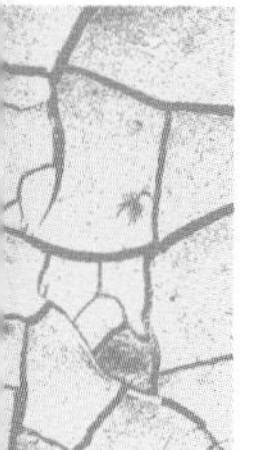

NGORONGORO CRATER

Formed by the dying eruption of an ancient volcano, the Ngorongoro Crater contains some 250 km² (96 sq miles) of grassland populated by African big game. In this giant arena predators and prey play out a struggle for survival.

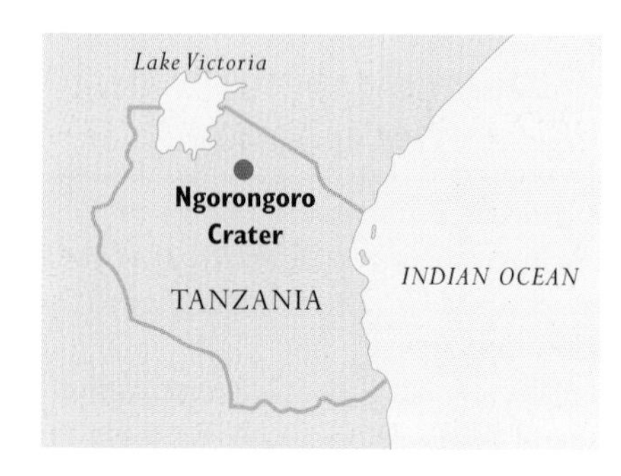

DUNG BEETLES ON PATROL

Africa's grasslands are kept fertile and clean thanks to the dung beetle, which tirelessly clears away animal droppings. Dung beetles use the droppings as food, both for themselves and their young. The adult beetles fashion the dung into table-tennis-sized balls, using their front legs and head as scrapers. They then roll these balls over the ground, before selecting somewhere to bury them.

During the breeding season, male and female beetles use one ball as a wedding banquet after they have mated. They then fashion several more, burying each one, and covering it up after laying a single egg. The ball acts as a nursery and a larder, and the grub inside it grows quickly once it has hatched. The adult beetle then crawls to the surface, ready to play its part in recycling nature's leftovers by gathering manure.

ON A ROLL *Dwarfed by their handiwork, these two dung beetles will push their dung ball onto soft ground, then bury it.*

HUNTERS IN THE NIGHT

The lions of Ngorongoro's grassy plains make most of their kills after dark, relying on their keen night vision. Lions cannot overtake zebras or antelopes running on open ground, so to hunt successfully, they need an element of surprise. Darkness is their greatest ally, allowing them to surround a group of animals before finally closing in. Their sense of smell is keen, but their eyesight is also sharp.

A lion's eyes can operate in very dim conditions because they contain a mirror-like layer, called a tapetum, just behind the retina. This structure reflects stray light back through the retina, so that the light-sensitive nerve cells have a second chance to pick it up. As a result, lions can easily detect their prey by moonlight, or even by the much fainter light of the stars.

Many carnivorous mammals have a tapetum, and it makes their eyes glow when they are caught in car headlights. In lions, this 'eyeshine' is normally red, but in foxes it is often green.

THE ZEBRA'S UNIQUE FINGERPRINT

For many years people have argued about why the zebra has distinctive black and white stripes. Some believe that the stripes act as camouflage, making zebra herds difficult to spot from far away. Another suggestion is that the stripes confuse predators when they stage an attack, making it harder for them to single

NOT QUITE BLACK AND WHITE *These plains zebras have faint brown stripes between their characteristic black and white bands.*

out an individual from the herd. A more plausible theory, based on the fact that no two zebras look exactly the same, is that the stripes work like identification labels, allowing zebras to tell each other apart. A foal, for example, will remember its mother's pattern from birth, so that it can tell her from others in the herd. Adults identify each other in the same way.

On the rare occasions when unstriped zebras are born, they tend to be shunned by the rest of the herd, confirmation at least that stripes are a passport to zebra society.

SMALL GAME HUNTER *The aardwolf's keen sense of smell means that it can easily detect the smelly defensive secretions of soldier termites.*

HYENA'S COUSIN TACKLES TINY PREY

One of Ngorongoro's strangest inhabitants is called the aardwolf, whose diet consists almost entirely of insects. Its favourite prey are harvester termites, which live on or near the surface of the soil. The aardwolf has exceptionally good hearing, and can track down the termites partly by sound, listening to the faint crackling noise they make as they haul dead grass blades underground.

Once the aardwolf has found a termite colony, it quickly sets about feeding. Using its sticky tongue, it laps up worker termites by the thousand, swallowing them with mouthfuls of soil. It has to work quickly because the soldier termites soon retaliate, squirting toxic chemicals to defend their nest. Even so, an aardwolf can eat up to 1 kg (2 lb) of termites a night.

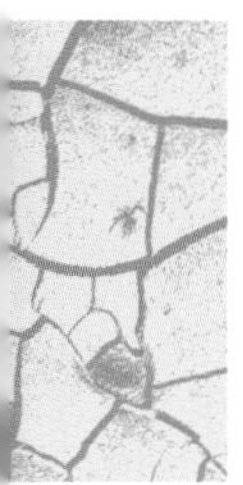

UNDERWATER WORLDS

The seas contain many creatures whose habits are poorly understood. Some form part of the plankton, the mass of life that drifts in the open sea. Others live in the depths of the oceans, in a habitat that remains largely unexplored.

CAMOUFLAGED CANNIBALS

Huge masses of sargassum weed drifting in tropical seas create a perfect hunting ground for the sargassum fish. This finger-sized cannibal is a master of disguise. Its lumpy body is covered with leafy flaps, breaking up its outline as it hides among the weed. It can also change colour, from mottled dark brown when it is deep in the weed, to light yellow-green when close to the surface in the sunshine. Camouflage is the key to its survival.

Even so, it needs to be vigilant. The sargassum fish's list of enemies is a long one, and it includes others of its own kind. This cannibalistic streak explains why sargassum fish are wary when courting their partners, and why they otherwise keep well apart.

THE FLOATING GENERATION

When they start life, some sea creatures look nothing like their parents. Barnacles are a good example. They hatch from eggs, and to begin with have feathery spines and a single eye. Smaller than a pea, they

HIDDEN DANGER *Lurking in a tangle of floating weed, a sargassum fish waits for unwary prey to come within reach of its camouflaged jaws.*

drift in the surface waters of the sea, feeding on other floating animals and plants. Eventually they settle on rocks, turn into adults and never move again.

TINY LONG-DISTANCE SAILORS

A tiny marine animal is specially adapted to make the most of the wind and water. Known as the 'by-the-wind sailor' catches the breeze with a tiny upright 'sail'. It uses wind-power to travel thousands of kilometres across tropical seas.

The by-the-wind sailor is closely related to jellyfish, but its body is round and almost flat, and contains small spaces filled with air. Its sail is permanently hoisted to make the most of the breeze. Although the sail is only 2 cm (less than 1 in) high, the animal is so light that it is easily blown along.

SAILING BY *Swarms of by-the-wind sailors in the Atlantic Ocean and Mediterranean Sea can be more than 1 km (two-thirds of a mile) long.*

In bad weather, by-the-wind sailors are sometimes washed up on beaches. But some of them always escape: their sails are set at two different angles, so that while half the fleet is beached, the other half is blown safely out to sea.

THE SEA WASP AND ITS TRAIL OF DEATH

All jellyfish have poisonous stings, and some inflict painful wounds. One species, the Australian box jellyfish, or sea wasp, produces such virulent poison that if a person brushes against its tentacles they could be dead within minutes.

The Australian box jellyfish, found off the continent's northern coast, gets its name from its rectangular body, which can be 25 cm (10 in) across. Beneath this hang trailing tentacles, up to 4 m (13 ft) long, armed with millions of lethal stinging cells and rows of tiny capsules containing a substance that works like glue. If any animal touches the tentacles, the glue is released, making the tentacles stick fast so their stings can do their work.

TORPEDO-SHAPED MONSTER THAT GLOWS AT A TOUCH

Imagine something that looks like an opaque test tube pumping its way through the sea. The tube is up to 10 m (33 ft) long, with an opening at the back. The organism has no head, no eyes, and no obvious sense organs, but if it is touched, it stops; light flashes along its body, and then it slowly swims away.

Though it looks like a single organism, it is in fact a floating colony of pyrosomes. These 1 cm ($^3/_8$ in) long tubular creatures travel through the seas together in their thousands. They siphon water up through their tubular bodies, expelling it in tiny jets which propel them along. As well as pumping the water, the pyrosomes filter out food from it.

The vast colonies have been taken for sea monsters and even torpedoes, a panic-filled reaction to some of the most harmless animals afloat.

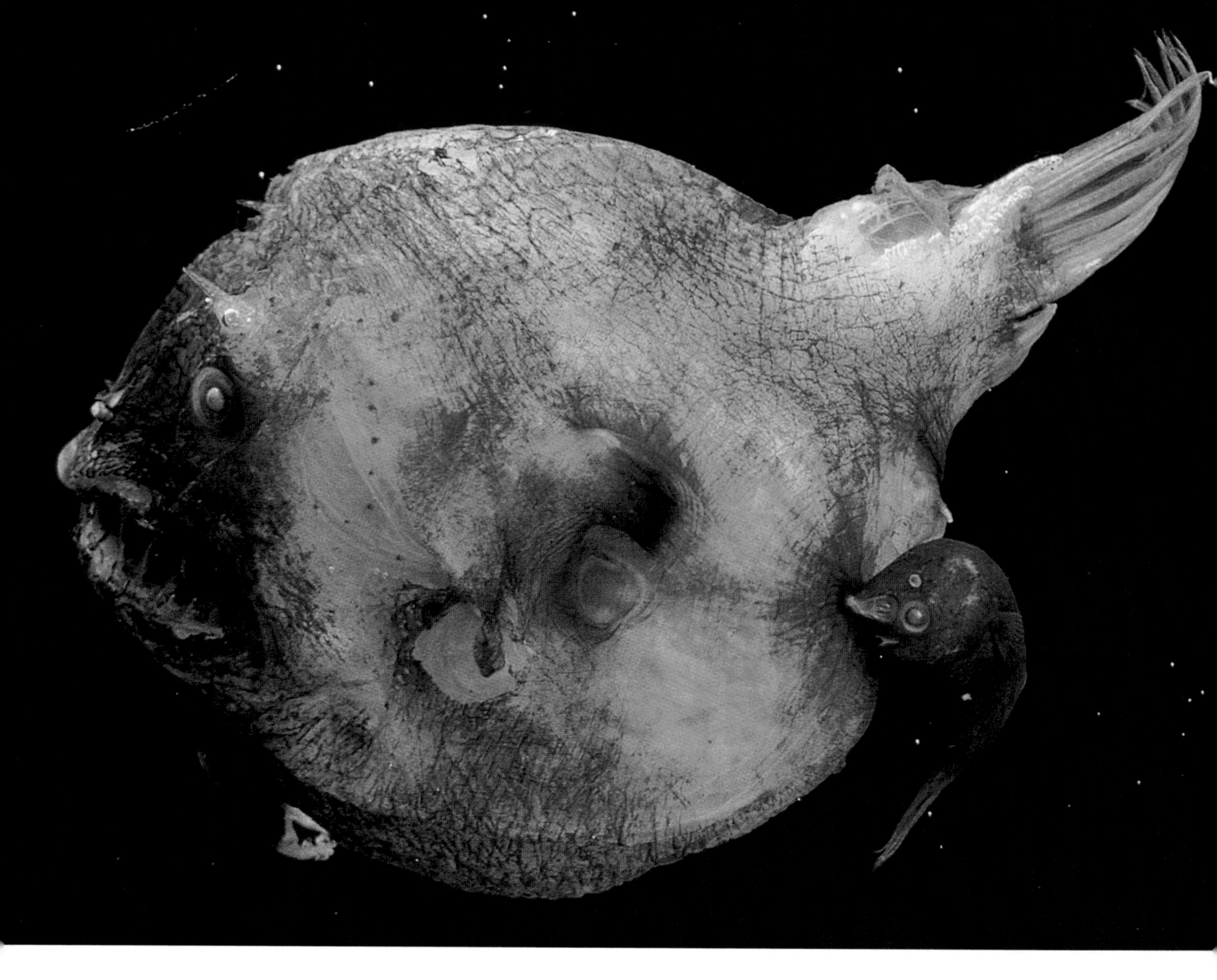

LOCKED IN A PERMANENT EMBRACE

It is not only food that is hard to find in the oceans' depths; mates can be even more elusive. The male bearded anglerfish makes sure that he never loses his partner in an unusual arrangement whereby he literally latches on to her.

Male bearded anglerfish are about 2 cm (3/4 in) long, about a quarter the length of the females. They do not feed, and spend all their time trying to track down the waterborne scent of a female, and to spot her alluring light.

As soon as a male locates a female, he sinks his jaws into her flesh. His mouth eventually fuses with her skin, and their blood supplies connect. For the rest of his life the male exists as a parasite. She supplies his food and he provides her with sperm.

INSEPARABLE PARTNERS *A male bearded anglerfish latches on to his much larger mate. Sometimes females attract two males at a time.*

HUGE MOUTHS AND GIANT APPETITES

In the vast expanses of the deep sea, encounters between animals are rare. When one fish meets another, it has to be ready to eat it, or it risks becoming a meal itself.

This rule of life has led to a deep-sea arms race, with fish evolving larger and larger mouths. The gulper eel, for example, has small fins and a slender tail, but huge jaws that make up more than a quarter of the length of its body. Like most deep-sea fish, it has an extremely elastic stomach, so it can swallow and digest fish almost its own size. To survive, animals like these have to be economical with energy, so their skeletons are light, and their muscles are small. For weeks at a time, they patiently float in the depths, waiting for that all-important moment: a chance meeting with a passer-by.

FISH UNDER PRESSURE

At the bottom of the ocean, the pressure of the water is intense. Just 10 m (33 ft) of water produces as much pressure as the atmosphere, and the pressure increases the farther down you go. At the oceans' deepest point, in the Mariana Trench (*see page 401*), it reaches about 1 tonne per sq cm (14,200 lb per sq in), and can crush submarines. But seabed fish and other animals survive in this pressure unscathed.

Fish can manage this because their bodies are made mostly of fluid and oily flesh which, unlike air, cannot be compressed. They spend the whole of their lives at the same pressure as the water around them, and because the inside and outside pressures are equal, they produce no ill effects.

Some types of deep-sea fish have air bladders, on-board buoyancy tanks that they use to float, and because the air bladders are also under pressure, the fish cannot survive changes in depth. These fish are rarely seen intact, because if they are pulled up to the surface, they burst.

ANGLERFISH'S DEADLY LIGHT

In a world of utter blackness far beneath the sea's surface, light is a precious commodity. It gets you noticed; it can get you a mate; but it can also get you killed. Deep-sea anglerfish are experts at using lights to attract unwitting prey.

Almost spherical, and with vicious teeth, anglerfish have a movable spine protruding from their foreheads. At the tip of the thread is what appears to be a glowing morsel, hanging close to the fish's mouth. If anything swims up to inspect the lure, the bait flicks up, the anglerfish's jaws spring open, and the animal is sucked inside.

Many deep-sea creatures use light as a recognition system, just as surface animals use their markings. In the sea's depths, these eerie displays bring together members of the same species, so that they can breed. Sometimes, as in the case of the deep-sea anglerfish, lights serve more sinister purposes. Most fish eat other fish, so approaching anything that glows can prove fatal.

ALL-ROUND KILLER *Most anglerfish are flattened to help them to hide on the sea floor. This species is a mid-water hunter.*

DEEP-SEA CREATURES THAT TRAVEL AT NIGHT

As night falls at sea, a ship's sonar detects something rising from the seabed, something so large that it dwarfs the ship itself. Eventually, it almost reaches the surface, but several hours later the mysterious object sinks again, and is back in the depths by dawn.

It sounds like science fiction, but this kind of sonar trace is recorded at sea throughout the world. The 'object' is a swarm of zooplankton, and the predators that feed on it, migrating towards the surface after dark.

Two facts explain why small sea animals commute upwards from the depths. Food is much more plentiful near the surface, because this is where tiny floating algae harness the energy in sunshine. During daylight, predators make the surface a highly dangerous place to be. By visiting it by night, these vertical travellers avoid daytime hunters, eating their fill before they return to the sanctuary of the depths.

LIFE IN MINIATURE *A microscopic web of life is adrift in the sea, made up of millions of zooplankton (animals) and phytoplankton (plants).*

NATURE'S MASTERPIECES

Rock gardens under the waves

Like underwater jungles sculpted in living stone, coral reefs are the largest and most spectacular structures ever built by living things. Here, nature's palette seems almost to overflow with colour, from the brilliant hues of the corals themselves, to the vivid markings of coral reef fish. This is a world full of nooks and crannies where animals can hide – one reason why coral reefs are home to a third of all marine fish species.

Corals often look rather like plants, but they are actually clusters of tiny animals called polyps. Rarely bigger than a fingernail, each one protects itself by making a hard case, and these cases build up to form the reef. Reef-building corals contain microscopic algae that harness light, and because of this partnership, the corals themselves need sunshine to thrive. Their beautiful shapes are designed to gather light, while at the same time withstanding the force of the current and the waves. Most corals need warmth, which is why reefs are found predominantly in tropical seas.

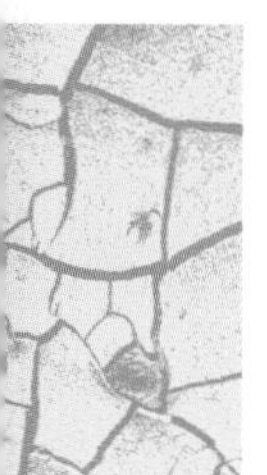

VOLCANIC VENTS

On the floor of the Pacific, near the Galapagos Islands, black, mineral-laden water gushes out of the Earth's crust at temperatures above 400°C (750°F). Yet even this unlikely environment teems with animals, many found nowhere else.

CREATURES THAT LIVE ON DANGER'S DOORSTEP

Seen in the beam of a powerful searchlight, deep-sea vents look like rocky chimneys pouring liquid smoke into the sea. The vent water is so hot it can make temperature gauges melt. Yet only centimetres away from the vent, the water temperature plunges to about 4°C (39°F) – the same as it is throughout most of the deep seabed. Near to this seemingly inhospitable environment vent animals such as giant clams thrive.

Vent animals cannot risk even an instant's contact with the vent water, but they need its dissolved chemicals to stay alive. As a result, they live in the narrow zone around the vent, close enough to use its nutrient chemicals, but not so close that they cook. In this rich micro-environment some animals grow to gigantic sizes. Clams the size of dinner plates gather around the chimneys, each anchoring itself with a fleshy foot. The clams' bodies harbour bacteria that provide them with food.

THE BIG-HEADED FISH THAT HUNTS IN DARKNESS

Since 1977, when deep-sea vents were discovered, more than 300 kinds of animal have been found in these remote habitats. They include sightless shrimps, pale-coloured octopuses, and several bizarrely shaped fish. One of them, the ventfish, from the vents near the Galapagos Islands, has a bulbous head and tiny eyes. But its most unusual feature is the way it reproduces. Instead of laying eggs, like most fish, it gives birth to live young. Unlike clams and giant tube worms, the ventfish does not harbour bacteria, so it has to find food to survive. It probably hunts other vent animals, or lives on their dead remains.

A NEW SEA VENT IS BORN

During a 1991 survey of deep-sea vents in the eastern Pacific, a research team witnessed the creation of a new vent. The sight that confronted the researchers was one of utter desolation: a huge eruption had obliterated most of the creatures that had previously been observed in the area. Dead tube worms and clams littered the sea floor, their carcasses covered in mineral sediment. The only signs of life were clouds of bacteria.

Two years later, the same vent was transformed. New colonies of tube worms were growing fast, while crabs and shrimps picked their way among banks of clams. Even though existing vents are often hundreds of kilometres apart, the animals had found their way to the new vent. Most vent animals produce tiny larvae that drift for long distances through the sea. The larvae develop if they reach a vent.

PARTNERS IN THE DEPTHS

Like a tangle of pipes with fleshy red cowls, giant tube worms are the strangest vent animals. Up to 3 m (10 ft) long, each one is topped by a cluster of about 200,000 bright red tentacles, which act as gills, collecting oxygen from the water. Remarkably, giant tube worms have no mouths, and no functioning digestive system. So how do they stay alive?

The answer lies in the billions of bacteria that thrive inside the worm's body. These process the hydrogen sulphide in volcanic vent water into food, and the worm gets a share in return for giving the bacteria a home.

WORMS WITHOUT MOUTHS *The tube that protects each giant vent worm is made of chitin, which also forms the hard covering of some insects.*

SECRETS OF THE SEABED

For 99.999 per cent of the world's living things, sunlight is the source of life. It provides the energy that makes plants grow, and plants in turn keep animals alive. If the Sun went out, nearly all plants and animals would die out within weeks. Most deep-sea life, fuelled by food that drifts down from above, would also disappear.

Yet around deep-sea vents, life would continue because vent life depends not on solar energy but on chemical energy from deep in the Earth. As long as vent water keeps flowing, bacteria can divide and grow and vent animals can survive. In theory, vent life could continue on its own until Earth cools down and its volcanic activity ceases.

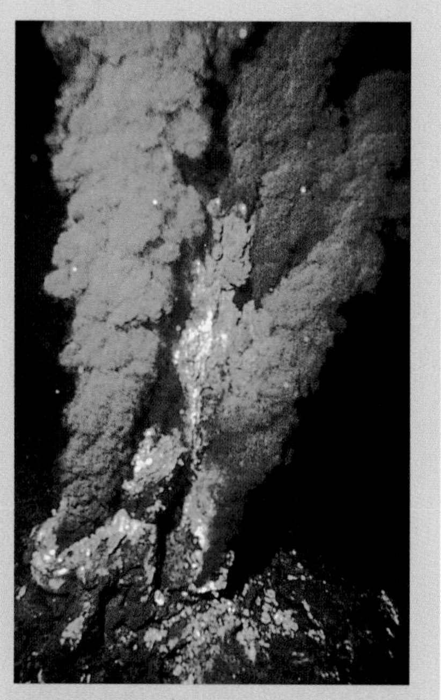

BLACK SMOKER *Mineral-rich water gushes from a vent.*

SURVIVAL ON THE BEACH

With its shifting sand, pounding surf, and swirling tides, the seashore is a demanding habitat. One big challenge is simply staying in place. Yet for some animals, the shore, with its constantly renewed food supply, is an ideal home.

SPEEDY GHOSTS RUN FOR COVER

There are few places to hide on sandy beaches, so small animals need rapid reactions to survive. On tropical coasts around the world lives a wary beach-dweller called the ghost crab. It avoids danger by fleeing and is one of the fastest runners on eight legs.

Always alert to the risk of attack from predatory sea birds, the ghost crab scavenges along the tideline for dead animals washed up by the waves. Compared to other crabs, this one has unusually good eyesight. It can detect small movements and shadows several metres away and is also very sensitive to vibrations in the air and on the sand – which could herald the arrival of a sea bird predator.

If a ghost crab senses trouble it reacts instantaneously. Soon it is speeding across the sand at more than 1 m (3 ft 3 in) per second, dodging and reversing to evade capture.

Ghost crabs often wander above ground for hours, but by memorizing surface features of the beach as landmarks, they always know exactly which way is home. This is important, because if a crab does enter another ghost crab's burrow, it risks getting abruptly ejected – with potentially fatal results.

GHOSTLY GATHERING *Jostling for a share of the feast, ghost crabs pick over the remains of a fish washed up by the tide.*

KEEPING TRACK OF THE TIDE

For a few nights each year, thousands of silvery blue fish can be seen wriggling over the beaches of southern California. The fish, called California grunion, often look as if they are stranded, but within a few hours, they have all disappeared back into the safety of the sea. This remarkable event occurs in the three or four nights following each extra-high spring tide.

The fish come ashore to lay their eggs where they will be safe from swimming predators. Each female lays up to 3,000 eggs, burying them just under the surface of the sand, and as she lays, several males writhe around

FISH OUT OF WATER *The grunion gathering on the Californian coast is quite a spectacle. Tourists are allowed to pick the fish up by hand.*

her, fertilizing the eggs with their sperm. Each fish is out of the water only just a few minutes, and all the egg-laying is completed before the tide retreats. The eggs lie in the sand for at least two weeks, until the next extra-high tide. When sea water moistens the eggs, they hatch in minutes, and the tiny fry swim away.

HOW DO MUSSELS GET A GRIP?

Imagine a glue that sets under water, turning into solid strands that not even the most powerful waves can break. Or one that works like cement, but which is much tougher, and which copes with the stormiest conditions on the seashore. These glues exist, and they are made by mussels and oysters – shore animals found worldwide whose survival depends on staying in one place.

Like many other molluscs, mussels and oysters anchor themselves to rocks. Here they filter food particles from the sea water, opening their shells when the tide is in, and closing them when it falls. They choose their homes as larvae, after an early life spent floating in the sea.

For adult mussels and oysters, getting a firm grip is vital: they cannot feed if they are swept away. Oysters use their 'glue' to fasten one of their two shells to a rock. The glue is so strong that if an oyster shell is broken away, pieces of rock sometimes come with it.

Mussels have a different technique: they secrete sticky threads from a special gland, like fine strands of toothpaste from a tube. When a strand touches a rock it sticks to it, and then hardens, locking the mussel in place.

SURE-FOOTED ROCK CLIMBERS' CHEMICAL 'MAP'

Limpets are famous for sticking to rocks, but they move about to feed. Twice a day, when the tide comes in, they crawl over rocks, grazing on algae.

Although they rarely stray more than 1 m (3 ft 3 in) from their base, limpets have an ingenious way of finding their way back home: they lay a chemical trail that does not wash away. When the tide starts to fall, the limpet simply follows the trail home.

Home, for a limpet, is a niche where its shell and the rock fit neatly together. Without any chinks in its armour, the limpet is safe from attack, and also protected from drying out. However, a limpet on the move can be dislodged from the rock by the tap of an oystercatcher's beak. The bird then feeds on the soft flesh inside.

SHORE PATROL *Hammering with its beak, an American oystercatcher tries to detach a limpet fastened to a seashore rock.*

HARDY THRIFT KEEPS A LOW PROFILE

On the world's stormiest temperate coasts, lighthouses and coastguard stations cope with gale-force winds and incessant salt-laden spray. The buildings do not always survive such harsh conditions, but some seashore plants manage to endure it almost every day.

One of the toughest of these seashore plants is sea pink, or thrift, which has flowers in round clusters, and slender leaves in cushion-shaped mounds. In sheltered places, sea pink can be up to 30 cm (12 in) high, but on coasts, its shape is changed to deal with much more punishing conditions. On clifftops and other rocky places, it grows close to the ground, held in place by woody roots that are like a ship's anchor at sea. By keeping a low profile, sea pink can cope with winds of up to 150 km/h (94 mph), high enough to rip most plants to shreds.

CLIFFTOP COLOUR *Clinging to a steep slope, sea pink creates splashes of colour above rocky shores. It stays in flower from April to July.*

SLIPPERY SURVIVORS' NATURAL WATER STORE

Seaweeds that grow on the shore often spend several hours each day out of the water. To survive, they need to stay moist, because once they dry out they die. One way seaweeds conserve water is by growing in shaded places, but kelps and other brown seaweeds also have a secret weapon: a substance called algin.

Algin is a building material, similar to cellulose, which is found in flowering plants. Cellulose is good at absorbing water, but algin is even better. A single tablespoon of it, added to a litre (1 3/4 pints) of water, is enough to turn the water as thick as honey. With algin in their fronds, brown seaweeds hold on to plenty of water, staying wet and slippery.

BETWEEN THE TIDES *Exposed by the falling tide, brown seaweeds show their tough, rubbery stalks – an adaptation for resisting the waves.*

UNDERSEA FLOWERBEDS

In temperate regions, shallow sea water is home to plants that actually flower beneath the sea. Called seagrasses, these plants are unusual in many ways. Unlike seaweeds, they have true leaves and roots, and they form huge underwater beds that are rooted in sand or mud.

There are hundreds of types of seaweed, but fewer than 50 seagrasses, and they are the only plants in the world that bloom and set seed at sea.

If you go swimming over a seagrass bed, do not expect to be dazzled by their flowers. Because seagrasses live in swirling salt water, they cannot stage a lavish display. Instead, their flowers are tiny, and are often tucked away in pouches inside their leaves.

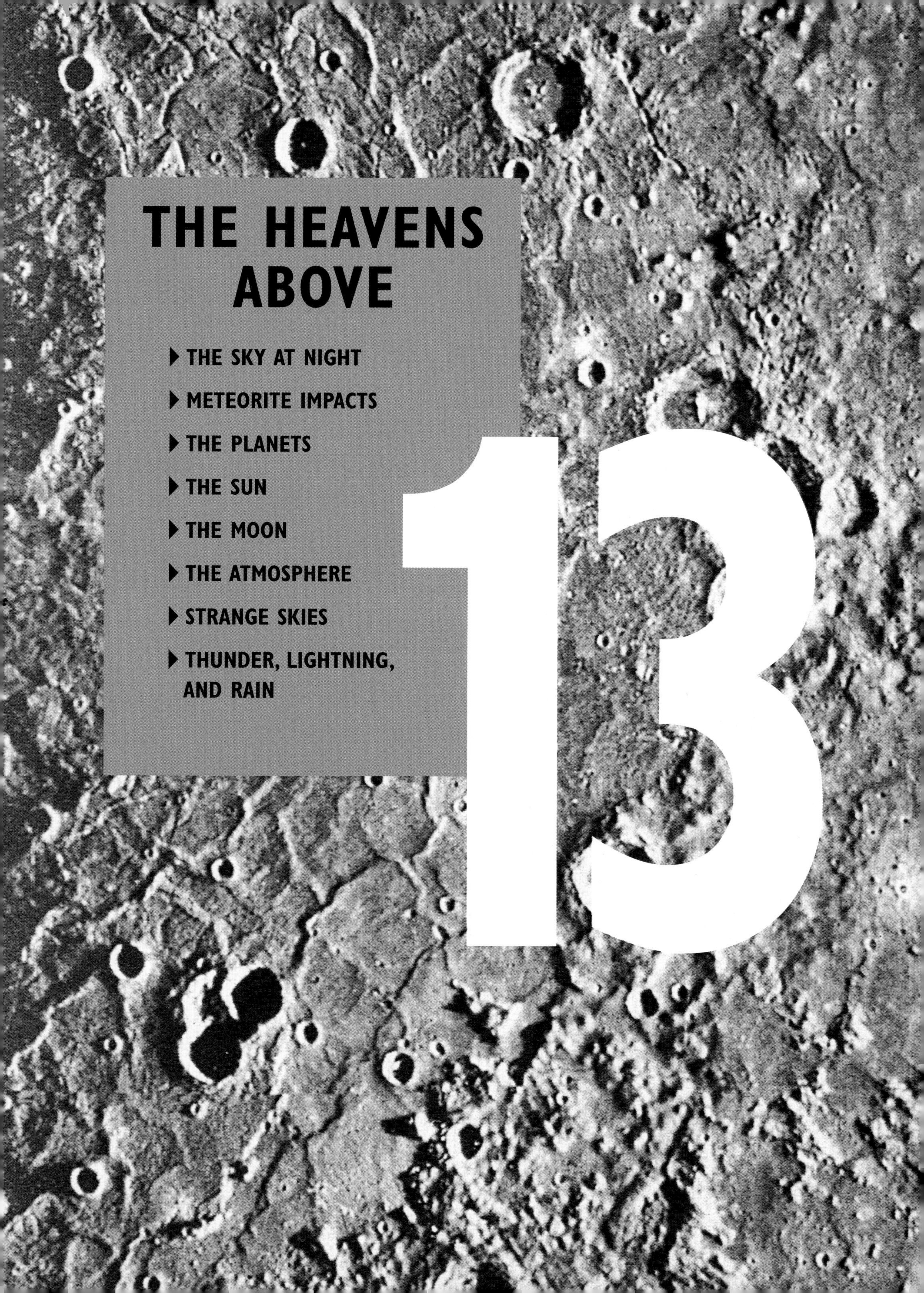

13

THE HEAVENS ABOVE

- THE SKY AT NIGHT
- METEORITE IMPACTS
- THE PLANETS
- THE SUN
- THE MOON
- THE ATMOSPHERE
- STRANGE SKIES
- THUNDER, LIGHTNING, AND RAIN

THE SKY AT NIGHT

Each night the sky is filled with a spectacular light show. Stars shoot, fireworks explode, and comets beam brightly. Many of these phenomena are almost as old as the Universe; some can be explained, others are a mystery.

SMALLEST CONSTELLATION LIGHTS UP SOUTHERN SKY

The Southern Cross is the most notable feature of the night sky south of the Equator. This is despite the fact that it is the smallest constellation in the sky, covering a mere 5 per cent of the area of the largest southern constellation, Hydra.

The Cross is highest on evenings in April and May and is a useful indicator for finding your way around the southern sky. Its long axis points to the south celestial pole directly over the Earth's South Pole, around which the stars appear to turn each night.

The view of the night sky differs between the Northern and Southern Hemispheres, because the Earth orbits around the Sun in a fixed plane. So, for example, the Orion constellation in the northern sky appears the other way up in the south.

Unlike the Pole Star in the northern sky, there is no bright star near the celestial South Pole. But close by is the Octans constellation, shaped like a navigator's octant. Centaurus is a large constellation in the southern Milky Way, depicting a mythical centaur. Its brightest star, Alpha Centauri, is the third brightest in the sky and the closest star to our Sun.

CROSS FROM EARTH *The stars of the Southern Cross are at different distances from our planet. Seen from anywhere other than Earth, they would form a different shape.*

REGULAR JOURNEY OF HALLEY'S COMET

Comets move in huge oval-shaped orbits at up to 20 km/second (12 miles/second). The first person to work out that comets moved to a regular pattern was the astronomer and mathematician Edmund Halley. In 1704 he calculated that one particular comet had passed by the Earth roughly every 76 years, and predicted that it would reappear again in 1758. Unfortunately he died 15 years before his comet came into view on Christmas Eve 1758. It has since been named after him.

SHOOTING STARS MAKE BIGGEST FIREWORK DISPLAY

Every August hundreds of shooting stars streak across the heavens in quick-fire bursts of light – as Earth passes through debris left in the wake of the comet Swift-Tuttle. These are known as the Perseid showers after the constellation Perseus from where they appear to come.

WHAT IS A COMET?

Imagine a giant snowball several kilometres wide and loosely packed with lumps of rock and ice. This is a comet, and there are billions of them tucked away in a ring beyond Pluto's orbit. Every now and again the gravity of a passing star nudges one of these comets deep into the heart of the Solar System, leaving it orbiting around the Sun in great wayward loops.

As a comet draws near the Sun it heats up, boiling off gas and dust in a large head that can be thousands of kilometres wide, and leaving behind a tail of debris, millions of kilometres long.

Even though it seems to be burning itself out in a brief blaze of glory, a comet has enough ice and rock to last for hundreds of journeys around the Sun, unless it is destroyed by crashing into a planet or some other object in space. Comets are among Earth's oldest neighbours in the Solar System. They are frozen relics from 5 billion years ago, before the Sun and planets formed from a large cloud of dust and gas.

These bodies moved through this primitive cloud until they were eventually forced out to the edge of the Solar System by the gravitational force of planets such as Jupiter, Saturn, Uranus, and Neptune.

TAIL BLAZE *Halley's Comet swings past Earth every 76 years or so in a blaze of burning ice and dust. Comets predate the formation of the Sun.*

RARE VISIT *Hale-Bopp was a rare and exceptionally bright comet when it was seen moving through the northern skies in 1997.*

Shooting stars are not, in fact, stars at all. They are meteors – the streaks formed as pieces of celestial debris, often from the trail of a comet, enter the Earth's atmosphere.

JUPITER'S HIDDEN POWER

In 1993 three amateur astronomers discovered a comet hurtling through the Solar System. Little did Eugene and Carolyn Shoemaker from the USA and David Levy from Canada realize that 'Shoemaker-Levy 9' – named for them – would end its days in the most cataclysmic explosion ever seen from the Earth. A year later the comet swung so close to the gravity field of Jupiter that it was torn into pieces and the debris sent round on one last orbit. Then in July 1994, the remains plunged into Jupiter and exploded with more energy than all the nuclear weapons on Earth detonated at once.

ANCIENT VISITOR MAKES SPECTACULAR REAPPEARANCE

When Hale-Bopp streaked through the northern skies in 1997, it was one of the finest comet displays ever seen. It had a brilliant head of dust and gas between 40 and 100 km (25 and 62 miles) across, and a pair of tails, one white and created from dust, the other blue from its burning gas.

Hale-Bopp is named after Alan Hale, a space scientist from New Mexico, and Thomas Bopp, an amateur astronomer from Arizona. By coincidence they both identified the comet at different locations on the same night in July 1995. It could be seen with the naked eye for the next 19 months, setting a record for the length of time a comet could be viewed from Earth in this way. Sightings of Hale-Bopp are rare: the Egyptians saw it last, 4,000 years ago.

STAR PATTERNS ARE SIGNPOSTS IN THE SKY

Long ago, people gave names to the constellations in the night sky that related to the patterns that their stars made. In the Northern Hemisphere, the easiest constellation of stars to find in the night sky is the Plough, or Big Dipper, which also resembles a saucepan with a bent handle.

The seven stars of the Plough are always above the horizon unless you are close to the Equator, and they make a wonderful signpost to other stars and constellations in the crowded night sky. They also form an important aid for navigation.

BIG DIPPER *The stars of the Plough are part of the constellation of Ursa Major, the Great Bear. They are one of the most easily distinguishable formations in the northern sky.*

DOG STAR OUTSHINES THE COMPETITION

Although not the biggest or most fiercely burning of stars, Sirius is the brightest star in the night sky. It is the leader of the Great Dog constellation otherwise known as Canis Major, and because it shines so strongly it is not difficult to distinguish. Sirius radiates 26 times more light energy than the Sun, even though with a diameter of 2.7 million km (1.7 million miles) it is only twice the Sun's size.

The apparent brightness of Sirius is due to it being so close to Earth, less than nine light-years away. Apart from Alpha Centauri at 8.7 light-years away, it is the closest of all the major stars and is approaching Earth at 8 km/second (5 miles per second).

Sirius also appears to twinkle, hence its Greek name, meaning 'sparkling'. This effect is caused by the Earth's unstable atmosphere.

Egyptians worshipped Sirius as the Nile Star, because its first appearance in the dawn sky every year marked the annual flooding of the Nile, on which crops in the fields depended.

LUNAR ECLIPSE MAKES THE MOON BLUSH

About every six months, a shadow passes over the Moon. Usually it only partially darkens the Moon, but sometimes it almost blots it out completely. This is a lunar eclipse. It occurs when the Earth passes between the Moon and Sun, blocking the path of light from the Sun to the Moon's surface.

One of the curious features of a lunar eclipse is that the Moon appears to turn red. This is caused by the Earth's atmosphere bending the rays of sunlight that are illuminating the Moon's surface – in the same way that a simple lens or prism bends light. Unlike a solar eclipse, when the Moon passes in front of the Sun for only a matter of minutes, a lunar eclipse can last for up to 2 1/2 hours.

MOON SHADOW *During a lunar eclipse, seen here in stages, the Earth passes between the Sun and the Moon, casting a shadow over the Moon's surface.*

BEAMING LIGHT ACROSS THE UNIVERSE

In the night sky, you can often spot very bright points of light. These are quasars, the brightest objects in the Universe. A quasar can give out the same energy as hundreds of galaxies, although it is not much bigger than our Solar System. This is equivalent to a torch shining as powerfully as all the lights in London.

How could something so small give out so much energy? It is thought that each quasar is a galaxy with an exceptionally massive black hole at its centre, giving out stupendous amounts of energy as it gobbles up millions of stars and clouds of gas. Quasars are some of the most distant objects in the Universe from Earth. Light from them takes billions of years to reach us, although they blast through space at enormous speeds, in some cases at more than 270,000km/second (168,000 miles per second), or 90 per cent of the speed of light.

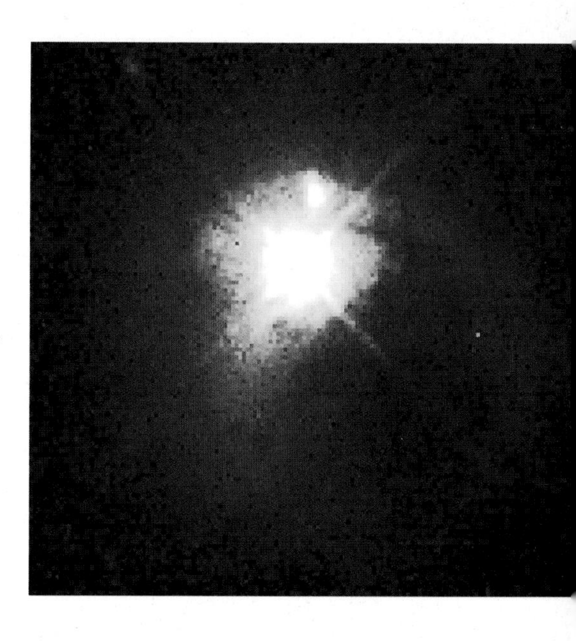

BRIGHT SPOT *Quasars were first identified in the late 1950s. Quasar is an abbreviation of quasi-stellar object: an object masquerading as a star.*

THE VAST COSMIC CLOUD THAT IS OUR HOME

On a dark night the Milky Way appears as a band of misty cloud, studded with more than 200 billion stars: this is our galaxy. Our closest star neighbours are four light-years away (their light takes four years to reach us); those at the end of the galaxy lie 2,500 light-years distant.

Some of the stars are moving across the Milky Way like a school of fish, clustered together in an embrace of gravity that was established after they formed from the same clouds of dust and gas billions of years ago. These clusters will eventually disperse, but other groups, formed before the Milky Way took shape, are permanently clumped in a halo.

METEORITE IMPACTS

Asteroids and meteorites are lumps of debris that blast through the Solar System like wayward bombs. When one hits Earth it can explode with immense force, producing clouds of dust that cause catastrophic climatic changes.

MISSILE BLASTS A GIANT HOLE IN THE DESERT

Just off Highway 99 in Arizona's Painted Desert lies a strange round pit nearly 180 m (600 ft) deep and 1.2 km (3/4 mile) wide. It was not created by an earthquake or quarried out by people – it was made by an extraterrestrial body. Some 50,000 years ago a meteorite from space penetrated the Earth's atmosphere and smashed into the ground, gouging out a crater.

Now known as Meteor Crater (although a more correct name would be Meteorite Crater, as the lump of rock that created it was a meteorite not a meteor), it is the best preserved example of its kind left on Earth.

The meteorite responsible for it was rich in the metallic element iron. Meteorites can also be stony, sometimes containing organic substances and even microscopic diamonds and rubies. They are formed from matter left over after the creation of the Solar System, or from bits of planets that have been blasted into space by collisions with comets or asteroids.

HUGE HOLE *Meteor Crater in the Arizona desert, USA, was caused by a meteorite colliding with Earth some 50,000 years ago.*

BALL OF FLAMES SENDS SHOCK WAVES AROUND EARTH

On June 30, 1908, a ball of flame, brighter than the Sun, was spotted over Siberia. Soon afterwards there was a massive explosion, which sent shock waves around the world. A meteorite or comet had exploded with the force of a nuclear bomb, about 10 km (6 miles) above the surface of the Earth. The explosion struck over uninhabited Siberian forest. The only damage was to the trees, which were flattened within a radius of 32 km (20 miles).

MASSIVE ASTEROID THAT KILLED THE DINOSAURS

There is evidence that some 65 million years ago an asteroid tore into the Earth. Travelling at more than 200 times the speed of sound, the asteroid – which was about 12 km (7 1/2 miles) wide – displaced 50,000 km^3 (12,000 cu miles) of rock.

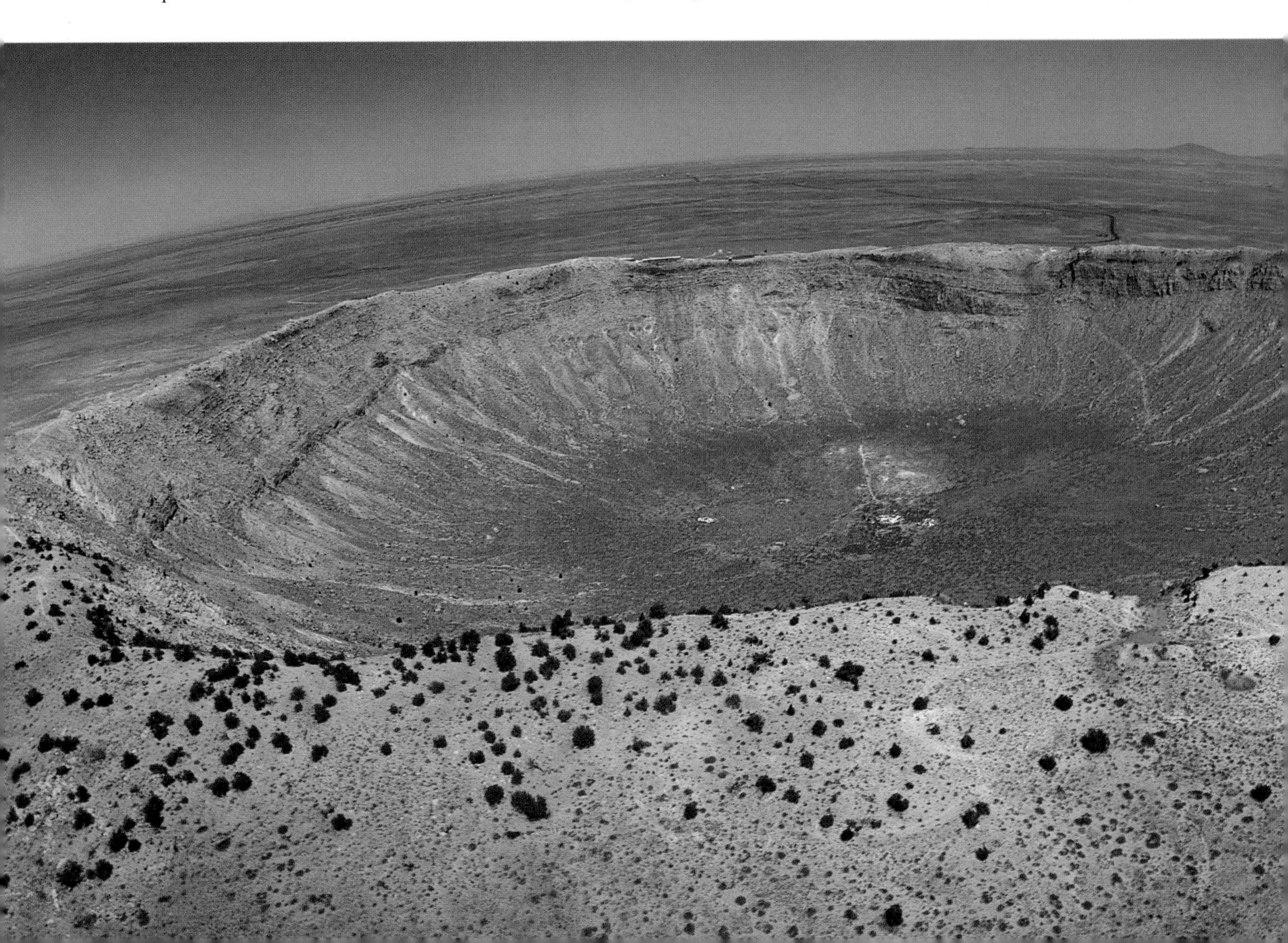

DEEP IMPACT *Mexico's Chicxulub Crater was created by an asteroid. The devastating effects of its impact almost certainly contributed to the demise of the dinosaurs.*

The scar that it left is now buried beneath sediment in the Yucatan Peninsula in Mexico, in the 195 km (120 mile) wide Chicxulub Crater.

This collision triggered a firestorm and a devastating fast-moving ocean wave called a tsunami (*see page 438*) that was several kilometres high. Billions of tonnes of debris blew round the world in a thick cloud, shrouding the Earth in darkness for a month. World temperatures fell and huge amounts of sulphuric acid in the air burnt the ground.

As the air cleared, temperatures soared, because carbon dioxide left over from the explosion set off an intense greenhouse warming effect which lasted for centuries. Plants withered and animals starved to death. Almost 90 per cent of all the animal species on Earth became extinct. The most famous casualties of this devastation were the dinosaurs.

THE BIGGEST METEORITE ON EARTH

Beside a dusty road in Grootfontein, Namibia, lies a lump of rock as big as a car. Only about a metre (3 ft 3 in) is visible above the ground; the rest is buried beneath. This is the largest known intact meteorite on Earth, composed mainly of iron and nickel, and weighing more than 60 tonnes. It lies where it fell thousands of years ago.

Since there is no trace of a crater, it is thought that the surrounding area has slowly eroded and the meteorite is the only evidence of the impact.

The world is constantly being bombarded with meteorites. Most of these missiles fall harmlessly to Earth as dust or small stones. It is rare for large meteorites to penetrate the atmosphere and hit the ground, and even rarer for one to survive intact without being obliterated on impact.

DOUBLE ASSAULT ON THE EARTH'S CRUST

NASA's Langley Research Center in Hampton, Virginia, sits on the edge of an underground crater 82 km (51 miles) wide. It was created 35 million years ago by a meteorite crashing into Earth in probably the most traumatic geological event to hit what is now North America.

It is thought that the missile sliced through water at 80,500 km/h (50,000 mph), instantly vaporizing the rocks and sediments up to 1 km (1/2 mile) below. Seconds later an enormous column of water and dust mushroomed like an atomic bomb blast. Dust particles from the eruption have been found as far away as the Indian Ocean. Everything within 1,000 km (600 miles) of the impact would have been destroyed.

Only a few thousand years later, another meteorite blast hit Popigai in northern Siberia. The two impacts created a blanket of dust in the atmosphere that blocked the Sun's warmth and triggered global cooling for the next 100,000 years.

SPACE ROCK *Just the tip of the world's largest meteorite can be seen in Namibia, south-western Africa. It hit the Earth around 80,000 years ago.*

THE SOLAR SYSTEM

The Solar System is like a tiny oasis in space, a rare planetary system in a galaxy where relatively few other planets are known. At the hub is our star, the Sun, around which the nine planets and their moons orbit. It is the powerful gravitational pull of the Sun which choreographs the system into a vast spinning whirl.

The Solar System may have been created from the debris of an old star which exploded, blasting out a spinning mass of gas and dust. About 4.6 billion years ago, this vast hot cloud of debris pulled together into the Sun and the leftovers spun round in a disc from which the planets formed.

The four innermost planets – Mercury, Venus, Earth and Mars – are small, solid and rocky. Jupiter, Saturn, Uranus and Neptune are gas giants, composed mainly of hydrogen and helium, each with a small solid core and a vast atmosphere of liquid or solidified gases. Pluto is the odd-man-out, a small frozen oddity that has a different orbit from the other planets. Some astronomers believe that Pluto may, in fact, be a rogue asteroid rather than a proper planet.

Lying between Mars and Jupiter there is a belt of tens of thousands of rocky asteroids, so large they are sometimes called minor planets by astronomers. The biggest one, Ceres, has a diameter of 947 km (605 miles).

Beyond Pluto lie billions of comets, circling on the outer fringes of the Solar System.

HEAVENLY BODIES
The largest planet in the Solar System is Jupiter, with an equatorial diameter of 143,884 km (89,408 miles) – just over 11 times that of Earth. The mass of all of the nine planets put together adds up to less than 1 per cent of that of the Sun.

THE PLANETS

All Earth's neighbours circle around the Sun. The planets closest to the Sun are deserts, those most distant are mainly giant balls of gas. Life exists on Earth because it has liquid water, but is there water and life on Mars?

IS THERE EVIDENCE OF LIFE ON MARS?

The planet Mars has much in common with Earth. It has a day of just over 24 hours, an axis tilted at nearly the same angle, and yearly seasons. Most intriguing of all, Mars also has water in the form of polar ice caps consisting of water and frozen carbon dioxide. On Earth, bacteria survive under deep glaciers in Antarctica, and so it is possible that primitive life could thrive on Mars.

Evidence provided by photographs taken by a spacecraft in 1999 shows highland terrain etched with a network of small channels, reminiscent of streams on Earth, which indicates that there could have been flowing water on the planet more than a billion years ago, when the climate would have been warmer.

WATERY PAST? *Water, the key to life, might once have flowed through these 'river beds' on Mars, and even now may still lie deep below.*

VENUS, PLANET OF WIND, FIRE AND VAST VOLCANOES

Venus is a planet of volcanoes with huge cones, collapsed craters and volcanic plains. In the early life of the Solar System, when the Sun was pumping out 25 per cent less energy than it is now, Venus was a cooler place with oceans and an atmosphere like that of Earth's today.

As the energy output of the Sun slowly increased, Venus began to heat up. Its seas evaporated into water vapour, which acted as a greenhouse gas, trapping the Sun's heat and pushing the temperatures up still further. The oceans boiled away, releasing their carbon dioxide and accelerating the greenhouse effect. Today the atmosphere on Venus is composed of carbon dioxide and yellow clouds of sulphuric acid. Winds roar around the planet at 350 km/h (217 mph) – three times the speed of a hurricane on Earth. No recognisable form of life could survive in these conditions.

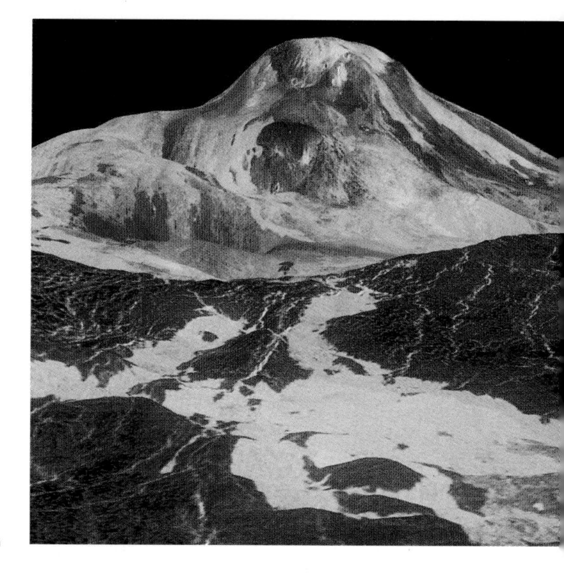

UNDER THE VOLCANO *Maat Mons, Venus, one of numerous volcanoes that dot the plains and pump out noxious clouds of sulphur.*

QUIET MERCURY HARBOURS A TRAUMATIC PAST

Rocky Mercury, the nearest planet to the Sun, is an inert planet now. But for its first 600 million years it was continually blasted by comets and asteroids. During this time, its interior melted and the planet grew. Volcanoes belched millions of tonnes of lava onto the vast plains around the craters.

Around 4 billion years ago an asteroid 100 km (62 miles) wide tore into its surface creating the Caloris Basin, which is 1,300 km (810 miles) across. It was one of the largest impacts in the history of the Solar System and sent such violent shock waves around Mercury that the crust of the planet buckled into hills and depressions. Temperatures here soar to 430°C (800°F) in the day at the equator and plummet to –185°C (–300°F) at night.

JUPITER'S VEIL OF MULTICOLOURED CLOUDS

As the Sun's light filters through the atmosphere of Jupiter it creates a stunning spectrum of reds, oranges, blues, and browns. Looking through the light and dark bands of dense clouds that cover the surface of the planet, certain areas or shapes are particularly noticeable: the most obvious of these is the Great Red Spot.

A colossal red oval larger than the Earth, the Great Red Spot has been observed for more than 300 years. It is the product of a vast high pressure system, sending out turbulent cloud patterns in its wake.

The vivid red colour probably comes from phosphorus, which lies close to the core of Jupiter, and is constantly being forced upwards into the atmosphere in vast plumes.

Other notable shapes among Jupiter's cloudy cloak are three large white ovals that drift slowly eastwards and some smaller white oval clouds; both are about 50 years old. Most of the other numerous cloud clusters or spots are short-lived, some lasting only a few hours.

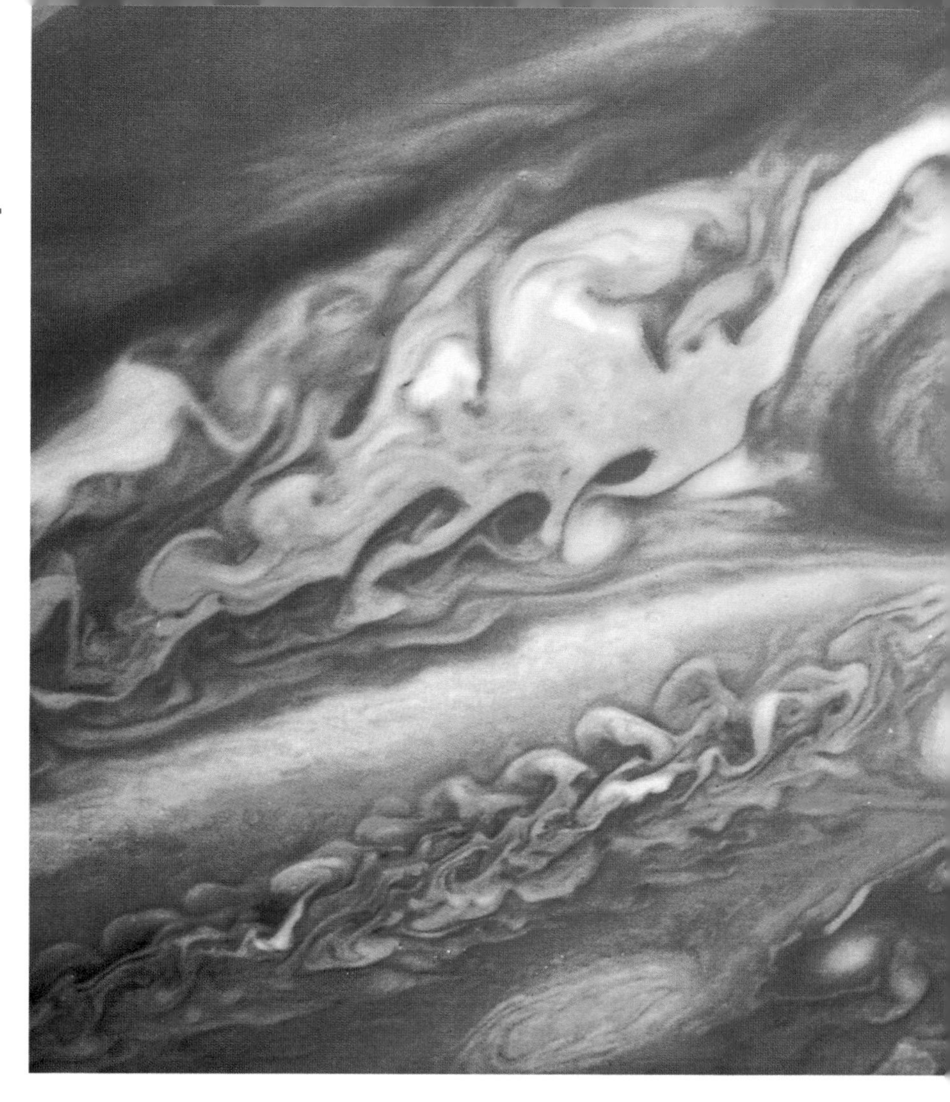

STORM CLOUDS *Jupiter's Great Red Spot is the biggest storm in the Solar System. In the eye of this continuous cyclone, winds can reach as much as 435 km/h (270 mph).*

PLUTO, THE FROZEN OUTCAST

An oddball among our planets, Pluto lies in the cold wastes of the outer Solar System. It follows an orbit unlike any other.

This small frozen rock rises and falls below the plane of the other planets and swings wildly through the Solar System in a great ellipse that for 20 years of its 248-year orbit around the Sun brings Pluto closer to the Sun than Neptune, its nearest neighbour.

Pluto's only moon is called Charon and is about half its size at 1,270 km (790 miles) wide. Charon is so large in relation to Pluto that some astronomers believe that the two bodies may, in fact, be a double planet, spiralling around each other and even sharing each other's thin atmosphere of nitrogen and methane.

SECRET OF THE CIRCLES THAT SWIRL AROUND SATURN

Saturn's glorious rings are made of nothing more than dust and numerous lumps of ice. Ranging in size from tiny particles to pieces about 8 m (25 ft) across, it is rather like having a mass of dirty snowballs whirling around the planet.

There are three main rings, extending to a diameter of 270,000 km (168,000 miles). They are only about 30 m (100 ft) thick and were created either by a cloud of wayward particles

BAND WIDTH *Saturn's rings measure 40,000 km (25,000 miles) from their inside to outside edges – more than three times the diameter of the Earth.*

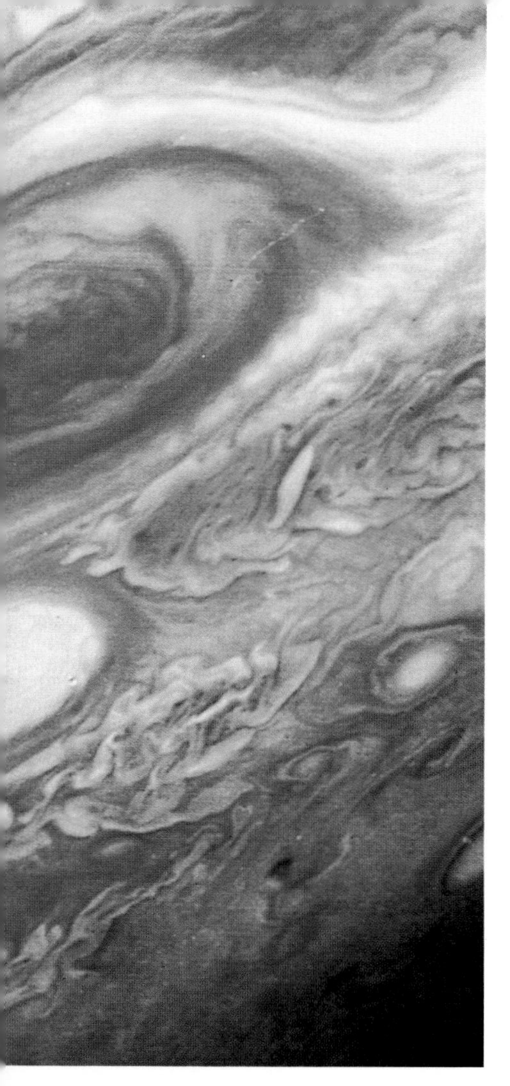

that came from the break-up of a moon, or from debris that did not coalesce to form a moon. Over time the cloud of colliding particles would have been bumped and shoved into a flat plane, forming a ring.

The Voyager 1 and 2 satellites of the 1980s revealed radial 'spokes' appearing and disappearing from the rings. It is thought that Saturn's magnetic field charges up dust particles with static electricity, causing dusty clusters to rise up temporarily, so creating the spoke effects.

THE BIG BLUE WORLD OF URANUS

Uranus appears blue-green from Earth because of light refracted by methane in its atmosphere. Temperatures in the upper atmosphere are so cold that methane condenses and forms a thin layer of clouds that lie above the other layers.

Miranda, one of the 11 moons of Uranus, has a landscape quite unlike any other in the Solar System. It features three enormous blots ranging from 200 to 300 km (125 to 185 miles) across of concentric circles and grooves, as well as a cliff plunging 15 km (9 miles). All these 'blemishes' may have formed when the moon was shattered by a mystery missile.

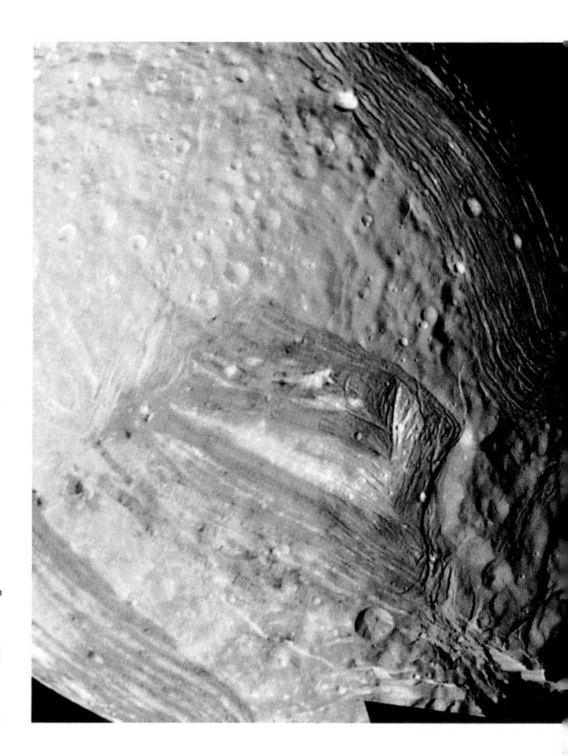

NEW MOON *Miranda, which orbits Uranus, may have been smashed to pieces in a collision then reassembled by gravitational attraction.*

NEPTUNE, PLANET OF THE WINDS

Howling bands of winds rip round Neptune's atmosphere at up to 2,000 km/h (1,200 mph). These storms and hurricanes, the most ferocious of all the planets, appear as huge dark spots. The powerhouse for this weather comes not from the Sun – Neptune receives only a tenth of 1 per cent of the sunlight that reaches Earth – but from heat deep inside the planet.

One of Neptune's moons, Triton, features extraordinary geysers that emit fountain-like plumes of nitrogen 8 km (5 miles) high. Triton's surface seems to be frozen with water ice. As the ice has melted on occasions it has collapsed, carving out an erratic landscape of pits and fissures crossed by ridges and vast smooth plains.

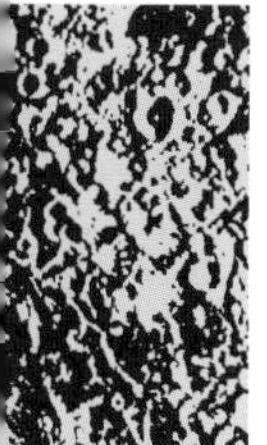

THE SUN

At the heart of the Solar System lies the Sun, a titanic furnace fuelled by a thermonuclear reactor. It is a star with sudden and violent tantrums, that can upset the delicate balance of life on Earth and cause chaos out in space.

THE GREAT NUCLEAR REACTOR IN THE SKY

The Sun makes a hundred million times more energy than all nine planets together. Yet it is only a huge, fiery ball of gas, composed mostly of hydrogen like other stars.

The secret to the Sun's power is a gigantic nuclear reactor buried deep inside its core, where atoms of hydrogen are squeezed together under tremendous pressure to form heavier helium atoms. This example of nuclear fusion turns 5 million tonnes of matter into energy every second, driving up the temperature in the Sun's core to 15,000,000°C (27,000,000°F).

The phenomenal power inside the Sun blasts out at its surface as visible light and heat and bathes the Solar System in energy. Even though Earth receives only a billionth of the total energy output of the Sun, it is sufficient to sustain life in the world.

SOLAR WIND *A stream of charged particles – the solar wind – flows from the Sun. Here, it is most intense where the image is brightest.*

SOLAR ERUPTIONS WREAK HAVOC IN SPACE

The Sun's surface can rip open and its innards burst out in torrents of electrical particles. These outbursts are called coronal mass ejections. They shoot out up to 10 billion tonnes of highly charged matter at 60,000 km/h (37,000 mph). These are the biggest explosions in the Solar System, approaching the power of one billion hydrogen bombs. When a solar explosion hits Earth's atmosphere it can disrupt power, radio, and satellite communications, and cause storms in the upper atmosphere, which create spectacular aurorae (*see pages 378–9*).

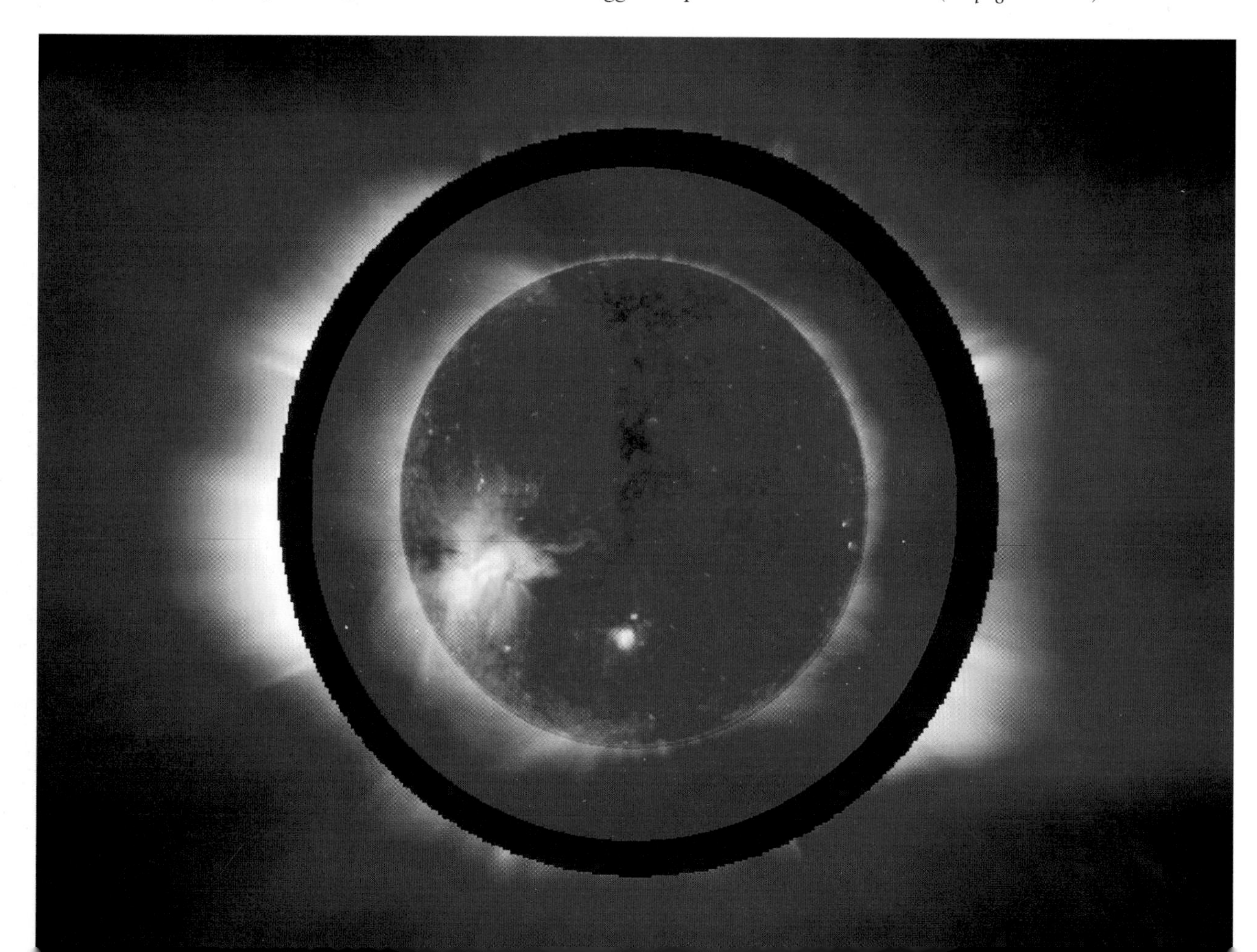

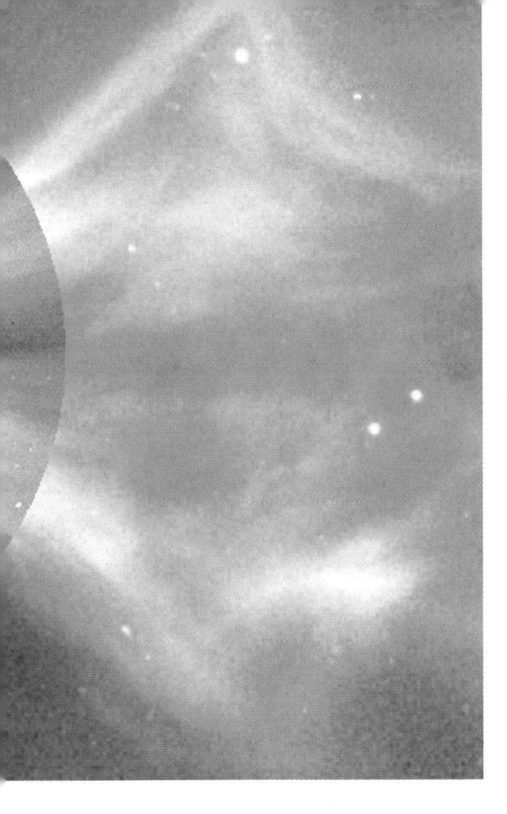

SUN BURSTS *This composite image of a large coronal mass ejection shows the blast of electrical particles (white) emerging from the Sun (blue circle).*

SOLAR BLEMISHES MAY AFFECT EARTH'S CLIMATE

Dark spots sometimes pass across the face of the Sun. Called sunspots, they are relatively cool areas, a mere 2,000°C (3,600°F) compared with 5,500°C (10,000°F) over the rest of the Sun's surface. Sunspots are created when magnetic lines just below the Sun's surface become twisted and block the energy surging up from within.

There is controversy over whether sunspots affect weather on Earth. From 1645 to 1715 sunspots virtually disappeared and this coincided with the Little Ice Age, when winters were bitterly cold. The high numbers of sunspots in recent years could explain the rise in global temperatures.

THE LAND OF THE MIDNIGHT SUN

Summer is like one endless day in the polar regions of Earth. This is the phenomenon of the Midnight Sun, a cause for celebration among the inhabitants of these regions and an antidote to the total darkness of winter; a time when flowers bloom in the tundra and animals breed.

The effect is caused by the position of the Earth in relation to the Sun. The Earth circles the Sun tilted at an angle of 23.5 degrees, so that when the Northern Hemisphere tips towards the Sun the extreme northern regions are continuously bathed in sunlight. At the same time, the extreme Southern Hemisphere is in continuous darkness.

BLOCKING OUT SUN TO PRODUCE DARKNESS BY DAY

The sky turns dark, stars and planets appear, and it seems that night has descended early. Yet a few seconds or minutes later daylight returns as if nothing has happened. This is a solar eclipse. Total eclipses occur every 18 months when the Moon passes in front of the Sun and casts a shadow across the Earth. At the point where the Moon blocks out the Sun, all you can see around the Moon's dark shadow is the Sun's outer atmosphere, known as its corona.

THE MOON'S SHADOW *A solar eclipse, when the Moon passes between Earth and the Sun, offers a chance to see the Sun's corona.*

THE SUN: FACT FILE

- DISTANCE FROM EARTH: 149,600,000 km (92,960,000 miles).
- TIME FOR SUNLIGHT TO REACH EARTH: 8 minutes 18 seconds.
- DIAMETER: approximately 1,392,000 km (865,000 miles) – 109 times greater than the Earth.
- CIRCUMFERENCE: approximately 4,373,000 km (2,717,000 miles).
- MASS: 2 billion billion billion tonnes – 330,000 times greater than that of the Earth. The Sun makes up 99.8 per cent of the mass of the Solar System (Jupiter accounts for most of the rest).
- ROTATION: about 25 days at Equator to 36 days near poles. This is a result of its gaseous composition.
- TEMPERATURE AT SURFACE: 5,500°C (9,900°F).
- TEMPERATURE AT CORE: 15 million°C (27 million°F).
- SURFACE GRAVITY: 38 times that of Earth.
- AGE: approximately 4.6 billion years.
- EXPECTED LIFE SPAN: probably another 5 billion years before it starts to die.
- CHEMICAL COMPOSITION: 92.1 per cent hydrogen, 7.8 per cent helium; the rest is made up mainly of oxygen, carbon, nitrogen, and neon.
- CLASSIFICATION: G2 star – relatively small, yellow.

THE MOON

Our closest neighbour in space, the Moon, was born out of a gargantuan interplanetary collision, and has been blasted by space rocks ever since. Recent exploration in space has started to throw light on many of the Moon's mysteries.

MYSTERY OF THE DARK SIDE OF THE MOON REVEALED

We only ever see one side of the Moon from Earth. The Moon's other face – known as the 'dark side' – is always turned away from our gaze.

The Moon spins on its axis as it orbits the Earth. In the early days after its formation, the Moon was spinning faster than it does now, but Earth's gravity was so strong that it slowed the Moon's spin to the point where its spin and its orbit round Earth became locked together. It now takes the same length of time – 27.32 days – to complete a rotation and to orbit Earth once. This keeps the same side facing Earth all the time. And because the same side always faces us, the Moon appears not to be spinning at all.

In October 1959, the dark side of the Moon was finally revealed when the Soviet spacecraft, *Luna 3*, sent back images of it. These showed a landscape with many more craters and fewer 'seas' than the side facing Earth.

TWO SIDES TO THE MOON *Spacecraft images of the 'dark side' of the Moon show a landscape with similar features as the side visible from Earth.*

HOW THE MOON WAS FORMED

After centuries of speculation, lunar rocks brought back to Earth by the six manned Apollo missions between 1969 and 1972 finally solved the mystery of the Moon's origin. The rocks were found to be surprisingly similar in age and content to those on Earth, indicating that the Moon must have once been a part of our planet.

It is thought that some 4.5 billion years ago a wayward planet, roughly the size of Mars, smashed into the Earth so violently that it blasted out a shower of molten rock and sent it spinning around Earth. This cooled and coalesced into a solid moon, which was held close by the pull of gravity.

That the colliding planet itself was totally destroyed in the process and its contents injected deep into the core of the Earth might also explain the high iron content found there. In fact, you could think of the Earth as two planets merged into one – with a bit left over to form the Moon.

LUNAR SECRETS *Seismic experiments on the Moon have revealed that it is solid to a depth of around 1,000 km (600 miles).*

THE LARGEST HOLE IN THE SOLAR SYSTEM

The deepest basin in the Solar System can be found near the Moon's southern pole. The size of Alaska, the Aitken Basin is more than 2,500 km (1,550 miles) wide and at least 12 km (7½ miles) deep. It was discovered in the mid 1960s by the Lunar Orbiter satellite.

The basin even contains a hint of water. The results of a recent satellite mission suggest that ice might be tucked away in its deep, dark shadows.

Craters pepper the Moon because, unlike Earth, it has no atmosphere to burn up incoming comets or meteorites before they hit the surface. The craters, once formed, remain intact, as there is no water or weather to erode them away.

THE ATMOSPHERE

The Earth's atmosphere is wrapped around the planet like layers in an onion. Without it life as we know it would be impossible. Each layer shields us from harmful cosmic forces such as the Sun's rays and meteorite bombardments.

WHERE THE WEATHER HAPPENS

If you climbed to the top of Mount Everest, you would be at the ceiling of the troposphere. This is the thin layer of atmosphere closest to the Earth where almost all our weather takes place.

The troposphere extends from 8 to 16 km (5 to 10 miles) above the Earth's surface and is composed largely of nitrogen, oxygen, and water vapour, with much smaller traces of other gases such as carbon dioxide.

All these gases help to trap heat around the Earth and keep the surface warm. The heat of the Sun sends the gases into turmoil. Warm and cold air battle across the globe in the form of winds and storms. Water vapour rises into the troposphere and condenses into clouds, then falls as rain, sleet, hail, and snow.

Increasing levels of carbon dioxide and methane in the atmosphere, caused by industrial pollution, are now trapping too much heat in the troposphere creating global warming.

THE EARTH'S MULTILAYERED ATMOSPHERE

The atmosphere that surrounds our planet is made up of several different layers. Above the troposphere (*see above*) lies the stratosphere extending up to 50 km (30 miles). This region holds the ozone layer which shields Earth from the worst of the Sun's ultraviolet rays. Next, the mesosphere extends to 85 km (53 miles) from the Earth's surface. It is the coldest part of the atmosphere at –93°C (–135°F). Here intense friction causes meteors to burn up into shooting stars.

The thermosphere extends to 193 km (120 miles). Temperatures here rise dramatically to 1,727°C (3,140°F). This region is also called the ionosphere because the Sun's rays break molecules into positively and negatively charged units called ions, that can bounce radio communications around the world.

The exosphere is the outermost layer of the atmosphere, reaching out 960 km (600 miles) from the Earth. It is the transition zone into space. The atmosphere is very thin in gases, because a lack of gravity allows molecules to escape into space.

ATMOSPHERE ANATOMY *Weather forms in the layer known as the troposphere, which contains most of the atmosphere's mass, including nearly all its water vapour and dust.*

STRANGE SKIES

The sky is often a theatre for spectacular natural light shows. There are rainbows, ghostly mirages conjured up by the clouds, and sometimes the stunning spectacle of a ring around the Moon, created by high-altitude ice crystals.

THE MULTICOLOURED ARC OF HOPE

The ancient Norse people believed that a rainbow was the 'bridge of the gods', connecting heaven to Earth. Many other cultures have regarded rainbows as a symbol of peace and hope.

Rainbows form when beams of sunlight break through a shower of rain. Each raindrop acts as a miniature prism, bending and slicing the white light of the Sun up into the colours of the spectrum – ranging from violet to deep red.

The great wonder is how countless numbers of raindrops can project the spectrum of colours into a vast arc across the sky. The answer is by reflection: once the sunlight is split into its constituent colours, the back of each raindrop reflects these, like a curved mirror would, into a 42 degree arc.

Occasionally a pair of rainbows occurs when sunlight is reflected twice inside the raindrops.

CHASING RAINBOWS *A pot of gold is said to lie where a rainbow meets the Earth but since it is an optical effect the spot is impossible to find.*

HEAVENLY HALOES ARE A SIGN OF RAIN

On fine, still days when the sky is filled with high, feathery clouds, the Sun appears to have a halo around it. The same type of halo sometimes appears around a bright full moon. Haloes are made by the high cirrostratus clouds that often precede rain or snow. Indeed, when the Zuni Indians of New Mexico saw the Sun 'inside its wigwam' they forecast rain.

Haloes are created if the clouds are high enough for their water to freeze and form ice crystals. Water crystals can take on various shapes and sizes, but if they form hexagonal cylinders, less than 0.1 mm (4/1,000 in) wide, they bend sunlight by 22 degrees. With countless tiny crystals lined up in the same direction, the combined effect is a bright ring around the Sun or Moon.

RARE SUMMER CLOUDS DRIFT HIGH IN THE SKY

Noctilucent, or night-shining, clouds are the highest and rarest clouds in the world. They sit about 80 km (50 miles) above the Earth's surface, more than seven times higher than any other cloud.

Appearing in summer only, these ghostly silvery white threads tinged with metallic blue can be seen around twilight at latitudes greater than 45 degrees – mostly in Canada, northern Europe, and Asia.

Noctilucent clouds are made either by burnt-out meteors leaving a trail of tiny smoke particles which collect ice crystals, or from methane, a product of pollution. This might explain why noctilucent clouds were first observed in 1885 and why they have become increasingly common.

SOOT CREATES A BLUE MOON

One of the most spectacular blue moons was seen across northern Europe on the night of September 20, 1950. It was caused by fine sooty particles in the atmosphere that had been swept over the Atlantic from a huge forest fire in Canada. The particles diffracted the bright moonlight, making it appear blue when viewed from the ground.

Blue moons and suns were also seen widely after the volcanic eruption on Krakatoa, Indonesia, in 1883. Depending on the size of the bits of soot, a full Moon can appear bright pink or orange, too.

SUNDOGS MAKE TOO MANY SUNS IN THE SKY

When the Sun is low in the sky and veiled in high clouds, a pair of bright spots often appears on either side. They look like ghostly suns and are known as sundogs. The name probably derives from the long horizontal ray of white light that sometimes sticks out at one end of them, like a dog's tail.

Sundogs, like haloes (*see page opposite*), are created when sunlight bends through ice crystals of a certain shape that form in high clouds.

The sight of what appear to be extra suns has been known to cause panic. In a battle during England's Wars of the Roses, in 1461, the Yorkist army was alarmed to see three suns in the sky. Their leader, Edward, Duke of York, took the sight as a good omen and rallied his troops to victory. When he was crowned Edward IV, he adopted the emblem of three suns in the sky for his personal coat of arms.

SECOND SIGHT *Sunlight diffracted by ice crystals in high clouds creates an image of a second or even third sun, known as a 'sundog'.*

SPECTRES OF THE MOUNTAINTOPS

Many of the first climbers on Brocken in Germany's Harz Mountains were dumbfounded to see huge ghostly images. These giant 'ghosts' became known as the Brocken Spectre, and were usually seen near the summit in the early morning mist, or sometimes just before sunset.

These mirages occur when shadows of the mountaineers are projected on to low clouds, then reflected back by the water droplets in the mist, each droplet acting like a tiny curved mirror. With the perspective of the light and cloud, the shadow looms up into a ghoulish giant. The droplets can also produce brilliantly coloured rings or haloes around the shadow.

BROCKEN SPECTRE *The giant ghoul playing over the mountaintop is a trick of light and cloud reflecting the climber's image.*

COLUMN OF LIGHT *A sparkling finger of light, known as a sun pillar, is created by ice crystals in the air reflecting light like pieces of tinsel.*

THE DIAMOND SHOWER THAT FALLS FROM THE SUN

On a bitterly cold, crisp day, you might be able to see what looks like a flickering shower of tiny diamonds plunging down from the Sun. This exquisite spectacle is known as a sun pillar.

A sun pillar is created when the temperature falls low enough for tiny ice crystals to form in the lower atmosphere. These perfectly horizontal crystals then gently drop through the skies and, as they do so, they reflect the sunlight into a single shaft of brilliant white light.

Often the vertical pillar of light can be seen above the rising or setting Sun, when it will be a blaze of yellow, orange, or red. It is best to view this sight when the Sun is hidden behind a building or a hillside so that the eye is not too dazzled.

WHEN THE AIR PLAYS TRICKS AND SEEING IS BELIEVING

In 1957, passengers aboard the liner *Edinburgh Castle* in the English Channel saw a row of ships on the horizon – upside-down. What they could see was an example of a superior mirage, where layers of warm and cold air behave like lenses or prisms, bending images upwards so that they seem to float in the air.

The classic example of shimmering water on the horizon of a desert is an inferior mirage, where an image of the sky is bent downwards.

Mirages, such as of mountain ranges, are especially common in the polar regions where warm air sits over extremely cold layers.

A FLASH OF GREEN FROM THE SETTING SUN

As the Sun sets, it can sometimes send out a brilliant beam of emerald green light. The French author Jules Verne wrote of it in *The Green Ray* in 1882: 'If there is green in Paradise it must be this green: the true green of hope!'

The green flash is made from layers of still air, which bend and split light in the same way as a glass prism. As the Sun sinks below the horizon, its last rays are teased apart into their component colours like a rainbow. The red tends to mask the green until, as the red dips below the horizon, a glint of green appears. This is most marked above the sea where the atmosphere is sharply divided into layers of different temperature and humidity. As sunlight passes through the layers it is split and amplified to produce the fabulous green flash.

GLOWING RINGS OF COLOUR ENCIRCLE THE SUN

In late 1883 people around the world began to report strange colours in the sky. In Hawaii one observer, Sereno Bishop, noticed a disc of luminous light, fringed with colour, encircling the Sun. The disc was 'of whitish haze with pinkish tint, shading off into lilac or purple against the blue'. The phenomenon became known as Bishop's Rings.

These beautiful visions were the product of the volcanic eruption on Krakatoa, Indonesia, in August, 1883. Dust in the atmosphere scattered the sunlight and created rings of colour around the Sun and Moon and intense, blood-red sunrises and sunsets.

THE GHOST OF A RAINBOW

A colourless, luminous bow, like the ghost of a rainbow, sometimes appears on the face of dense banks of cloud or mist. Known as a fogbow, it is composed of water droplets that behave just like raindrops by reflecting sunlight into an arc. The difference is that these droplets are so small – a hundredth of the size of a raindrop – and create such a feeble spectrum that we cannot see the colours in it.

A PYRAMID OF LIGHT FROM THE COSMOS

In the twilight a faint glow sometimes rises up from the horizon in a large cone of milky transparent light. It is easy to mistake it for the glow of the setting Sun, but this is no ordinary light. It comes from space, where dust floating in the Solar System reflects the last of the Sun's rays. In the tropics, the cone of light is almost perpendicular to the horizon. Farther north and south, it makes only a small angle with the horizon for much of the year.

STAR SHINE *A faint pyramid of light glows in the twilight. This phenomenon is seen in the Northern Hemisphere in spring and autumn.*

THE AURORAE

The night offers its most fantastic light show in the coldest parts of the world. Around the Arctic and Antarctic Circles, the aurora borealis and aurora australis, known as the northern and southern lights respectively, are a truly amazing sight.

Colossal storms raging on the Sun's surface hurl out streams of charged particles that react with the upper atmosphere, exciting the swirling molecules of the various gases that are there. These then glow in a blaze of colour that lights up the sky. Aurorae are our window on the violence swirling around at the margins of the Earth's atmosphere. The most spectacular aurorae are seen in the polar regions where the Sun's particles are drawn towards the Earth's magnetic poles, but with the most intense aurorae, lights can sometimes be seen in Scotland or Florida Keys. The best time to view them is at the peak of the Sun's 11-year sunspot cycle, when the Sun is at its most tempestuous.

THUNDER, LIGHTNING, AND RAIN

Thunderstorms produce some of the weather's most spectacular violence, unleashing giant electrical sparks, ferocious winds, bullets of ice, and even showers of frogs – and all from just three ingredients: heat, air, and water.

A GIANT AMONG CLOUDS

The thundercloud, cumulonimbus, is the king of all clouds. It can tower up to 18 km (11 miles) into the stratosphere where its top freezes into ice. A thundercloud 5 km (3 miles) wide carries half a million tonnes of water and packs enough energy to power a small town for a year.

All thunderstorms begin life as cumulus clouds, feeding off warm air near the ground. Huge updraughts feed more warm, humid air into the cloud, turning it into a thunderstorm with gusts of wind reaching 160 km/h (100 mph). It crackles with electricity that eventually bursts out in giant sparks of lightning and claps of thunder. When the cargo of water in the cloud becomes too heavy the cloud releases it as rain – or sometimes hail.

CLOUDBURST *Lofty cumulonimbus clouds, seen here lowering above the Florida Keys, USA, feed off warm air from the ground.*

ILL WIND THAT STRIKES IN THUNDERSTORMS

On June 24, 1975, a jumbo jet approaching John F. Kennedy Airport, New York, was struck from above by an intense blast of wind. The aircraft was smashed into the ground, killing 113 people on board. The plane had been flying through torrential rain from a thunderstorm, but what caused the tragedy was a downdraught – a wind that bursts out from the bottom of a thundercloud as torrential rain drags cold air down with it. Downdraughts can hit the ground at speeds of up to 100 km/h (62 mph). This is the reason why aeroplanes try to avoid landing at airports during thunderstorms.

RAINING CATS AND DOGS – OR FROGS AND FISH?

Since ancient times, there have been reports of freak showers of frogs, fish, eels, periwinkles, crabs, jellyfish, and other small wildlife. Why do these extraordinary events occur? They are probably caused by powerful tornadoes that pass over rivers, lakes, and sea, sucking up small animals as they go.

Tornado winds travel at speeds of up to 511 km/h (318 mph) and are capable of carrying small animals aloft for kilometres. Eventually the cargo of creatures and rain grows too heavy for the cloud, which drops it all to the ground in a shower.

THE GREATEST ELECTRIC SHOW IN THE SKY

A lightning strike can carry up to 100,000 amps at 1 million volts. Travelling close to the speed of light, it heats the air to 30,000°C (54,000°F) causing an explosion which we hear as a clap of thunder.

Thunderstorms are born as hot, moist air rising from the ground hits cold air high above and condenses into cauliflower-shaped cumulus clouds. These grow into mighty cumulonimbus thunderclouds, which darken as they become increasingly dense with water droplets. As the droplets and tiny ice particles in the cloud collide with each other they create electricity.

The electric charges turn the thundercloud into a huge puffy battery, with a positive charge at the top and a negative charge at the bottom. The electricity grows so intense it has to escape somehow, and charges at the bottom of the cloud seek out the earth in a gigantic spark.

Sometimes a discharge takes several paths. This is known as forked lightning. Lightning strikes inside or between clouds are called sheet lightning and are seen from the ground as flashes.

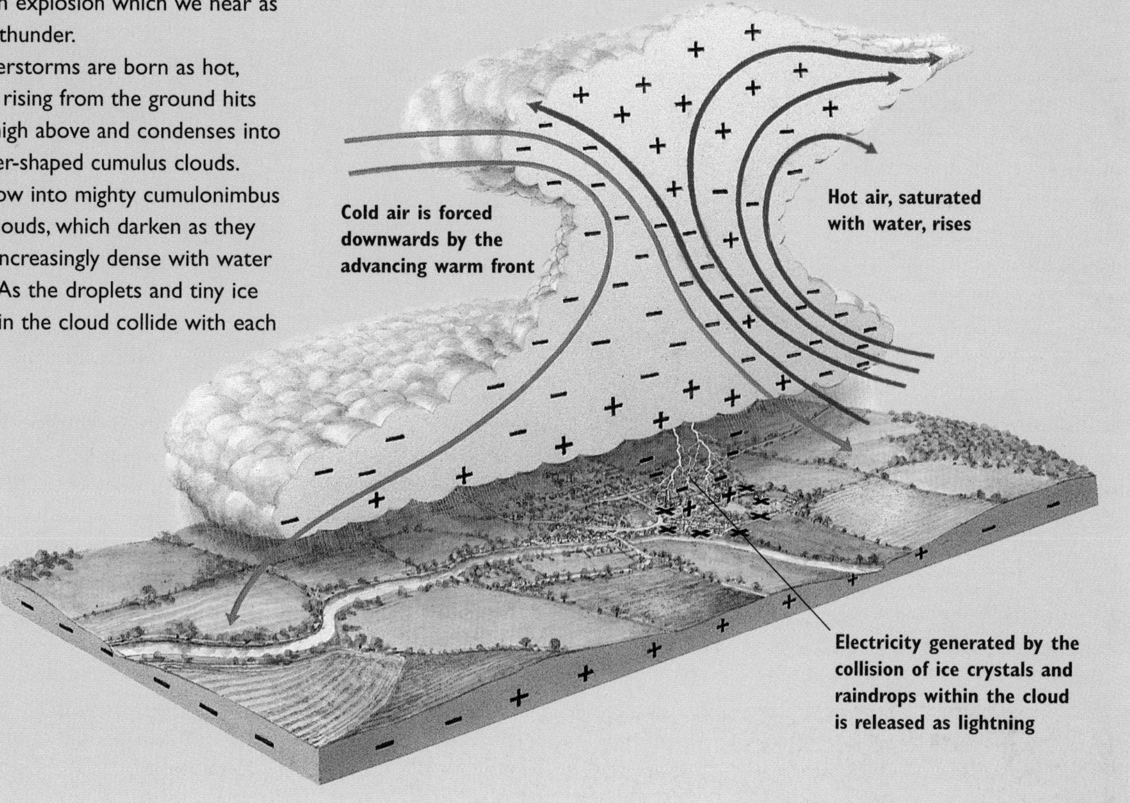

LIGHTNING THAT STRIKES UPWARDS

Lightning has been known to strike upwards from thunderclouds as well as downwards. In 1993, coloured lights were recorded bursting out of the top of some thunderclouds – orange 'jellyfish' with blue tendrils, puffs of blue light, and giant red blobs.

These lightning phenomena are called sprites. They shoot up to 95 km (60 miles) high in the sky, either striking out on their own or in swarms. Rarer sights include a blue beam of light shot up at more than 95 km/h (60 mph), known as a jet, and exploding discs of light called elves. All are created in the intense electrical fields above thunderstorms.

MYSTERY OF THE GREAT BALLS OF LIGHT

In June 1996 a ball of blue and white light the size of a tennis ball flew into a factory in Tewkesbury, England. The workers watched in amazement as it bounced around inside the roof, spun along girders, sent sparks flying, and finally exploded with an orange flash and a tremendous bang, disabling the company's telephone switchboard.

This was a dramatic example of ball lightning, a rare form of lightning that usually glides through the air, glows with the power of a 100-watt bulb, and lasts only about 15 seconds. It is a mysterious phenomenon, thought to be produced by electrical energy given off by a thunderstorm.

FLASH LIGHT *Ball lightning is capable of bouncing into buildings without mishap, although it has been known to explode into flames.*

AMERICA'S STORM OF THE CENTURY

March 1993 began with a heatwave along the eastern coast of the United States. Then suddenly the balmy spring calm was wrecked by a storm racing north with freezing temperatures, high winds, and surging tides. America's 'storm of the century' affected 26 states and about half the nation's population, killing 270 people. It left a 3,000 km (1,800 mile) trail of destruction. The snowfall was so heavy that roofs collapsed, and power pylons buckled and fell.

The storm was triggered by an unusual pattern of high-level jet-stream winds bringing abnormally warm air from the south into a collision course with bitterly cold air from the Arctic. As the warm air was thrust over the cold the result was a devastating weather front.

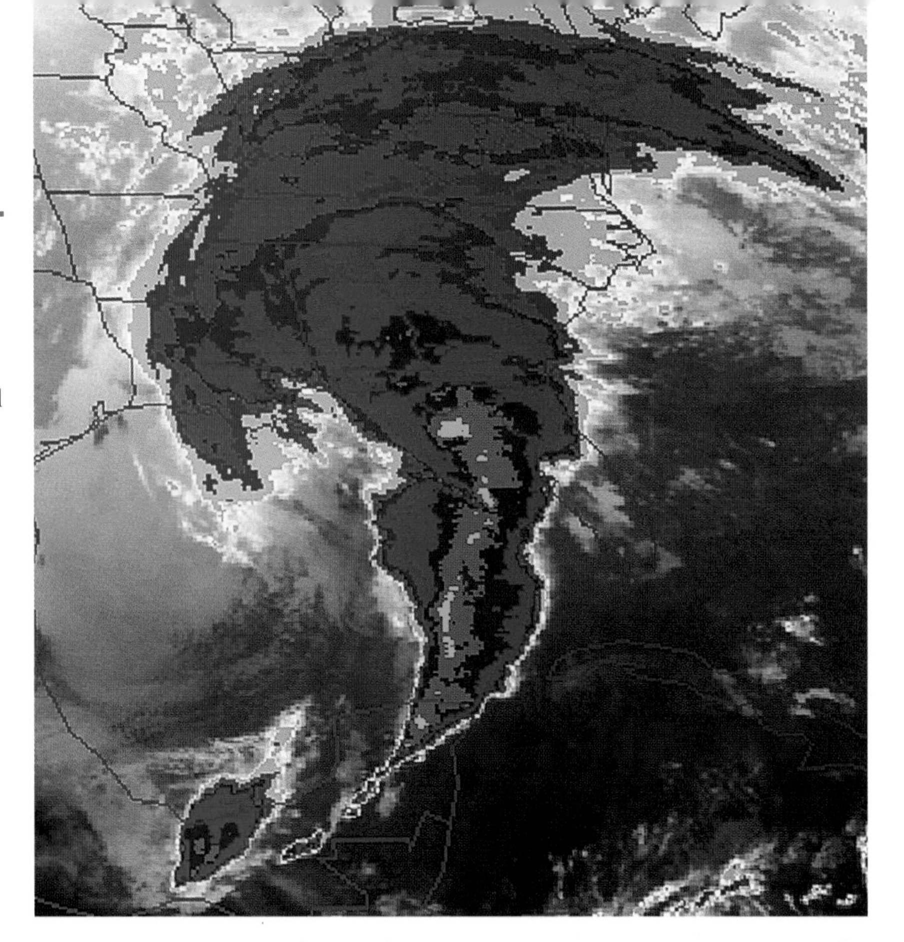

STORM FURY *Jet streams contributed to America's massive storm of 1993. These strong winds circle the globe at a height of 10 km (6 miles).*

GHOSTS OF LIGHTNING LEFT BEHIND

A lightning bolt burns through the air at 30,000°C (54,000°F), so it can inflict serious damage when it hits the ground. It also leaves its mark below the Earth's surface. If the soil is sandy, the lightning's heat melts the silica, converting it into a piece of glassy rock called a fulgurite. Fulgurites look like a knobbly set of small tree roots. They give an idea of how a lightning bolt tears across the ground, sometimes electrocuting people and animals several metres away from the strike point.

LIGHTNING RELIC *A fulgurite is the glassy imprint left by a lightning strike as it melts through sandy soil.*

HOW HAILSTONES GROW BIGGER THAN TENNIS BALLS

Thunderstorms can unleash a vicious and potentially lethal weapon – hailstones. In a furious storm over Munich, in Bavaria, on July 12, 1984, hailstones the size of tennis balls smashed a quarter of a million cars and caused damage totalling US$1 billion. In 1986, hailstones weighing up to a record 1 kg (2 lb 4 oz) killed 92 people at Gopalganj in Bangladesh.

So how are such hailstones formed? Updraughts fling raindrops to the top of a thundercloud, where they freeze, plummet down, collect more water, and are flung to the top again. This continues until the hail grows so heavy that it showers down in a hailstorm.

HOLY GLOW OF THE ELECTRIC STORM

One thundery day in 1983, a group of policemen in Swindon, England, were suddenly swathed in a spooky glowing light. This strange aura was due to an electrical glow called St Elmo's fire after the patron saint of sailors in the Mediterranean.

Although harmless, it can open up a path for a lightning strike. It is made by a thundercloud pulling electrical charges up off the ground – they often stream up tall objects, such as trees or ships' masts. If this electrical current meets charges coming down from the clouds then a lightning bolt instantly fires to earth.

FREEZING STORMS OF ICE AND SNOW

A single snowstorm can deposit 40 million tonnes of snow, carrying energy equivalent to 120 atom bombs. But what conditions are necessary for snow?

Basically, a snowflake can form if there is water and a low enough temperature. The air temperature high up in the atmosphere is below freezing point and turns minute water droplets into ice. Different shapes of ice crystals form at different temperatures. At around –1°C (30°F) ice crystals grow into thin plates; at –9°C (16°F) they form hollow columns; and at –15°C (5°F) the conventional snowflake with its delicate star patterns takes shape.

Snowflakes are clumps of snow crystals. When the temperature is near or slightly above freezing, the flakes become wet and stick to each other, creating the snow we see falling from the sky. Through a microscope these flakes look like irregular clusters of crystals. Only about 1 in 100 displays the beautiful symmetry we usually think of as snow crystals.

As the crystals fall to earth they float through air of different temperatures and change shape again, but each grows in its own way. No two crystals are ever the same because they never travel through an identical blend of temperatures.

Sometimes snowflakes grow to extraordinary sizes. At the end of the 19th century, flakes the size of small saucepans measuring 38 cm (15 in) across by 20 cm (8 in) thick fell over Fort Keogh in Montana, USA.

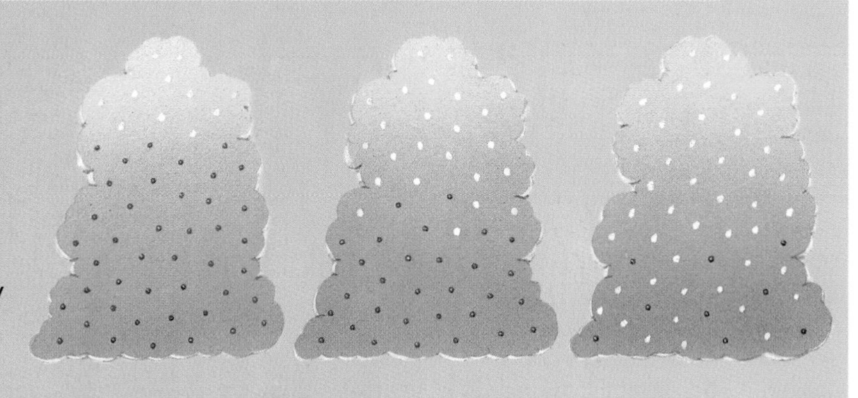

STORM CYCLE *A cloud grows as rising air lifts water particles upwards. Rising and falling air currents then cancel each other out, and snow falls. Finally, only downdrafts remain and snowfall decreases as the storm ends.*

The world record for the heaviest snowfall was during the winter of 1971-2 at Paradise, Washington State, USA. Some 31 m (102 ft) fell – enough to engulf a 10-storey apartment block. The greatest snowfalls in the world tend to occur where moist air from the sea is cooled as it climbs over mountains.

IMPERFECT SYMMETRY *The hexagonal feathery pattern that we associate with snowflakes is actually rare – most flakes have irregular shapes where they have been melted as they fall through the air.*

HOW LIGHTNING AFFECTS THE WORLD

Even though lightning can kill, set buildings ablaze, and blow up electrical equipment, it can also be a force for good. By discharging electricity from the air to the ground, lightning helps to disperse the enormous electrical charges in the atmosphere.

Lightning also fertilizes soil. The tremendous heat of lightning turns nitrogen and oxygen in the air into nitrous oxide and nitrogen dioxide, natural fertilizers that wash down with rain into the ground.

Worldwide each year, lightning produces up to 15 million tonnes of nitrogen fertilizer, a quarter of the natural world's production of nitrates. Trees and woodlands also benefit. Lightning sets fire to swaths of forests each year. These fires turn vegetation into mineral-rich ash, which fertilizes the soil. They even stimulate some seeds to germinate. In dense forests the fires clear the ground and help to regenerate the woodland.

DRAMATIC EFFECT *Lightning strikes the ground 100 times every second across the world. It is most prevalent in the tropics.*

WATER COURSE *Although it rains almost daily on Mount Wai'ale'ale, Hawaii, floods are rare: the rain forest below soaks up the water.*

RAIN FALLS MAINLY OFF THE PLAIN

Mount Tutunendo in Colombia is the wettest mountain in the world. Its average annual rainfall is 11.77 m (38 ft 6 in) – the height of a three-storey house.

The world's wettest areas tend to be where there is a combination of moist air, often coming off the sea, and mountains. The moist air rises over the mountain and cools, the water condenses into clouds, and so it rains.

It rains more often on Mount Wai'ale'ale, on the island of Kauai in Hawaii, than anywhere else. On average, rain falls 360 days a year. The mountain's summit is shrouded in almost permanent mist as the prevailing moist trade winds blow in from the Pacific.

Mount Wai'ale'ale means the 'Mountain of Overflowing Water' in Hawaiian, and an average of 11.18 m (37 ft) of rain each year cascades in huge waterfalls down the slopes.

Cherrapunji, in the Indian foothills of the Himalayas, has the world's most intense rainfall, averaging 2.7 m (9 ft) rain each June. The rain comes from the monsoon wind. The air is thick with moisture picked up from the Indian Ocean. When it reaches Cherrapunji it cools, condenses into black clouds, and rains in torrents.

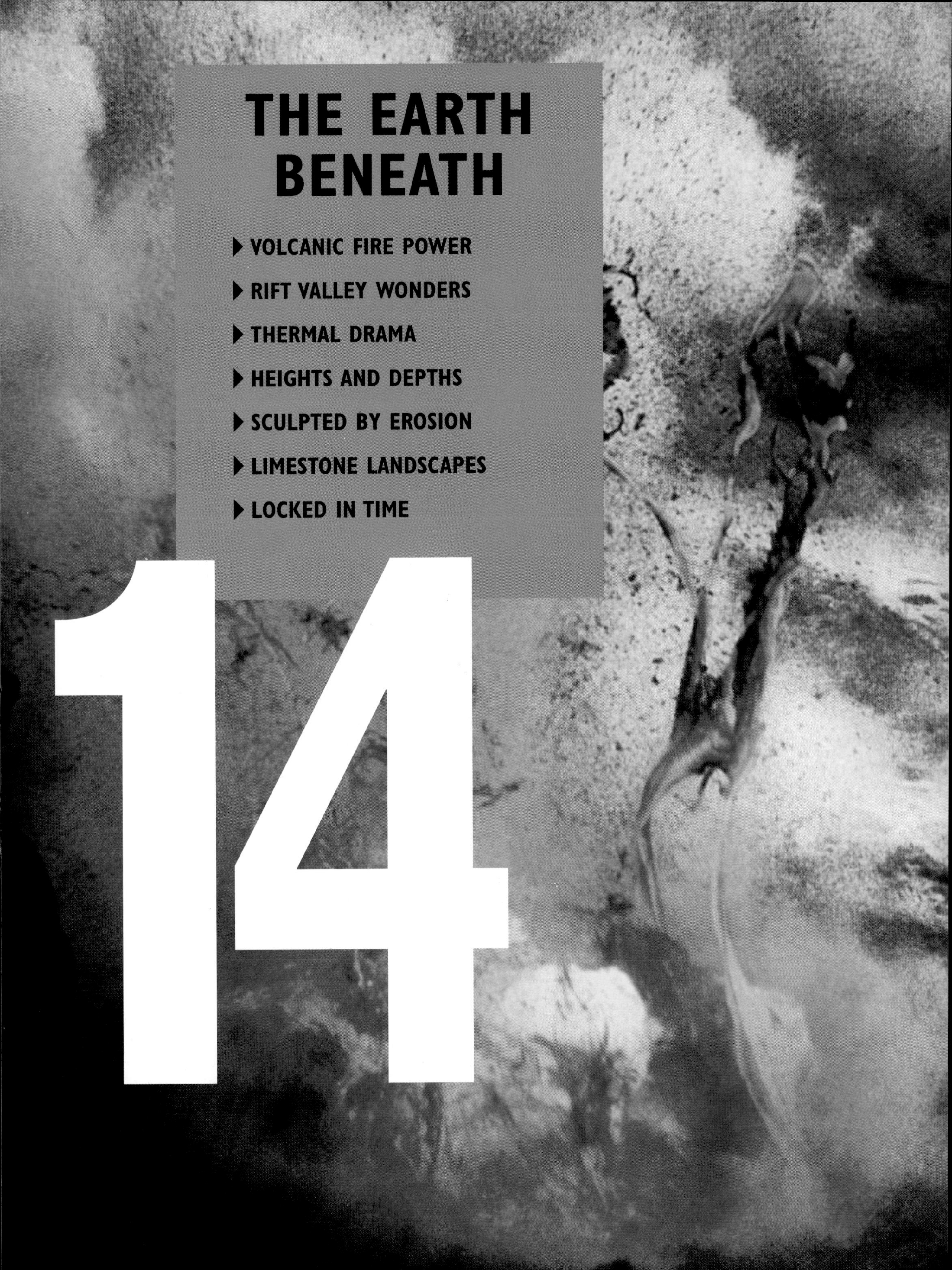

THE EARTH BENEATH

- VOLCANIC FIRE POWER
- RIFT VALLEY WONDERS
- THERMAL DRAMA
- HEIGHTS AND DEPTHS
- SCULPTED BY EROSION
- LIMESTONE LANDSCAPES
- LOCKED IN TIME

14

VOLCANIC FIRE POWER

Fire raging in the belly of the Earth erupts through the ground as volcanoes, spewing out red hot lava, poisonous gas, and rocky debris. The results, such as radical changes in the climate, can affect life all around the world.

RESURRECTION AFTER DEVASTATION

At 8.32am on May 18, 1980, a great earthquake rocked the volcano of Mount St Helens in Washington State, USA. It triggered a massive landslide. Gas and steam inside the volcano split its side open and hurled out debris with an explosion 1,300 times more powerful than the Hiroshima atom bomb.

The devastation was immense – an area of 595 km² (230 sq miles) was scorched by fires and blasted by the shock wave. A landscape of lush conifer forests, clear streams, and lakes was transformed into a barren, grey wasteland. Sixty people and thousands of animals died.

Yet within several weeks, signs of life appeared: tree shoots sprouted from roots or severed trunks buried beneath ash and pumice. The following year fresh vegetation created a blaze of green. It took a few years for insect life to return, although colonies of ants survived the eruption in logs or underground. Fourteen species of small mammal also managed to survive. By 2000, the blast area looked almost fully recovered, except for the conifer forests, which will not mature to their former splendour until about the year 2200.

THE YEAR WITHOUT A SUMMER

On April 7, 1815, Mount Tambora, on the Indonesian island of Sumbawa, exploded with the power of 100 atom bombs. It was one of the largest eruptions in 10,000 years and its effects were felt worldwide. Only 26 people survived out of a population of 12,000.

Some 150 km³ (36 cu miles) of rock were blasted out. Dust and sulphur dioxide were shot into the upper atmosphere and spread out in a veil across the world, blocking the Sun and sending global temperatures tumbling. The following year was called 'the year without a summer', when frosts in June led to crops withering across Europe and North America, causing widespread famine.

EARTHQUAKE BENEATH THE GLACIER

A series of earthquakes rocked the Vatnajökull glacier in Iceland on September 29, 1996. Two days later, a gaping hole opened up in the top of the ice and thick black

VIOLENT ERUPTION *One of the greatest explosions in modern history happened on Mount St Helens, USA, in 1980.*
1 *The volcanic blast flattened forests across a wide area.*
2 *Fifteen years later vegetation had regrown, but the tree cover will take 200 years to mature.*

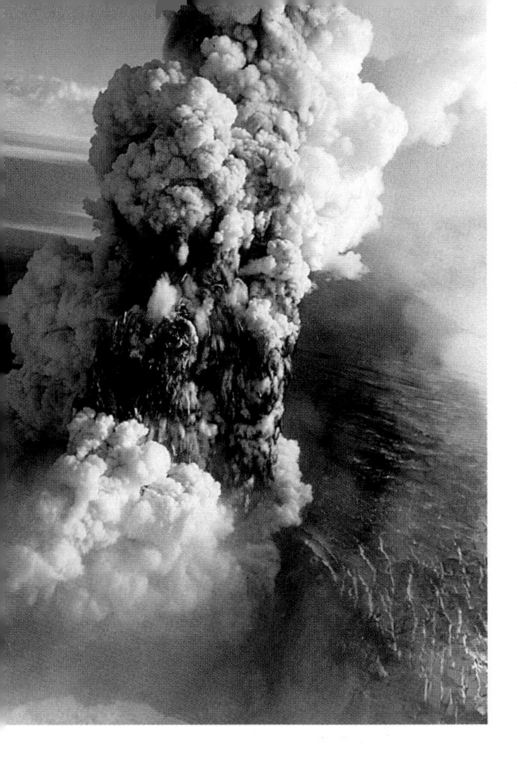

WATER POWER *The eruption of Grímsvötn in Iceland, 1996, caused meltwater to burst through a glacier and cascade down an ice sheet.*

clouds of ash billowed out. The Grímsvötn volcano had erupted under Vatnajökull, melting a vast chamber through the ice.

Several weeks later and tens of kilometres away, the meltwater burst out of the glacier, surging down the ice sheet in a river of water, mud, and stones. It swept away bridges and tore up roads and power lines, but bypassed inhabited areas.

THE VIOLENT BIRTH OF A NEW ISLAND

Thick black smoke belching from the sea on November 14, 1963, signalled the birth of the island of Surtsey off Iceland's south coast. For the next four years Surtsey erupted, smoked, and spewed lava until eventually it stood 169 m (555 ft) above the sea and covered almost 2.5 km^2 (1 sq mile) of new land.

The first creature to appear on the island was a fly, then a seagull, and by 1987 the landscape was carpeted with plants carried there by birds. Surtsey also became a favoured resting place for thousands of birds migrating between Iceland and Europe, and seals came to relax on its volcanic beaches.

THE STORY BEHIND VOLCANIC ERUPTIONS

The ground we stand on is a very thin solid crust resting on a mass of hot soft rock – like an eggshell on a soft-boiled egg. Under the crust there are pockets of intensely hot molten rock called magma, heated by pressure, friction, natural radioactivity, and the intense heat of the Earth's core.

The magma is lighter than the surrounding rock, causing it to rise up, and if it finds a weak spot in the crust it collects in a bubble called a magma chamber. If enough pressure builds up in the chamber, the magma bursts out through the ground as a volcano.

Most active volcanoes occur at the edges of the Earth's tectonic plates – which carry the oceans and continents – as they grind against each other (see *page 400*). The most devastating eruptions occur when an oceanic plate sinks under its neighbour. Sea water mixes with the magma underground, trapping gas bubbles. If that magma forces itself to the surface, the pressure of the gas makes the magma explode – like taking the cork out of a champagne bottle. The volcano then erupts with gas, ash, rock, and lava.

Volcanoes can produce an avalanche of ash and gas as hot as 700°C (1300°F), that races along the ground at up to 200 km/h (125 mph) for 25 km (15 miles), eradicating everything in its path.

Exploding volcanoes are usually tall, steep cones made of layers of rock and lava. Typical examples are Vesuvius, Etna, St Helens, and Fujiyama.

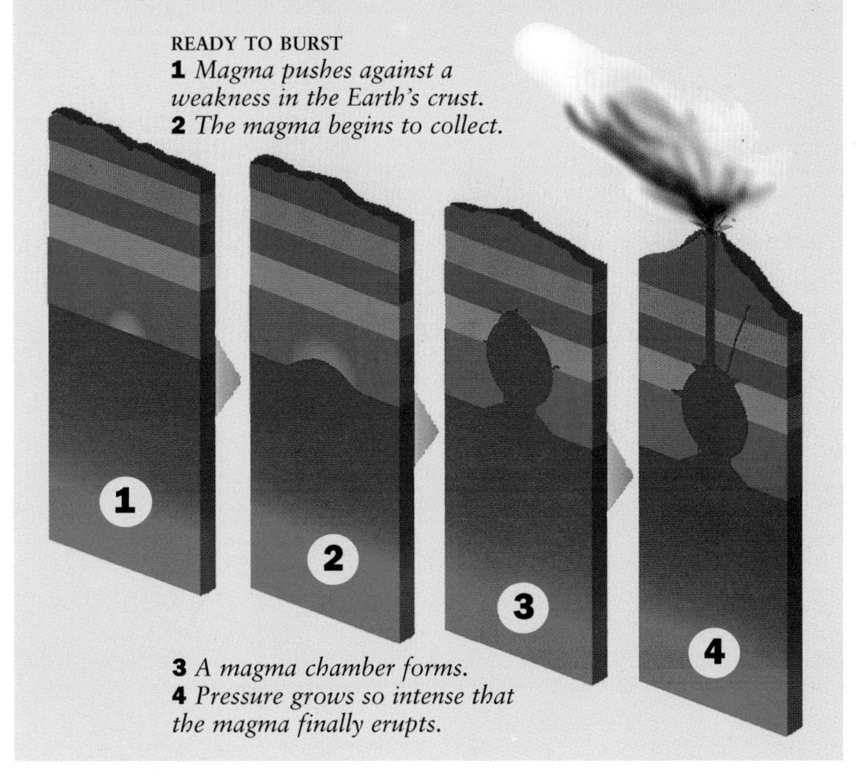

READY TO BURST
1 *Magma pushes against a weakness in the Earth's crust.*
2 *The magma begins to collect.*
3 *A magma chamber forms.*
4 *Pressure grows so intense that the magma finally erupts.*

THE PACIFIC'S RING OF FIRE

Volcanoes erupt and earthquakes occur with great frequency along the Pacific Ring of Fire. Stretching 48,000 km (30,000 miles), from the tip of the Andes, to Alaska and the Aleutian islands, Japan, the tropical islands of Indonesia and Papua New Guinea, and on to New Zealand, the ring comprises three-quarters of the world's volcanoes. Some of the most violent eruptions have occurred along it, such as that of Mount St Helens (*see opposite page*).

Seismic activity along the Ring of Fire also generates powerful earthquakes, including those that regularly shake Japan and California.

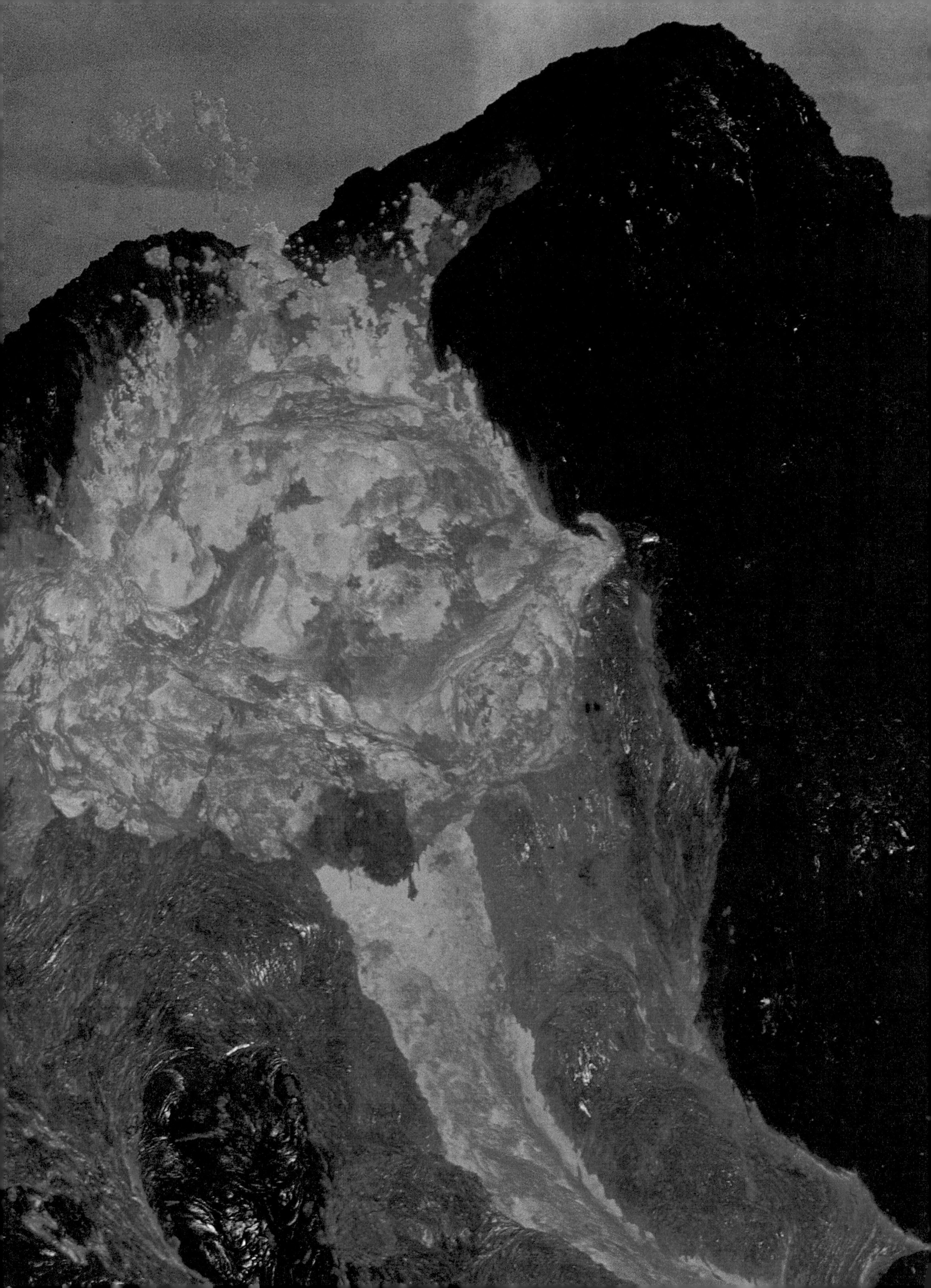

MAJOR ERUPTIONS OF THE 20TH CENTURY

- **Mount Pelée, Martinique, 1902**
30,000 people killed by a pyroclastic flow (avalanche of hot gas and debris) that swamped the port of St Pierre.
- **Santa Maria, Guatemala, 1902**
About 2,000 killed by a pyroclastic flow.
- **Mount Katmai, Alaska, 1912**
A pyroclastic flow covered 750 km² (290 sq miles) of remote parts of Alaska.
- **Mount Lamington, New Guinea, 1951**
3,000 killed by a pyroclastic flow.
- **Nyirangongo, Zaire, Africa, 1977**
72 killed by lava flowing at up to 40 km/h (25 mph).
- **Mount Etna, Sicily, 1979**
Nine people killed by flying rocks.
- **Mount St Helens, USA, 1980**
About 60 people killed in the side-explosion of debris and rocks.
- **El Chichón, Mexico, 1982**
About 2,000 killed by a number of pyroclastic flows.
- **Nevado del Ruiz, Columbia, 1985**
21,000 killed by a lahar (river of mud) that destroyed the town of Armero.
- **Lake Nyos, Cameroon, West Africa, 1986**
1,887 killed by a carbon dioxide eruption spilling out from the lake.
- **Mount Pinatubo, Philippines, 1991**
About 1,000 killed in pyroclastic flows and lahars, which were compounded by typhoon rains.

WHY VOLCANOES OCCUR OVER HOT SPOTS

Hawaii sits on a hot spot in the North Pacific Ocean, thousands of kilometres from the major volcano zones. Its 20 islands have been formed by a chain of volcanoes that have risen from the seabed.

The hot spot occurs where there are isolated magma vents in the Earth's plates. A great plume of hot liquid rock, or magma, rises as far as 160 km (100 miles) from the depths of the Earth and breaks through weak spots in the crust to create a volcano. If one of the Earth's plates moves over a hot spot, a chain of volcanoes, such as the Hawaiian or Galapagos islands, forms.

HOT ROCKS BENEATH THE ICY CONTINENT

Antarctica might look like a continent trapped in a deep freeze, but it has five volcanoes. The most dramatic of these is Mount Erebus, 3,794 m (12,447 ft) high, the world's most southerly active volcano. It was erupting violently when the explorer Captain James Ross discovered it in 1841. Erebus has been active again more recently. In 1984 it threw bombs of hot rock 610 m (2,000 ft) above the rim of the crater in a mushroom cloud of black ash, and it still pumps out steam regularly.

Antarctica's volcanoes have been created along the margins of a rift system, rather like the East African Rift Valley, which is very slowly spreading outwards.

ISLAND UPRISING *Hawaii, like Tristan da Cunha and the Canary Islands, was born from molten magma bursting through the Earth's crust.*

AN EXPLOSIVE CHANGE IN THE WEATHER

When Mount Pinatubo in the Philippines erupted in 1991, it caused the biggest shake-up in the world's climate in recent history. Mount Pinatubo shot out huge amounts of dust into the stratosphere, which then spread round the world, reaching as far as the Antarctic. The dust blanketed the sun and reduced global temperatures by about half a degree Celsius. No volcanic eruption had had such far-reaching effects since Krakatoa in 1883.

The eruption of Mount Pinatubo also created an acid cloud containing 6 million tonnes of sulphuric acid aerosol – a fine mist of tiny airborne particles. Like a heat shield, the aerosol helped to bounce the Sun's heat back into outer space and cooled the planet even further.

The acid cloud also attacked part of the ozone layer in the stratosphere, and is thought to have increased ozone destruction by a half over the tropics, where it is most concentrated.

VOLCANIC AVALANCHE *In 1991, in the Philippines, Mount Pinatubo released an avalanche of gas and debris that engulfed communities.*

RIFT VALLEY WONDERS

Africa's Great Rift Valley is so immense it is the only geological feature on Earth that can be seen from the Moon. The valley harbours volcanoes, mountains, and lakes with some of the most inhospitable climates

THE GREAT RIFT DIVIDING A CONTINENT

The Great Rift Valley is the biggest scar on the surface of the Earth. It stretches 6,400 km (4,000 miles) from Lebanon in the Middle East to Mozambique in Southern Africa.

The Rift Valley began to form 20 million years ago, when the Earth's crust started to tear apart. Earthquakes rocked the fault lines, and magma squeezed through the gaps and erupted into volcanoes. Today they are among the world's highest dormant volcanoes and include Mount Kilimanjaro in Tanzania. Water filled the vast canyon, creating some of the largest lakes on Earth, such as lakes Victoria (*see page opposite*) and Malawi.

The widening rift tore Saudi Arabia away from Africa and made the Red Sea. Today the Rift Valley is still opening up, and eventually the Horn of Africa will split off from the rest of the continent to become a huge island.

THE SNOWY PEAK ON THE AFRICAN PLAINS

You would not expect to find snow in the middle of tropical Africa, just 3 degrees south of the Equator. Yet Mount Kilimanjaro in Tanzania is so high that it is permanently capped with snow, frozen in temperatures of –20°C (–4°F).

Kilimanjaro is an ancient volcano towering 5,895 m (19,340 ft) above sea level. It is the highest peak in Africa and the highest isolated mountain on Earth. The volcano lies on a fault line crossroads in the Earth's crust where steam and sulphur still puff out of vents in the ground.

The summit of Kilimanjaro was covered by a glacier 100 m (330 ft) deep, but global warming in the past four decades has shrunk the ice and it could vanish altogether by about 2050.

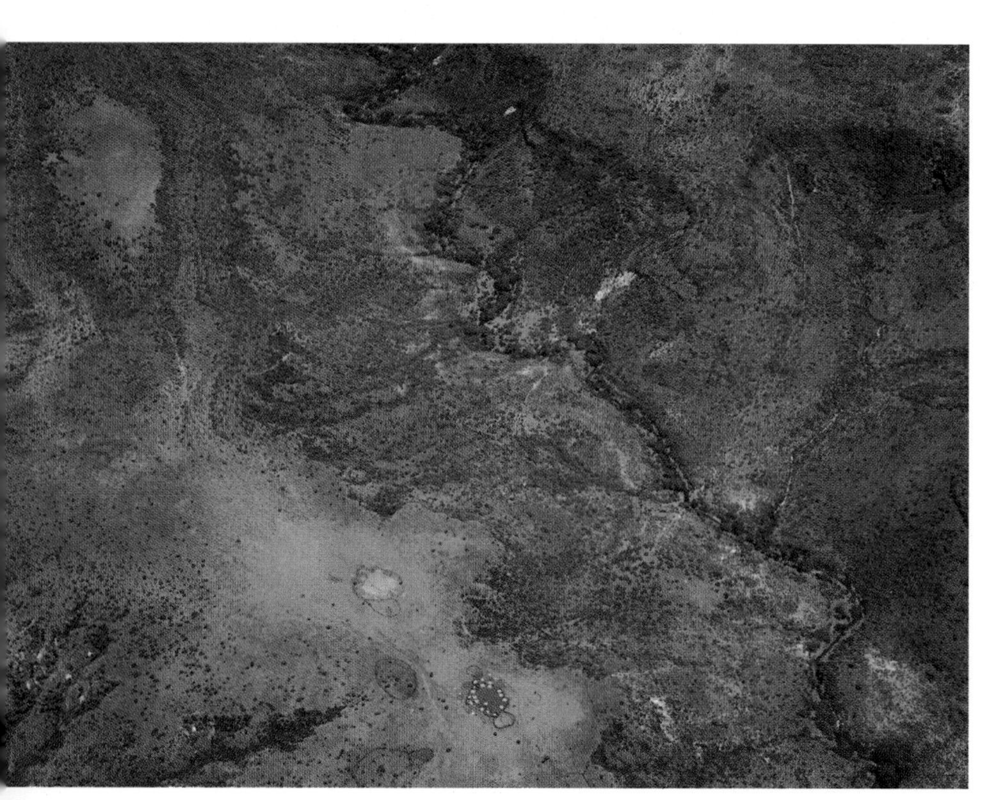

FAULT LINE *A river snakes through the plains of the Great Rift Valley in central Kenya. The valley averages 65 km (40 miles) wide.*

A SEETHING CAULDRON ON THE DEVIL'S MOUNTAIN

The Danakil Depression in Africa is an astonishing landscape of volcanoes and boiling lakes. The valley floor lies 220 m (720 ft) below sea level, and is the hottest place in the

world with the temperature averaging 34.5°C (94°F). In this hostile landscape in Ethiopia rises Devil's Mountain, Erta Ale, a volcano that has been erupting for more than 90 years.

At the summit of Erta Ale a lake of lava cooks at 1,000°C (1,830°F). Crusts of cooler, silvery lava float on the molten lake, broken by jets of fiery lava and gas bubbles. Onlookers here have to wear fireproof clothing.

LIFE IN THE SODA LAKES

A red and blue lake of alkaline soda baked under the searing sun sounds like a scene from another planet. Lake Natron, in northern Tanzania, is one of several East African soda lakes. Soda salt is washed down from surrounding volcanoes and makes the lake shimmer with bright red blooms of salt-loving algae. The soda crust is so thick you can drive a car across it.

Each year, hundreds of thousands of flamingos migrate to breed in this desolate environment, attracted by supplies of small invertebrates and algae. But they risk their lives. If the birds stand in the water for too long their legs become caked in salt and they may never fly again.

SALT CRUST *The soda lakes of East Africa are a feature of local volcanic activity. Soda is flushed from craters and baked under the tropical sun.*

LAKE THAT GIVES RISE TO THE WORLD'S LONGEST RIVER

Lake Victoria is Africa's largest lake and the world's second largest freshwater lake. It covers 68,400 km^2 (26,400 sq miles), and is the chief source of the Nile.

Nestling in a hollow between the Western and Eastern Rift Valleys, the lake is bordered by Uganda, Kenya, and Tanzania. It supports 400 species of fish, including the shallow-feeding cichlids that brood their eggs in their mouths. But many species are now at risk from industrial pollution, predation by the commercially introduced Nile perch, and the prolific floating water hyacinth that chokes the lake.

THERMAL DRAMA

The heat from the heart of the Earth bursts through its surface as fountains of steaming water, clouds of toxic gas, and lakes of boiling mud. In some places this energy has been harnessed as a source of natural warmth and power.

CASTLES AND WATERFALLS CAST IN CALCIUM

A strange natural fairytale castle hangs on the side of a cliff in the foothills of south-western Turkey's Cokelez Mountains. Pamukkale, meaning Cotton Castle, with its dazzling white ramparts, basins and frozen cascades, is made entirely of calcium.

The top of the cliff gurgles with springs of hot water full of calcium carbonate seeping out of the limestone bedrock. As the water drips down the steep slopes it evaporates and deposits its cargo of calcium, laying it down as travertine – better known for making stalagmites and stalactites in caves. This brilliant white mineral has gradually built up into hundreds of basins, some 20 m (66 ft) tall and jutting out 100 m (330 ft) from the cliff. As they fill with warm water the basins overflow and drip with white waterfalls. Where the cliff face is shallower, the travertine solidifies into terraces of steps.

FAIRY STEPS *Hot springs and calcium salts have created an extraordinary castle-like formation at Pamukkale, overlooking Turkey's Curuksu plain.*

VOLCANIC GATEWAY TO THE UNDERWORLD

Just 16 km (10 miles) west of Vesuvius, near Naples in Italy, lies a range of volcanoes fuming with steam and sulphurous vapours. The ancient Romans thought this place was the gateway to the Underworld. One of the most active of these volcanoes is Solfatara, named after the yellow sulphur that crystallizes out of the hot gases from its smoking craters.

As magma rises to the surface it heats the rocks and gases and pumps up more sulphur dioxide fumes. Some volcanic vents give off sulphur gases so acidic that they corrode the rock as they rise through it, creating a great cauldron of hot mud. The mud pools in Solfatara are 60°C (140°F). Some are cool enough to wade in, and are prized as beauty treatments for skin.

Sulphur from volcanic vents is used in rubber tyres to make them more durable – a process called vulcanization after the Roman fire god, Vulcan.

THE HOT SPRING CAPITAL OF THE WORLD

One of the world's greatest geothermal landscapes is at Rotorua, North Island, New Zealand. The eruption of nearby Mount Tarawera, in 1886, left a legacy of seething hot springs, emerald green and orange lakes, boiling mud pools, steaming vents, and smoking craters.

There are dozens of geysers, many surrounded by terraces of sparkling crystalline silica, built up from minerals dissolved underground by the geysers' superheated waters. Currently, the largest geyser is Pohutu, which shoots twin jets of steaming water 30 m (100 ft) into the air for up to 40 minutes at a time.

STEAMING LAND OF MULTICOLOURED POOLS

When Mount Tsurumi in Kyushu, Japan, erupted in 867 more than 3,500 geysers, hot springs, and steam vents burst forth. The volcano is now extinct, but the springs around the spa town of Beppu still spout hot water, geysers, and bubbling pools in a dazzling palette of colours.

The pools all contain the same ingredients of alkali, sulphur, carbon, and iron, but in different proportions, which give them their varied colours. The most spectacular pools are called *jigoku* ('burning hell'). *Chinoike-jigoku* ('blood-pond hell') is deep red from the red clay stirred up in the boiling waters; another is a cobalt blue pond of steaming water surrounded by tropical plants; a third spurts colourless water from the ground that turns milky white.

ICY HOME OF THE ORIGINAL GEYSER

In 1294 a huge jet of hot water erupted from the magma welling beneath Iceland on the edge of the Arctic Circle. Those who saw it called it Geysir, 'The Great Gusher', and thought it was boiling water from hell. It became so famous that the name geyser was adopted to describe all such hot springs. Geysir performed regularly for the next 600 years, its column of water shooting more than 60 m (200 ft) into the air. But since the early 20th century its power has waned and now it spurts into life only occasionally. It has been eclipsed by neighbouring Strokkur, 'The Churn', which erupts every few minutes, spurting jets 20 m (66 ft) high.

NATURAL FOUNTAIN *Iceland, and its hot springs such as the mighty Strokkur, sits on a crack in the mid-Atlantic floor where the Earth's crust is hot from upwelling magma.*

THE COUNTRY THAT RUNS ON THERMAL ENERGY

For an island bordering the Arctic Circle, Iceland is full of hot water. When Ingólfur Arnarson, the first settler, set up home in the 9th century he called it Reykjavik, or 'Smoky Bay', after the plumes of steam gushing from hot springs.

Iceland is a volcanic land where the Earth's crust is so thin that magma lying close to the surface heats the rocks and underground water. Every home in Reykjavik is heated with this natural hot water, and power stations are driven by steam from the ground.

The hot water is also used to heat greenhouses for growing tropical fruits, and bread is baked on hot rocks in subterranean caverns.

Triple-headed geyser

Fly Geyser in the Black Rock Desert of Nevada, USA, is possibly the most freakish spout of water in the world. Water trickles down from its three heads into muddy pools. The bizarre conical shapes are the result of calcium in the water slowly solidifying to produce the mineral travertine. The geysers were created from man-made wells and gush 24 hours a day, all year round. The rocks and shallow pools close to the geyser are too hot to walk on without shoes. However, pools farther away are cool enough to allow visitors to bathe in them. The geysers are fed by underground streams and lakes, and much of the Black Rock Desert itself consists of ancient dried-up lakebeds. The region embraces a designated wilderness area that was a lake about 2 million years ago. It is so huge and flat that it is possible to see the curvature of the Earth's surface on the horizon.

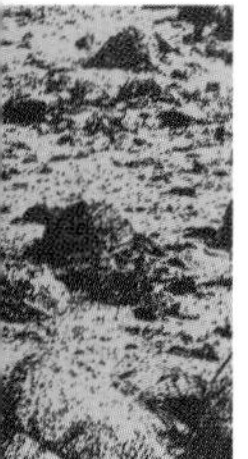

HEIGHTS AND DEPTHS

From the highest mountain to the deepest ocean, the Earth is a planet of record-breaking extremes. There are caves that can house small cathedrals and gaping canyons that are wide enough to fly a plane across.

THE LAKE A QUARTER OF A MILE BENEATH THE OCEAN

The Dead Sea, which lies between Israel and Jordan, is the lowest point on the world's land surface. Its shores lie 397 m (1,302 ft) below sea level. It is also the world's saltiest lake. The mineral-rich water contains eight times more salt than the Mediterranean, which makes it exceptionally buoyant – human beings float effortlessly like corks.

Fed by the River Jordan, the Dead Sea has no outlet: instead, water is lost through evaporation. In summer, when the temperature soars to 50°C (122°F), hundreds of islands appear in the lake as evaporation leaves the tips of pillars of salt protruding above the water. Huge shallow pools have been built to crystallize the salts and mine them for fertilizers and industry. Such high levels of salt are toxic for most life-forms, and until recently the Dead Sea was thought to contain no living creature. We now know that halobacteria thrive in the briny water (*see page 342*).

SALINE SOLUTION *Water evaporates from the low-lying Dead Sea in the heat of summer to expose a multitude of sculptural salt pillars.*

THE LARGEST HOLES IN THE WORLD

The world's biggest cave is the Sarawak Chamber in Borneo. Measuring 70 m (230 ft) high, 700 m (2,300 ft) long, and averaging 300 m (980 ft) wide, it is three and a half times as long as St Peter's in Rome and twice as wide. It could hold eight jumbo jets noise-to-tail with plenty of room at the sides. The cave was discovered in 1984.

The Reseau Jean Bernard cave system in the Haute-Savoie region of France, at 1,602 m (5,256 ft) deep, held the record for the world's deepest cave until 1998, when a Polish caving team in the Austrian Alps near Salzburg discovered a shaft at the bottom of a cave called Lamprechtsofen, 1,632 m (5,354 ft) deep. Then in January 2001 a Ukrainian party in the Western Caucasus Mountains of Georgia discovered a cavern called Voronja, or 'Crow's Cave', 1,710 m (5,610 ft) underground, a little deeper than the deepest part of the Grand Canyon.

AMERICAN LABYRINTH BREAKS ALL RECORDS

Kentucky, USA, is home to the world's longest cave system. Mammoth Cave has more than 555 km (345 miles) of caves and passages that have been explored – equivalent to the entire London Underground system.

The caves were carved out during the past 30 million years as the Green River worked its way through layers of sandstone and limestone. Because the water was slightly acid, it dissolved the limestone and etched out underwater passages. By a million years ago, the river had carved out a huge gorge that drained away all the water and left many passageways high and dry. Today, the harder sandstone shell acts as a roof that preserves the cave system.

A RIVER AND THE WORLD'S BIGGEST HOLE IN THE GROUND

The Grand Canyon is the world's largest gorge, stretching 444 km (276 miles) along the Colorado River in Arizona, USA. The canyon, which spans 29 km (18 miles) at its widest and plunges 1.6 km (1 mile) at its deepest, was created by the Colorado River cutting through the surrounding rocks. The land tilted up over millions of years, allowing the river to cut even further into the rock.

As the river sliced deeper into the ground, it uncovered layers of rock that date back through geological time like a vertical calendar. The deepest and oldest rocks are from 4 billion years ago – a time when the planet was still being formed. The desert climate has ensured that the steep slopes of the canyon have not been washed away.

TIME MACHINE *The Grand Canyon cuts a swath up to 1.6 km (1 mile) deep through Arizona, USA, providing a clear record of the Earth's history.*

WHERE THE EARTH MEETS THE SKY

The highest mountain on Earth is Mount Everest on the Nepal-Tibet border in the Himalayas. It is 8,850 m (29,035 ft, almost 5 1/2 miles) above sea level and its summit was first reached on May 25, 1953, by Sir Edmund Hillary and Tenzing Norgay.

Because Everest is so high in the troposphere – the layer of atmosphere that contains most of our weather – it has very little oxygen, which poses a serious problem for climbers (sometimes killing them). The height of the Himalayas also interferes with high-level jet-stream wind, sending it into waves that can affect the weather over China. The mountains act as a barrier to rain from the Indian Ocean reaching the Gobi Desert to the north.

A LAKE AS DEEP AS THE GRAND CANYON

Lake Baikal in Siberia is the deepest lake on Earth. At 1,637 m (5,371 ft) in depth it is, by coincidence, as deep as the Grand Canyon. Although Lake Baikal is only the size of Belgium, it holds a fifth of the world's freshwater reserves – more than all five of North America's Great Lakes combined.

Baikal has existed for at least 20 million years. A thousand or so species of animals and plants are unique to the lake, including the world's largest flatworm, 40 cm (16 in) long. Baikal is also home to one of the smallest seal species in the world (*see page 336*).

TOP OF THE WORLD *The world's highest mountain is named after Sir George Everest, Surveyor General in India between 1830 and 1843.*

THE FASTEST-GROWING MOUNTAINS ON EARTH

When India crashed into Asia, about 50 million years ago, the Earth's crust was pushed up to form the Himalayas. The mountain range is still growing.

Each year the Indian plate moves about 2 cm (3/4 in) northwards, pushing the Himalayas up about 5 mm (1/4 in) higher. If they continue to grow at this rate they will be a kilometre (3,274 ft) taller in another 200,000 years.

Yet the Himalayas are also the fastest-eroding mountains in the world. The slightly acid rainfall etches away the rocks and flushes huge quantities of sediment down into the Brahmaputra, Ganges, Yangtze, Mekong, and Indus rivers, creating enormous river deltas.

ICY COVER TO THE WORLD'S HIGHEST CONTINENT

The Antarctic has the thickest ice sheet in the world. The ice is 4 km (2 1/2 miles) thick in places, and holds 70 per cent of the world's fresh water. If all the ice melted, sea levels around the world would rise by between 37 m and 91 m (120 ft and 300 ft) – enough to engulf Miami, New Orleans, Bangladesh, and many other low-lying places.

The Antarctic is also the world's highest continent, averaging about

1,830 m (6,000 ft) above sea level, with mountain peaks soaring to 5,800 m (19,000 ft), but buried under ice. The height of the land is one reason why Antarctica is the coldest place on Earth: the average winter temperature is –60°C (–76°F).

It is possible that life may exist deep under the ice. Vostok, in the interior of Antarctica, conceals a lake from the time when the continent had a mild climate. Although the lake was frozen over 30 million years ago, it is thought that bacteria could have survived, and research is in progress.

BEST PLACE ON THE PLANET TO REACH FOR THE STARS

The snow-capped peak of Mount Chimborazo in the Andes of Ecuador is the point farthest from the centre of the Earth. Chimborazo stands 6,267 m (20,561 ft) above sea level. It is more than 2 km (1¼ miles) shorter than Mount Everest, but is 2,150 m (7,054 ft) farther from the Earth's centre than Everest because of an outward bulge at the Equator. This bulge is caused by the spin of the planet and makes the Earth's radius 21,000 m (68,900 ft) greater at the Equator than at the poles – Chimborazo is just 158 km (98 miles) south of the Equator. Even the beaches of Ecuador are farther from the centre of the Earth than the summit of Everest.

THE TALLEST MOUNTAIN IN THE WORLD

Mauna Kea, in Hawaii, is the tallest mountain in the world. It is almost 2 km (1¼ miles) higher than Mount Everest, although only 4,205 m (13,796 ft) is above sea level. The mountain rises 10,205 m (33,480 ft) from the seabed. Mauna Kea is one of five volcanic masses making up Big Island in Hawaii. Its summit is always above the clouds, providing good visibility for astronomical observatories.

MOUNTAIN FACTS AND FIGURES

- The greatest land mountain range is the Himalayas-Karakoram. It contains 96 of the world's 109 peaks that reach above 7,317 m (24,005 ft).
- The longest mountain range is the Andes in South America, which extends for 7,564 km (4,700 miles).
- The tallest mountain, from base to summit, is Mauna Kea, which forms part of Big Island in Hawaii. Only 4,205 m (13,796 ft) is above sea level (*see above*).
- The world's largest mountain range, the Mid-Ocean Ridge, is almost entirely hidden by the sea. It extends for 64,374 km (40,000 miles) from the Arctic Ocean, peaking above sea level in Iceland, through the North and South Atlantic, around Africa, Asia, and Australia, under the Pacific to the west coast of North America. It reaches 4,207 m (13,800 ft) above the ocean floor.
- The point farthest from the centre of the Earth is the summit of Mount Chimborazo in Ecuador, at 6,267 m (20,561 ft) high (*see left*).

HOW MOUNTAINS ARE MADE

If you sliced through the Earth you would see that the solid crust is relatively thin and made up of plates that slip and slide over a hot, putty-like mantle. Movement of these plates causes mountain formation.

Some mountains are formed when the Earth's plates crash into each other. About 70 million years ago the Himalayas were created in this way. Other mountain chains are made as ocean plates slip under neighbouring continental plates and lift the lighter land up into mountains. The edge of the continent crumples as the oceanic crust is forced under it, often creating mountains along the coast such as the chains of North and South America.

Block mountains are formed when two crustal plates tear apart, as in the Great Rift Valley of East Africa. The tearing compresses the margins of the plates, cracking and lifting up the crust to form mountains.

Mountains also form under the sea where an ocean plate has cracked open. Magma wells up and cools into fresh rocks which often rise up to form mountain ranges.

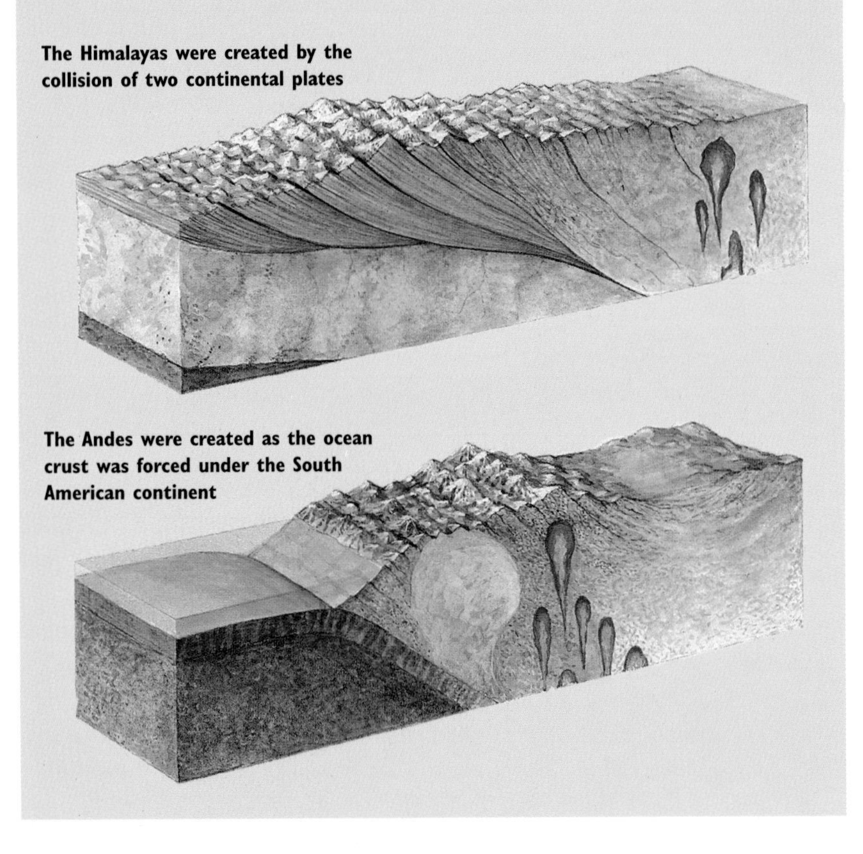

The Himalayas were created by the collision of two continental plates

The Andes were created as the ocean crust was forced under the South American continent

THE ROOF OF THE WORLD

The Tibetan plateau north of the Himalayas is the highest plateau on Earth. Averaging 4,500 m (14,760 ft) above sea level, it is higher than all the peaks in the Alps except for Mont Blanc and Monte Rosa, and well above the summits of most mountains in the United States.

Tibet originally lay under the sea. Then, about 50 million years ago, the continental plate carrying India crashed into Asia. As it did so, the land crushed, folded, and cracked, creating the Himalayas.

The collision was so severe that the sea behind the Himalayas was also compressed and lifted up to make the Tibetan Plateau. So Tibet is made from sea-floor rock.

RIDING HIGH *Movement in the Earth's plates lifted the sea floor 4,500 m (14,760 ft) high to create the lofty, barren Tibetan plateau.*

THE WORLD'S HIGHEST LAKE

High in the Andes between Bolivia and Peru lies Lake Titicaca, the world's highest navigable lake. Located at 3,810 m (12,500 ft) above sea level, the lake was formed 10,000 years ago and is the remnant of an ancient ocean which became trapped between the Cordilleras and the Andes. It is also one of the longest lakes on Earth, 177 km (110 miles) in length, and the second largest lake in South America.

The intense sun and strong winds evaporate so much water that the lake is quite salty and that restricts the range of wildlife. However, an

incredible variety of frogs lives on the shallow muddy bottom. There are more than 754 species of amphibian in the Andes, 95 per cent of which live nowhere else in the world.

THE FJORD THAT IS DEEPER THAN IT SHOULD BE

Scoured out by a glacier, Skelton Inlet, by McMurdo Sound on the coast of Antarctica, is the deepest fjord in the world. It plunges down 1,933 m (6,341 ft). The 1,450 m (4,760 ft) thick Skelton Glacier is thought to be carving out the fjord as it scrapes along the sea floor, grinding out the rocks like a giant chisel. But there is a mystery: the fjord may be too deep to have been carved just by ice. Ice would become buoyant at the deepest point in the fjord, and lose much of its scraping power. Yet what would be capable of digging so deep into the sea floor remains unknown.

GREAT CANYONS IN THE OCEAN FLOOR

Imagine a hole so deep you could sink Mount Everest in it with room to spare. This is the Mariana Trench in the western Pacific, the deepest place on the face of the Earth, 11 km (7 miles) below sea level. The Mariana Trench is a crescent-shaped canyon stretching 2,500 km (1,500 miles) across the floor of the Pacific. It averages 70 km (43 miles) wide. If you could fly down the trench it would look like a canyon flanked by vast, steep-sided mountains.

The trench is so deep because it sits on a geological fault line where the Pacific Ocean floor is plunging under the neighbouring Philippine crust at more than 11 cm (4 in) a year.

Challenger Deep is the deepest part at 10,925 m (35,843 ft). It was named after the British survey ship *Challenger II*, which pinpointed it off the Mariana Islands in 1951. The depths of the trench have been plumbed only twice – by unmanned submersibles.

SCULPTED BY EROSION

Over millions of years, water and wind have chiselled the Earth's rocks into magnificent sculptures; glaciers have carved sharp mountain peaks and deep fjords; rivers have dug canyons, and the sea has clawed columns from cliffs.

TIME AND ELEMENTS PUT ROCK ON A PEDESTAL

In the hills of Wyoming, USA, east of the Rockies, spheres of rock lie strewn across the ground. A few are balanced on stone pedestals at precarious angles, almost as if carefully placed there by a pair of gigantic hands.

Called concretions, the spheres were formed at the bottom of a shallow prehistoric sea. Each grew from a piece of bone or shell, which was slowly surrounded by layers of sand and the mineral calcite.

Today, the ancient seabed is more than 1,000 m (3,280 ft) above sea level, and the concretions are free of the softer sandstone that once encased them. Carved out over millennia by wind and rain they remain intact, each testament to a life lived millions of years ago.

THE BIGGEST PYRAMID IN EUROPE

It is not only the Matterhorn's 4,505 m (14,780 ft) height that inspires onlookers and mountaineers, but its spectacular shape: a pyramid carved by glaciers. Some 70,000 years ago an ice-sheet measuring up to 4 km (2.5 miles) thick grew across much of northern Europe, crushing the Alps. The glaciers slowly slid down over the mountains, chiselling, scraping and digging into the landscape like a sculptor carving a piece of stone.

On the Matterhorn, four glaciers all slipping and eroding on different sides of the mountain left a pyramid shape with knife-edges. When the world warmed up 10,000 years ago, the glaciers retreated revealing the towering pinnacle of the Matterhorn that now straddles the border between Switzerland and Italy.

THE MANY HUES OF ULURU

Deep in the Australian outback stands Uluru (Ayers Rock), the largest sandstone monolith in the world. One of the most surprising features of Uluru is the way it changes colour with the setting sun, turning orange, crimson, then brown and finally grey. On rare occasions when it rains, the rock takes on a silvery sheen. The light show is produced by marble and other stones in the rock, which reflect the light into different colours.

Rising 384 m (1,260 ft) above the plains and standing 2.5 km (1 1/2 miles) long and 9 km (5 1/2 miles) round, Uluru can be seen for tens of kilometres in all directions. It is the

TEE OFF *Balanced like a golf ball ready to be struck, this boulder shielded the sandstone beneath it from erosion, creating a pedestal.*

BLAZING ROCK *Reflecting the changing light, Uluru (Ayers Rock) in Central Australia is regarded as a sacred site by the Aborigine people.*

remains of an ancient mountain in a long-gone sea from 500 million years ago. As the Earth's crust gradually lifted, the sea drained and the mountain was left exposed. Millions of years of wind have slowly sculpted its rounded shape.

ITALY'S BALANCING BOULDERS

Spires of rock with boulders perched on their peaks like hats make an extraordinary sight in a valley in the Italian Alps. They are known as the Ritten Earth Pillars and are the work of glaciers.

When glaciers were on the move during the last Ice Age they picked up enormous cargoes of rocks and soil which were dumped when the ice melted. In some places a thick clay was laid down with boulders resting on top. As rivers swept through the valleys they carved out ridges in the clay, and rain caused further erosion. Eventually the rocks, which were more resistant to the rain, were left balancing on top of clay spires.

DID GIANTS PLAY IN THE MATOPOS HILLS?

In the Matopos Hills of Zimbabwe thousand-tonne boulders balance on stone columns, rocky outcrops, or even on top of one another. They look like the building bricks of some monstrous creature. The hills are what is left of a huge flow of lava that turned into granite mountains, which were then worn away by rain into craggy peaks. These relics cracked under daily heating and cooling, leaving boulders, piles and pinnacles.

THE WORLD'S LARGEST NATURAL BRIDGE

In the desert of Utah, USA, is a stone bridge 86 m (282 ft) high, which is almost tall enough to fit the Statue of Liberty beneath it. It is the largest natural bridge on Earth with a span of 84 m (275 ft), and is called Rainbow Bridge because of its striking arc shape.

The bridge was created when water flowing off a nearby mountain meandered across the local soft sandstone and dug out a canyon. The river looped tightly at one point around a wall of rock, then later took a shortcut and drove a hole right through the wall leaving a bridge of rock overhead. The river carried on with its new course, digging deeper and wider under the bridge. In time the river will dig so deep and wide that the bridge will collapse.

STILL STANDING *These pillars of limestone, known as the Twelve Apostles, have been carved by the thrusting waves and searing winds.*

APOSTLES CRUMBLE IN THE RELENTLESS SEA

In southern Australia, several great columns of rock can be seen standing in the sea, like a giant's fingers rising up from the water. These are the Twelve Apostles, a collection of sea stacks carved out of the crumbling limestone cliffs west of Melbourne by the unforgiving power of the waves.

The stacks no longer live up to their name, because some of the 12 have already collapsed. The sea continues to claw away at the cliff face, sculpting the rock and carving grottoes and arches. In time the coast road will fall victim to the constant battering.

STEPPINGSTONES FOR AN IRISH GIANT

The hexagonal columns of the Giant's Causeway in Northern Ireland are so precisely geometric they look man-made. For centuries they were thought to be the work of the legendary Finn MacCool, a giant who stepped across the stones and invaded Scotland. In fact, the causeway is an entirely natural phenomenon, the result of an eruption of lava 60 million years ago.

The lava spewed out through cracks in the ground and created a lake which hardened into a plateau. Some of the lava cooled so fast it contracted and formed 37,000 regular hexagonal columns. As wind, rain, and waves lashed the rocks, the columns eroded at different rates, leaving a higgledy-piggledy terrace of steps, and so the Giant's Causeway was made.

CHIMNEYS AND SPIRES GRACE AMERICAN DESERT

The deep red sandstone peaks of Monument Valley on the Utah-Arizona border in the USA are the result of millions of years of erosion. Wind, frost, and intense cycles of heat and cold have shattered the rocks into monoliths, pillars, chimneys, and spires. The almost surreal landscape of flat desert plains and steep red cliffs against deep blue skies has long been used by Hollywood as the setting for such classic Westerns as *Stagecoach*.

About 7,000 years ago the Anasazi people arrived in the region and began to live in homes dug out of the cliffs. The weather could well have been wet enough then to allow some sort of farming, but 750 years ago the climate suddenly turned very dry and the Anasazi vanished.

LIMESTONE LANDSCAPES

There is no other rock on Earth like limestone. It is pale, porous, and peculiarly sensitive to the acid in rain. Rainwater eats into the rock, opening up fissures and carving pinnacles, pillars, caves, canyons, and deep holes.

RAZOR-SHARP NEEDLES IN THE PINNACLES

Half-way up Gunung Api in Mulu National Park, Borneo, stands a stone forest of limestone needles. Known as The Pinnacles, some of the points rise to 45 m (148 ft) high.

They are the result of rain eating into vertical faults in the rock and causing large cracks to open up. Gradual erosion has produced these deeply incised, razor-sharp pillars.

In this challenging environment, carnivorous pitcher plants (*see pages 238-9*) overcome the poor soil by growing bowls from their leaves to collect rainwater and drown insects, in order to obtain the food they need.

MEDITERRANEAN PLANTS IN IRISH LIMESTONE

On the wet and windy west coast of Ireland, a landscape of flat limestone supports a collection of Mediterranean plants. The region, known as The Burren, covers an area of 300 km² (115 sq miles).

With plenty of rain and a warm breeze blowing from the Gulf Stream, the climate is surprisingly mild. The rain has eaten into weak fault lines in the limestone and opened up cracks where just enough soil collects to let the plants grow and shelter. Alpines and miniature trees normally found in warmer climates can flourish here.

JAMAICA'S HIDDEN LIMESTONE WORLD

The island of Jamaica features a hilly rain forest almost untouched since it was found by Columbus more than 500 years ago. Called The Cockpit Country, it is an uninhabited limestone landscape, complete with sink holes, underground rivers, waterfalls, and caves, covering 1,300 km² (500 sq miles). It formed about 20 million years ago when the shallow seabed around Jamaica lifted up out of the Caribbean.

Limestone is made from the compacted skeletons of sea creatures, and today the maritime origin of The Cockpit Country is evident in its rocks, which are rich in the fossils of sea animals.

The Cockpit Country's isolation makes it an important refuge for plants and animals, including 500 species of fern. It is also home to the endangered giant swallowtail butterfly, the largest in the Americas and second largest butterfly on Earth.

MADAGASCAR'S LIMESTONE WARRIORS

Rising up from the west coast of Madagascar a huge plateau of limestone towers above the island. On one side are gently rolling hills – peaks called *tsingy* – and to the south a bewildering array of knife-edged pinnacles standing parallel to each other like an army of warriors.

The Tsingy de Bemaraha is an impenetrable wilderness, and much of the vegetation is unique to the area, including a wild banana and a particular type of ebony tree. Desert plants such as the succulent aloe are perfectly adapted to the dry soils.

The inhospitable stone forest has protected native wildlife such as the red forest rat, a spiny chameleon, and the Madagascar grey-throated rail bird from human incursion.

CLIFF HANGING *The bizarre limestone cliffs of the Tsingy de Bemaraha National Park in Madagascar, a World Heritage Site.*

DESERT TEETH *The Pinnacles of Western Australia were long thought to have been carved by ancient man, but they are entirely natural forms.*

A CRAZY LANDSCAPE OF LIMESTONE TEETH

They look like neolithic stone pillars, but the thousands of tooth-like blocks in the desert of Western Australia are natural. Known as The Pinnacles, these limestone rocks form part of Nambung National Park. They are made from sand that was washed ashore tens of thousands of years ago. Among the sand were broken sea shells. Over time these were dissolved by rainwater and turned into calcium carbonate, eventually producing limestone rock that was buried under sand dunes. In the wet season rainwater seeped into the limestone, carving it into underground pillars.

Scattered all around The Pinnacles are fossil roots – a clue that the dunes were once wet enough to support vegetation. The roots would have helped the passage of water underground. About 25,000 years ago the climate became so dry that the sand dunes were blown away, revealing the pillars of limestone. The exposed rocks were further chiselled by the wind to create today's jagged shapes.

CHOCOLATE DROPS OR GIANT'S TEARS

Legend has it that the dome-shaped Chocolate Hills of Bohol in the southern Philippines are the hardened tears of a giant. He is said to have shed them when his lover died. There are 1,776 symmetrical hills, standing about 30 m (100 ft) tall, which appear chocolate brown in the dry season as the vegetation dies back.

They are the limestone relics of an earlier erosion. Some geologists think that the area was once entirely submerged and the hills were formed on the sea floor by volcanic eruptions and then eroded into domes by underwater currents.

Others believe that the hills were formed by the weathering of a marine limestone formation over an impenetrable clay stone.

THE GORGE CARVED OUT BY A HIDDEN RIVER

The Gorge du Verdon, in south-western France, formed when the roof of an underground cavern collapsed after thousands of years of erosion. The gorge is 20 km (12 miles) long with grey limestone cliffs rising to 700 m (2,300 ft).

The River Verdon, which flows in its valley, is a greeny turquoise colour – which is strange because there is almost no soil or organic debris in the water. The colour comes from the Verdon's source in the Alps where the glaciers erode the underlying rock and wash away fine particles that create a green-blue colour in sunlight.

SPAIN'S TOWERING LIMESTONE GATEWAY

Monte Perdido, on the Spanish side of the Pyrenees, is Europe's highest limestone peak at 3,352 m (11,000 ft). The huge bulky limestone mass towers over the Ordesa Canyon, its slopes covered with plants and animals. During the

upwards. The largest cave is the Carlsbad Cavern, which is more than 477 m (1,565 ft) deep and contains the cathedral-sized Big Room. It is decorated with stalactites, stalagmites, and thick crusts of minerals.

Lechuguilla Cave, also in the complex, has astounded researchers with its community of 1,200 different types of microbe. This is the world's greatest collection of bacteria living on rock.

The microbes draw their energy from sulphur, manganese, and iron – an interesting parallel to the microscopic organisms that feed off the hot mineral springs that vent from the ocean beds (*see page 354*).

THE LEGENDARY GORGES OF THE YANGTZE

The most dramatic stretch of China's mighty Chang Jiang (Yangtze) is the 200 km (120 miles) of gorges through Sichuan Province. In places the limestone gorges narrow to 100 m (330 ft) across, squeezing the river into a fierce torrent in the rainy season. The steep cliffs, up to 1,200 m (3,937 ft) high on both sides of the gorge, were born 70 million years ago when the Earth's crust shifted, thrusting up the land through which the Chang Jiang river carved its channel. But soon the beauty of the gorges will be lost to a hydroelectric scheme.

height of the Ice Age, the Pyrenees were a refuge for many plants and animals which have since evolved into organisms found only in these mountains. A few species are found only on Perdido, such as a small alpine saxifrage, a mountain frog, and the Pyrenean chamois, a type of goat.

A DESERT CAVE ADORNED WITH SEA CREATURES

In the dusty, dry Chihuahuan desert in New Mexico, USA, an ancient tropical reef decorates a massive complex of caves. The reef, which is unusual because it is made almost entirely of sponges and algae instead of corals, was created around 230 million years ago.

Since then the reef has been heaved up from the seabed and left high and dry as mountains of limestone. Inside the caves, it can be viewed like a walk-in aquarium exhibit.

Unlike most limestone caves, which are etched out by weak carbonic acid in rainwater, these caves were eroded by sulphuric acid made from underlying oil and gas as it seeped

ACID ATTACK

As rain falls through the atmosphere it absorbs enough carbon dioxide to make carbonic acid. When the slightly acidic rain falls on limestone, formed mainly from the calcified skeletons and shells of sea creatures, it eats into the rock and sculpts out some interesting formations. The limestone landscape that results takes thousands of years to form.

First small cracks form in a lattice pattern, exposing microscopic lines of weakness in the rock. As the cracks grow they open into gulleys, leaving regular blocks of rock exposed on the surface. The gulleys open wider into shafts or potholes leading into caves. Streams and rivers travelling over the limestone disappear down sinkholes, then flow through underground caves and reappear kilometres away.

Sometimes a cave roof collapses, leaving a valley such as the Gorge du Verdon in France. Gorges can also form as a result of rivers, like China's Chang Jiang, slicing through the limestone.

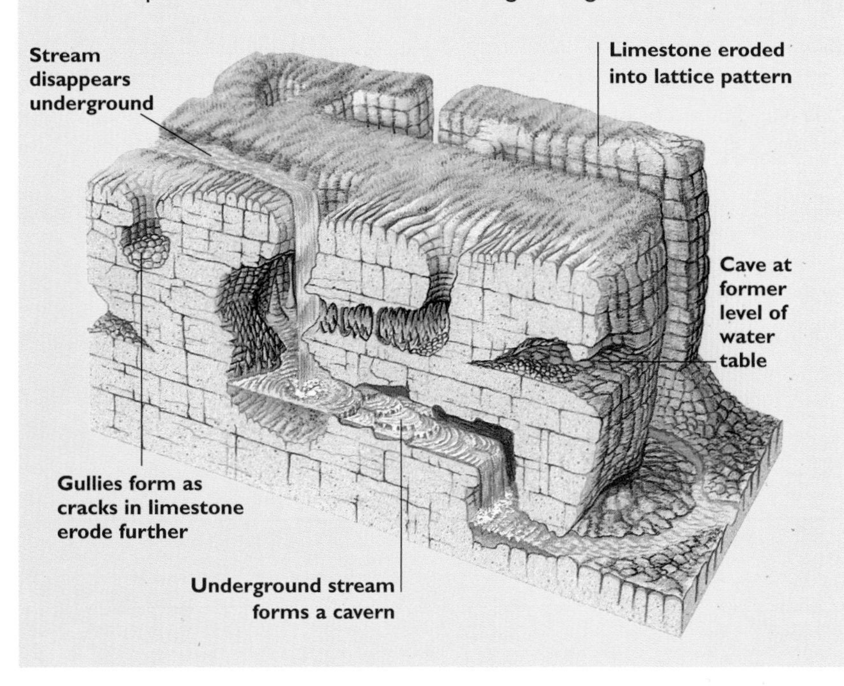

Nature carves a menagerie

Guilin in China is a limestone wonderland of hills and caves thrusting up into massive pinnacles, cones, and domes. Some formations look like factory chimneys, others resemble galloping horses, and many have taken names from their weird shapes, including the Camel, the Climbing Tortoise, and the Elephant Trunk. The Folded Silk Hill is where several peaks rise closely next to each other.

The limestone peaks protrude from the flat plains on the Li river like giant's teeth, soaring up to 100 m (330 ft) high and stretching along the river for 50 km (30 miles).

These heavily eroded hills were originally part of an ancient ocean floor that was gradually heaved up into dry land, exposing thick layers of limestone to the elements. Vast amounts of softer, spongy limestone were eroded away, leaving harder plugs of limestone towering above a plain of bedrock.

LOCKED IN TIME

The history of life on Earth is recorded in fossils, the remains of animals and plants preserved in rock. By piecing together the different ages of rocks and fossils it is possible to chart the evolution of life over the past 3 billion years.

THE PREHISTORIC COOK AND HIS CHINESE BARBECUE

At Zhoukoudian, south of Beijing in China, is a series of hill caves where in 1929 the skull of an early human was discovered. The skull belonged to *Homo erectus*, nicknamed Peking Man, who lived half a million years ago. The cave also held charred animal bones. The remains of his barbecue is the earliest evidence of our human ancestors using fire.

Peking Man looked in many ways like humans today. He was intelligent enough to make stone tools by hammering and using an anvil. But it was the mastery of fire that allowed early humans to survive in the cold and dark so that they could expand into more northerly regions, colonizing Asia and Europe.

THE ANCIENT FOREST THAT TURNED TO STONE

Deep in the Painted Desert of Arizona, USA, thousands of chopped-up tree trunks lie strewn about. The trees seem somewhat out of place in the hostile desert conditions until you realize that they are in fact the fossilized remains of trees that lived millions of years ago.

According to legend, a goddess tried to light a fire with the logs but they were too damp. She was so furious that she turned them into stone, ensuring that they would never burn again. In so doing she made the Petrified Forest.

The forest is made up of fossilized araucaria trees, similar to pine trees, that once stood up to 60 m (200 ft) tall. Millions of years ago a huge flood washed the logs many kilometres downstream until eventually they were

FALLEN GIANTS *The fossilized remains of a devastated primeval forest appear brightly pigmented. The vivid colours are due to the presence of iron, manganese, copper, and lithium in the original wood.*

all dropped into a shallow basin. They were buried so quickly under silt that they did not have time to decay. Over millions of years, minerals in the water slowly replaced the woody tissues and turned them to rock. Some of the minerals crystallized into valuable gemstones, such as amethysts, but most were plundered by fortune-seekers more than a century ago.

GIANTS UNEARTHED IN ARGENTINA

On the northern frontier of Patagonia in Argentina lies one of the largest dinosaur fossil sites in the world. The exposed rocks around Plaza Huincul provide a record of dinosaurs during their last 35 million years on Earth.

In 1993 the largest predatory dinosaur ever found was unearthed there. This killer beast, which weighed 6–8 tonnes and measured 14 m (45 ft) long, was named *Giganotosaurus carolinii* (giant southern lizard). The creature overshadows the Northern Hemisphere's most notorious carnivore, *Tyrannosaurus rex*.

An even more staggering find at Plaza Huincul was *Argentinosaurus*, the biggest animal that ever walked the surface of the Earth. It was a giant plant-eating dinosaur, 45 m (148 ft) tall, the height of a five-storey building, and weighed 100 tonnes.

FOOTPRINTS FROM THE PAST

In Laetoli in Tanzania, East Africa, three sets of footprints mark the first solid evidence of human-like creatures walking on two legs. They were discovered by the British anthropologist Mary Leakey in 1978 and provide a clue to the missing link between apes that walked on all four legs and humans that walk upright on two.

The footprints were made some 3.6 million years ago, after a volcanic eruption. Ash was scattered over the ground, then a rain shower turned it into a sort of wet concrete. Three of the creatures strode through the sludge, and their footprints were preserved.

SEA MONSTER FOUND AT COASTAL RESORT

In 1812 a young British girl astonished the world when she found the skeleton of a 'sea dragon', known as an ichthyosaur. The child, Mary Anning, made the find at Lyme Regis on the south coast of England, and hers was the first complete specimen of the fish-like reptile to be discovered in Britain.

The cliffs of Lyme Regis are packed with the fossilized remains of marine creatures from the Jurassic period. This is because 200 million years ago the limestone and clay that make up the cliffs were at the bottom of a tropical sea. When the sea creatures died they were covered in silt and eventually turned into fossils. After a storm, sections of cliff often collapse, spilling out their precious treasure.

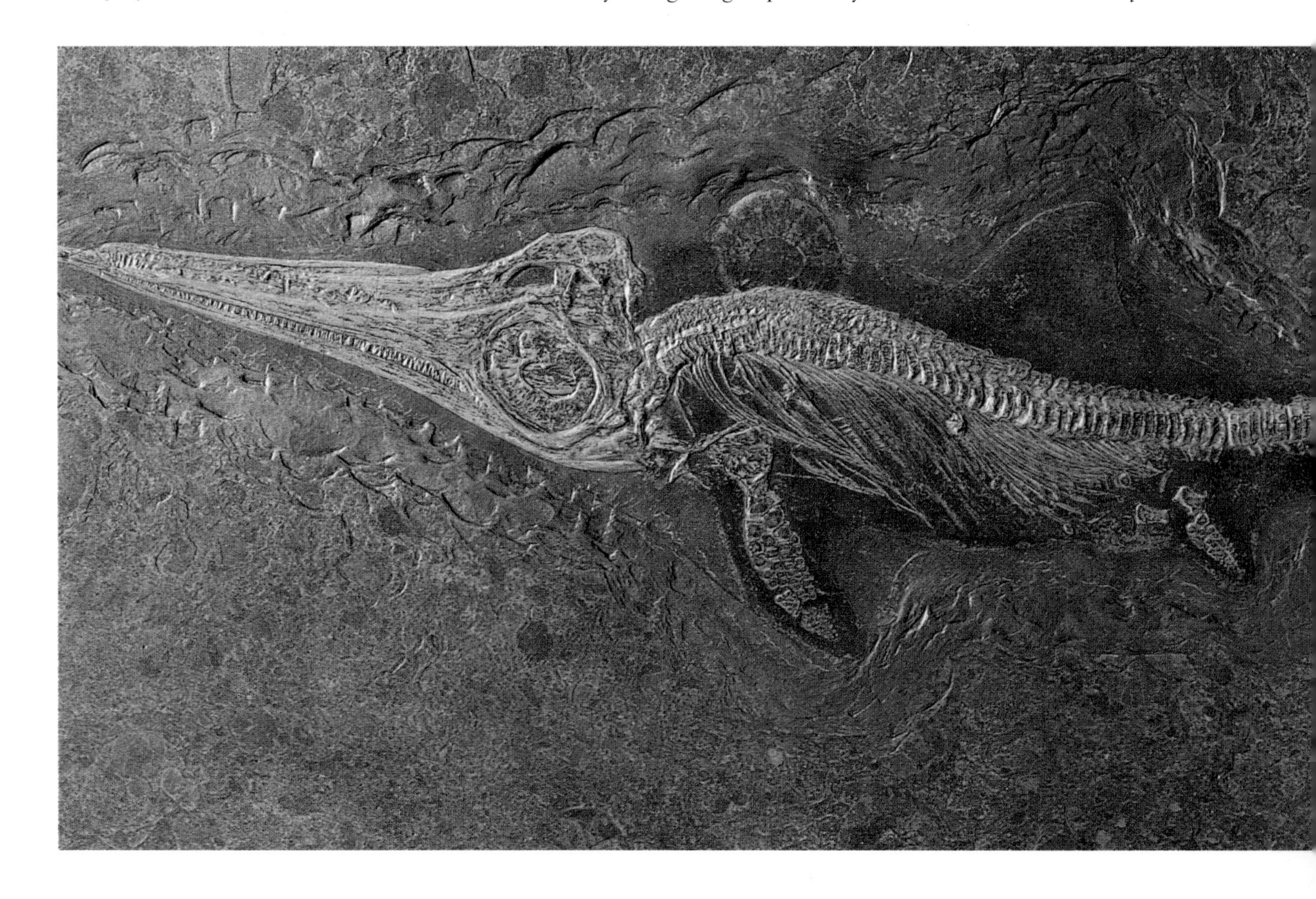

JURASSIC MONSTER *The ichthyosaur, first unearthed in Britain in 1812, was a 12.5 m (40 ft) long sea dragon that preyed on ammonites.*

NESTING SITE *Every year the winds and harsh winter weather of the Gobi Desert uncover a valuable haul of dinosaur and mammal fossils.*

GOBI DINOSAUR NEST SHOWS LINK TO BIRDS

In the Gobi Desert of Mongolia in 1923, palaeontologists discovered a fossilized nest which harboured toothless meat-eating dinosaurs. These creatures, with parrot-like beaks, are known as oviraptors. It was a ground-breaking find since it implied that dinosaurs looked after their young like modern-day birds.

The events that led to the creation of this important fossil occurred more than 80 million years ago. A group of meat-eating creatures was nesting in the shelter of a vast sand dune when a torrential downpour of rain caused the dune to collapse, completely swamping the nest and its occupants. Eventually the dunes were buried by sediments which turned to sandstone and the bones were fossilized.

In the 1920s, the same desert yielded the bones of a *Velociraptor* –

a carnivore that could run around on two powerful hind legs. Later finds have uncovered some of the finest examples of early mammals including the earliest marsupials (pouched mammals).

FOOTPRINTS LEAD TO FIRST MODERN HUMANS

Just over 100,000 years ago a 1.5 m (5 ft) tall woman walked down a steep dune after a storm and left her mark in the chalk. Sediment quickly washed over her footprints and fossilized them in rock. Today we have that footprint, the imprint of a heel, arch, and toes of a modern human, found in a lagoon near South Africa's Cape coast.

All the evidence points to modern mankind, *Homo sapiens sapiens*, coming out of South Africa. Not far away, at Klasies River Mouth, modern humans sheltered some time between 60,000 to 120,000 years ago and left behind prehistoric rubbish: mussel shells, bones of seals and penguins, and the ashes of fires. Yet another site in the region features 40,000-year-old prehistoric animal bones and stone tools. Here early people butchered meat, and cut marks made in the bones by knives can be seen.

REVEALING THE AGE OF THE EARTH

Today, it is possible to determine the age of rocks very accurately using radiometric dating, which measures the change of radioactivity over millions of years. But up until the 20th century geologists had to rely upon fossils to help them to estimate the age of sedimentary rocks (rocks originally laid down by ancient rivers, lakes, and seas).

By prising apart the layers in sedimentary rocks the fossils of dead plants and animals are exposed. These not only help to date the rocks, but also give some idea of past climates, and researchers have put together a history of the Earth dated from their fossils.

The dating of rocks began in 1828 when the British geologist Charles Lyell realized that young rocks contain many fossils which are similar to living life-forms, and that progressively older rocks have fewer and fewer species similar to today's organisms.

Era	Period	Epoch	m. yrs
CENOZOIC	QUATERNARY	HOLOCENE	0.01
		PLEISTOCENE	1.6
	TERTIARY	PLIOCENE	5.3
		MIOCENE	23
		OLIGOCENE	36.5
		EOCENE	53
		PALAEOCENE	65
MESOZOIC	CRETACEOUS		135
	JURASSIC		205
	TRIASSIC		250
PALAEOZOIC	PERMIAN		290
	CARBONIFEROUS		355
	DEVONIAN		410
	SILURIAN		435
	ORDOVICIAN		510
	CAMBRIAN		570
PRECAMBRIAN ERA			4600

AGES OF THE EARTH *The Earth is around 4,600 million years old, although fossils are rarely found in rocks over 570 million years old.*

THE EARLY MAMMALS' OILY GRAVEYARD

The Messel Pit near Darmstadt in Germany is the richest site in the world for early mammals. Since the first fossils were discovered there in 1875, 40 different animal species have been identified including a marsupial opossum, anteater, ostrich, and tiny ancient horses the size of a domestic cat. These were so well preserved that fossilized leaves were found in their stomachs. Even the metallic colourings of insects were preserved in the oily pit.

These are relics of the Eocene era, 57 to 36 million years ago, when the dinosaurs were long dead and mammals were coming into their own.

The fossils were preserved so well because the pit used to be a muddy lakebed. About 50 million years ago the land subsided and filled with water. In the hot tropical climate a thick layer of algae turned into sludgy mud, mixed in with sands and clay silts, to make an oily rock in which thousands of plants and animals were fossilized.

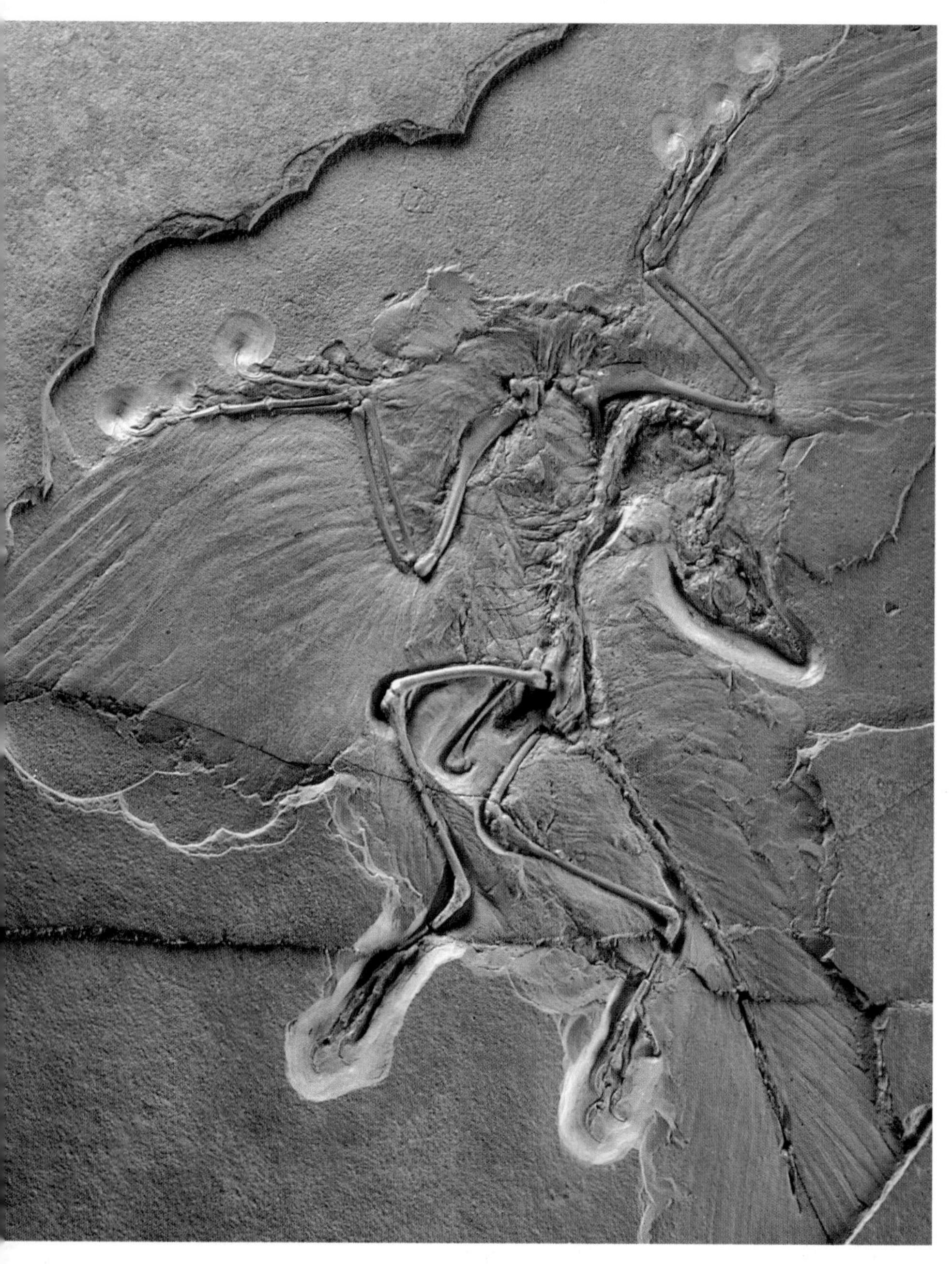

AMAZING FIND *The fossilized skeleton of* Archaeopteryx *appears to carry the faint impression of feathery wings and a tail.*

ARE FEATHERED DINOSAURS AVIAN ANCESTORS?

The origin of birds has long been one of the biggest mysteries of evolution. But a stunning collection of fossils from Liaoning Province in China suggests that today's birds could be the descendants of feathery dinosaurs. The region was covered by volcanic clouds of ash in the early Cretaceous period, making a fossil-rich lakebed. Local farmers digging in the rocks discovered a fossil the size of a chicken with fine indentations that suggested the thin hollow shafts of feathers. Since that initial discovery, other dinosaurs with feathers have been revealed – so well preserved that every detail of their skeletons, claws, beaks, and feathers can be seen. They were meat-eating theropods that lived 120 million years ago. Although covered with downy plumage, they could only leap using their tails for balance: they could not fly.

RESURRECTING MAMMOTHS FROM THE DEEP FREEZE

In a sensational experiment a Russian team are trying to recover DNA from frozen mammoths to create a mammoth-elephant hybrid. The frozen carcasses of mammoths, which became extinct about 10,000 years ago, are found all over Siberia. One mammoth was so well preserved that there was still grass in its teeth and stomach from its last meal.

To find good quality mammoth DNA the animal would have had to have been frozen rapidly after death. It is hoped that sperm extracted from the mammoth's gonads will be able to impregnate a living elephant to make a mammoth-elephant hybrid. Many doubt that this technique will be successful, let alone whether the resulting hybrid could survive.

Mammoths had to eat huge amounts of plants to survive, and their appetites led some of them too close to the edges of deep bog pools where they fell to their deaths.

CANADA'S DINOSAUR GRAVEYARD

Dinosaur Park, in the badlands of Alberta, Canada, is so named for its wealth of fossilized remains. Many types of dinosaur from the late Cretaceous period 75 million years ago have been unearthed here, including 38 species of dinosaur and 150 complete dinosaur skeletons.

Although Alberta's badlands now form an arid landscape of tabletop mountains, rocky pillars, and canyons, in the Cretaceous period this part of Canada was hot and subtropical. Hadrosaurs – duck-billed herbivorous dinosaurs – roamed around in lush forests and swamps. The fossils found today are mainly the remains of dinosaurs that fell into swamps or rivers and whose bones were quickly covered with layers of sediment. Over millions of years, as the sediments turned to rock, the dinosaurs were fossilized. In the Ice Age, 15,000 years ago, the rocks were scraped by moving ice, exposing the fossils.

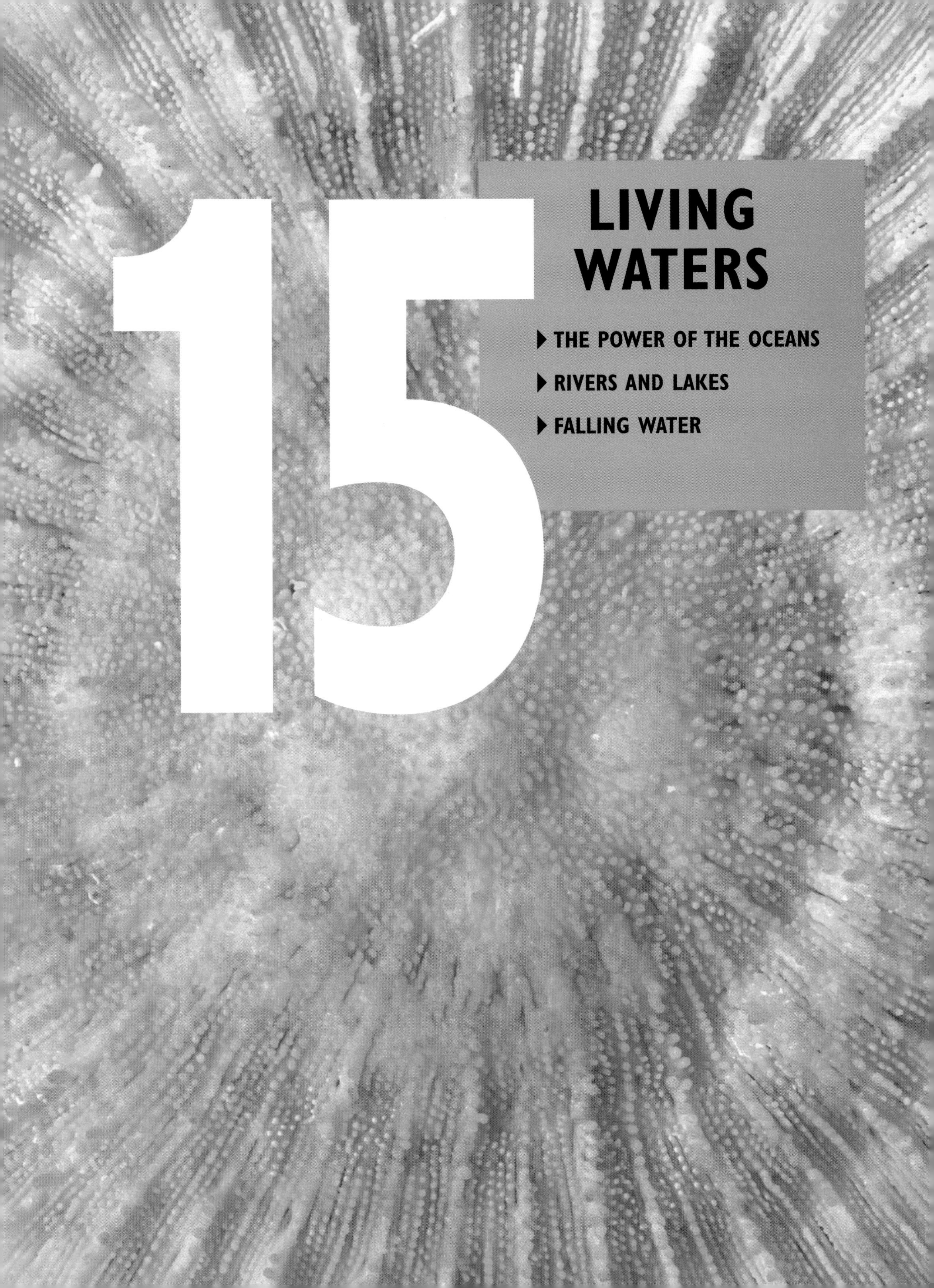

15 LIVING WATERS

- THE POWER OF THE OCEANS
- RIVERS AND LAKES
- FALLING WATER

THE POWER OF THE OCEANS

Seen from space, the Earth is a blue planet covered mostly by water. These vast oceans harbour a hidden world of volcanoes, giant gas bubbles, and strange mineral deposits. They also play a vital role in controlling the world's climate.

THE LONELIEST AND MOST TEMPESTUOUS OF SEAS

The Roaring Forties of the Southern Ocean are the most violent seas in the world. Here, around the Antarctic, sailors venture at their peril. Waves as tall as houses are not uncommon.

The Southern Ocean is the only ocean in the world uninterrupted by land, flowing in a complete ring. Its current carries more water than any other in the world, each day pumping 165 million tonnes and reaching depths of 3 km (2 miles). It is one of the world's most important places for marine life, where plankton are the staple diet for a food chain of krill (a prawn-like creature), fish, whales, seals, penguins, and a host of sea birds.

RAGING SEA *The savage Southern Ocean carries 150 times more water around Antarctica than the flow of all the world's rivers combined.*

MINERAL RICHES ON THE SEA FLOOR

Deep at the bottom of oceans are dark brown, potato-shaped balls that are rich in minerals, especially manganese. Iron, copper, nickel, and cobalt are also present. It is not known what created these manganese nodules. Possibly, liquid rock oozing up out of cracks in the ocean floor rolled up into balls which were then swept over the seabed.

The nodules are only 2.5–15 cm (1–6 in) across, but there are billions of them and they represent a major world resource. Manganese is an important industrial element used in steel, aluminium, and cast iron, but the economic and technical problems of plumbing such great depths have so far been impossible to overcome.

THE GULF STREAM: A GLOBAL HEATING SYSTEM

With the heat equivalent of 20 million nuclear power plants, the Gulf Stream brings warmth to north-west Europe. This great ocean current carries warm water from the Caribbean to the colder waters of the Arctic Circle via the Atlantic. It is a phenomenal force: every hour 90 trillion tonnes of water pass through Cape Florida.

As the Gulf Stream travels through the Atlantic the heat keeps north-west Europe several degrees Celsius

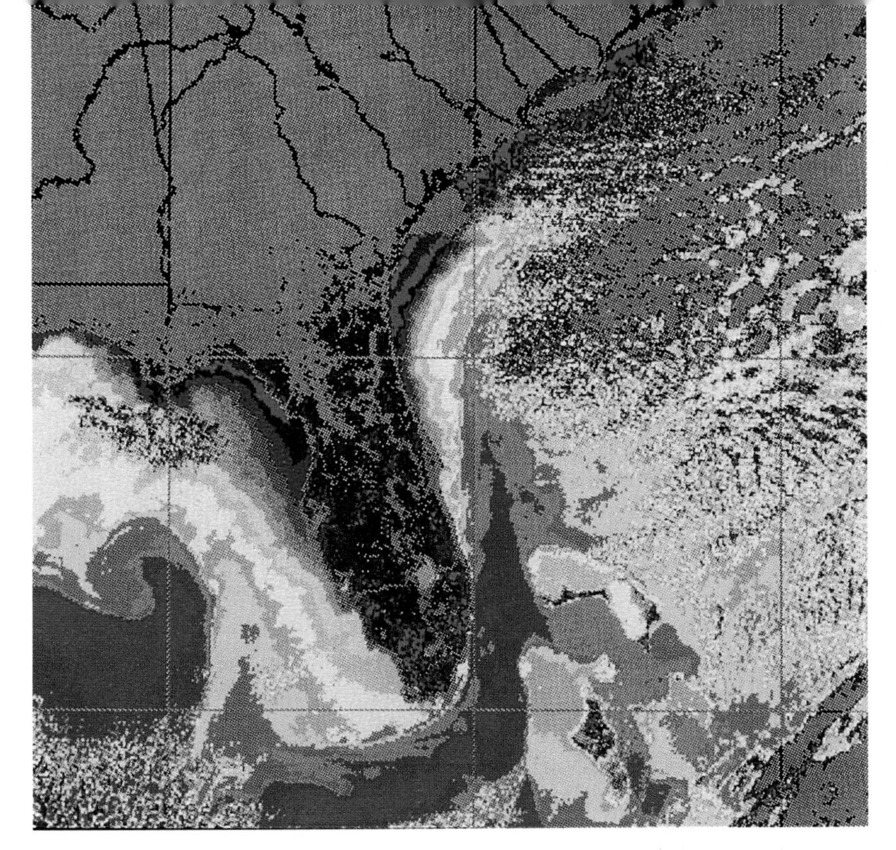

HEAT SOURCE *The Gulf Stream sweeps past Florida up to northern Europe, carrying warm water that has a major effect on the weather.*

warmer in the winter than equivalent latitudes in Siberia or Canada. Where the full force of the Gulf Stream hits County Kerry in the south-western corner of Ireland, it has nurtured a natural subtropical paradise of plants and trees.

When the warm current reaches the Arctic, it cools, turns more salty and sinks to the ocean floor. As it does so, it sucks more warm surface water northwards from the Gulf of Mexico.

MAGNETIC MYSTERY OF THE BERMUDA TRIANGLE

If you believe all the stories, the area of sea bordered by Bermuda, Florida, and Puerto Rico is the most mysterious place on Earth. This is the Bermuda Triangle, where over the past 100 years more than 50 boats and 20 aeroplanes have disappeared for no apparent reason. For example, on December 5, 1945, five US Navy planes vanished on a flight between Florida and Bermuda. A search plane sent to find them also disappeared without trace. Supernatural forces have been blamed, but there are possible scientific explanations.

The Bermuda Triangle is one of the few places on Earth with local magnetic anomalies in the bedrock of the ocean that make compass needles point to the true north instead of the magnetic north pole (an anomaly of about 20 degrees), and that could easily confuse navigation.

The Triangle is close to the Gulf Stream, a swift current with eddies that can sink boats. Also, there are pockets of methane gas lying under the seabed in the area. The methane could escape in violent bubbles that might destroy the buoyancy at the sea's surface and sink even large ships.

DEVASTATING FORCE OF THE SURGING RIVER WAVE

The greatest river bore in the world is found on the Tsientang Kiang in China. At spring tides a sudden wall of water up to 3 m (10 ft) in height can surge upriver at 25–35 km/h (15–20 mph), leaving a trail of fallen trees, dead animals and toppled boats in its wake.

Bores happen on rivers which have a large height difference between low and high tides, a bottleneck mouth, and face out to sea. On the Tsientang Kiang, bores occur when exceptionally high tides are squeezed upriver at two to three times normal speed.

WONDROUS FACTS ABOUT WATER

The vast majority of people in the world depend on the oceans in some way or another, whether for food, transport, oil, or gas.

- 80 per cent of the Earth's surface is covered by water.
- 97 per cent of the Earth's water is sea water.
- The volume of the world's oceans is 11 times the volume of the land above sea level.
- The Earth's oceans cover about 360 million km^2 (140,500,000 sq miles) and contain almost 1,370 billion km^3 (329 billion cu miles) of water.
- There is enough groundwater in North America to cover the continent with a sheet of water almost 30 m (100 ft) thick.
- The largest ocean in the world is the Pacific at 166 million km^2 (64 million sq miles).
- Every 24 hours about 1,000 km^3 (250 cu miles) of water evaporates from the sea and land.
- The pressure down in the depths of the ocean is more than 1,044 kg/cm^2 (2,300 lb/sq in).
- Sea water is about 1,035 times heavier than fresh water because it contains 35 kg (77 lb) of salt in each 1,000 kg (2,205 lb) of water.
- The world's seas contain 50 quadrillion tonnes of dissolved salts.
- Sea water freezes at –2°C (28°F).
- The ocean is blue because when the sun shines through it, it hits tiny particles and blue light bounces back.

CUTTING EDGE *The action of the tides and rising temperatures in spring cause ice to sheer off Greenland's glaciers and float away.*

HIDDEN DANGER OF THE ICEBERGS

Each year up to 15,000 icebergs break away from Greenland's west-coast glaciers. They can be hundreds of metres high. The tallest known Arctic iceberg, seen in 1967, towered 170 m (557 ft) above the ocean, slightly less than half the height of the USA's Empire State Building.

Most icebergs melt away before they reach the North Atlantic, but a number slip into the shipping lanes and pose a big threat, particularly since two-thirds of the iceberg will be below the surface of the water.

Antarctic icebergs are much larger than those found in the Arctic. They can be found as far as 1,600 km (1,000 miles) north of the pack ice.

THE NOTORIOUS NORTHWEST PASSAGE

In 1999, a pair of icebreakers escorted a vessel right through the hitherto frozen Northwest Passage for the first time. For more than 500 years mariners such as Sir Francis Drake and Vasco de Gama had tried in vain to find a way to the east around the northern edge of North America. Recently the ice has thinned dramatically due to global warming, making this epic voyage possible.

Over the past 30 years the waters of the western Arctic have risen by 3°C (5°F) and almost 40 per cent of the pack ice has disappeared in the last decade alone. It is predicted that the Arctic pack ice could vanish altogether by the end of the 21st century. The route would cut 8,000 km (5,000 miles) off sea journeys between Europe or the east coast of North America and China.

ICY THREAT *Icebergs are a hazard for shipping. In 1912 the* Titanic *hit an iceberg off Newfoundland and sank with the loss of 1,513 lives.*

TREACHEROUS POOLS OF WHIRLING WATER

The Maelstrom off the coast of northern Norway is the world's largest whirlpool at 6 km (3¾ miles) wide. It whirls at up to 6 m (20 ft) a second, as currents flowing in opposite directions set off the whirlpool like two fingers spinning a toy top. Many fishermen have lost their lives to this monster of the sea.

Other whirlpools are created when currents are squeezed through narrow straits, such as in the Gulf of Corryvreckan in western Scotland. Some can be so violent that they actually make a noise. The Old Sow whirlpool sounds like a pig grunting as it rushes around in the straits between islands in the Bay of Fundy in Canada.

A CONGREGATION OF VOLCANOES

In the mid-eastern Pacific Ocean lies one of the world's densest concentrations of active volcanoes. Discovered in 1993, over a thousand volcanoes are gathered in an area about the same size as England, or New York State. The volcanoes were found with the help of satellite measurements of changes in ocean height caused by the underlying topography followed by sonar scanning. The cones rise 1,800–2,100 m (6,000–7,000 ft) from the sea floor.

The Pacific is the largest and deepest ocean on Earth, covering more than a third of the planet's surface. Finding active volcanoes beneath the surface is no great surprise as the region is renowned for its many volcanic islands, some of which are long dormant while others remain active.

VOLCANIC PEAK *Hawaii is part of an isolated string of volcanoes that stretches across the Tropic of Capricorn in the Pacific Ocean.*

A TROPICAL OCEAN OF ROTATING SEAWEED

The Sargasso Sea is so named because of its vast floating community of sargassum seaweed. The seaweed grows in this region of the North Atlantic around Bermuda, because the sea is rich in iron, thought to be blown on dust from the Sahara. The Sargasso is a giant lens of clear water, about 5 million km^2 (2 million sq miles) in area, which is set in motion by the Earth's rotation.

Freshwater eels from Europe and the eastern USA migrate thousands of kilometres to breed in its waters (*see page 183*).

SEABED MAGNETISM HOLDS KEY TO EARTH'S HISTORY

In 1963, geophysicists Fred Vine and Drummond Matthews were reading magnetic traces across the seabed when they noticed a strange phenomenon: the rocks on either side of the North Atlantic mid-ocean ridge exhibited identical stripes of magnetism, alternating between north-south and south-north polarity.

When dated, the bands were progressively older the farther they were from the central ridge. Iron particles in magma emerging at the ocean ridge had aligned themselves to the Earth's magnetic field and the alternating polarity showed that the Earth's magnetic field has reversed several times in geological history.

THE FROZEN SEA

The Arctic Ocean is the smallest and shallowest of all the world's oceans, at 1,000 m (3,300 ft) deep. In winter sea-ice covers 15 million km^2 (6 million sq miles), though it shrinks to half that in summer.

It is thought that global warming is the reason why the polar ice is melting at an accelerated rate. If the ice continues to melt, the effect on the rest of the world could be devastating, affecting global and regional weather systems in an unpredictable way.

EL NIÑO AND LA NIÑA

In 1997 the Pacific Ocean was gripped by a powerful force. Known as El Niño, it was the biggest climate disaster of the 20th century. It killed more than 4,000 people, caused $20 billion-worth of damage, and ruined the livelihoods of millions of people.

For reasons not fully understood, every few years a massive bulge of warm water, almost 30 cm (12 in) high and covering an area larger than the USA, builds up in the Pacific off the coast of South America. It signals a new El Niño, 'The (Christ) Child', so named because Peruvian fishermen found that it tended to occur off their coast at Christmas. El Niño is a mysterious lurch in climate, in which the Pacific 'does a somersault'. The cold waters off Latin America become warm, and on the other side of the world the warm waters from Australia to Indonesia grow cold.

That flip in the Pacific currents creates havoc. The coasts of California, Peru, and Chile, which are usually dry, are subject to storms, floods, and landslides as the warmer sea results in massive rainfall on the land. Meanwhile, the monsoon rains of Indonesia, Papua New Guinea, and northern Australia are shunted off course, sending the regions into drought.

La Niña (El Niño's little sister) has the opposite effect – the seas off the Americas turn colder, bringing drought, while those off South-east Asia turn warmer, bringing torrential rain and floods.

SEA CHANGE *Floods in Kenya were just one disaster caused by the El Niño of 1997–8, which altered the temperature of the sea currents.*

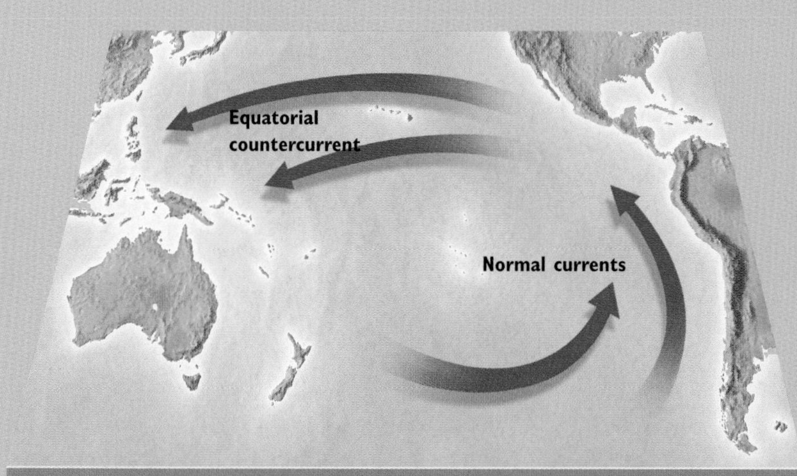

AGAINST THE FLOW *During El Niño the normal flow of the Pacific currents is reversed, causing weather chaos in the surrounding countries.*

RIVERS AND LAKES

Fresh water is vital for life on land. Yet only 2.5 per cent of the world's water supply is fresh water, and two-thirds of that is locked up in ice. The remainder runs off the land into lakes and rivers and eventually flows out into the sea.

THE MIGHTY FORCE OF THE AMAZON

The River Amazon in South America carries a fifth of all the water that runs off the surface of the Earth. From the deep gorges and rapids of the Andean foothills to the pounding surf of the Atlantic Ocean, this is the mightiest river in the world. It is fed by more than a thousand tributaries and discharges 800,000 million litres (170,000 million gallons) per hour into the Atlantic. The force of this torrent surging out of the river's mouth is so immense it pushes back the Atlantic sea water for more than 160 km (100 miles).

Most of the River Amazon is flanked by flood plains and at many points along its course it is so wide you cannot see from one side to the other: at its mouth it is 140 km (87 miles) wide. During the height of the annual floods, water levels can rise as high as 13 m (43 ft).

The Amazon is home to alligators, monkeys, parrots, sloths and the world's largest snake, the anaconda. Piranha and the 2.5 m (8 ft) long pirarucu, or arapaima, are among the many species of fish that inhabit its waters.

FLOODED FORESTS *Rainfall over the Amazon annually tops 2.5 m (8 ft). Forests are regularly inundated, creating flood plain ecosystems.*

LIFELINE WINDING THROUGH THE DESERT

The River Nile is the world's longest river. It is 6,648 km (4,132 miles) in length, which is greater than the distance from New York City to San Francisco, or from London to Lagos. Stretching from Burundi and Lake Victoria in Central Africa, winding through the Sahara Desert, and emptying into the Mediterranean, the Nile drains the largest area of any river in Africa, covering 3,349,000 km^2 (1,293,000 sq miles), about one-tenth the area of the continent.

Rich sediment washes down the river, fertilizing the flood plains that helped to nurture the Egyptian civilization about 5,000 years ago. The ancient Egyptians flourished for more than 3,000 years, longer than most other civilizations in world history, thanks in part to the bountiful Nile, which provided them with water, food, and a means of transportation.

They worshipped the sacred ibis, a bird which has not been seen on the

PICTURE OF THE PAST *Three young sacred ibis investigate a Nile crocodile. Today both species occur together only in sub-Saharan Africa.*

Nile for 100 years, although its close relatives, the red-billed ibis and glossy ibis, still feed on the waters of the Nile. The river teems with fish including the Nile perch and the tilapia, which holds its young in its mouth. Also lurking in the waters is the fearsome Nile crocodile, looking like an alligator, and reaching 3 m (10 ft) in length and up to 680 kg (1,500 lb) in weight.

A SACRED RIVER

The Ganges is India's sacred river, named after the Hindu goddess Ganga, the daughter of the mountain god Himalaya. It starts as a trickle of water melting off a glacier in the Himalayas and stretches for 2,506 km (1,557 miles).

The volume of water and silt pouring out of the river is so enormous that it dilutes the salt content of the Bay of Bengal and stains the water a muddy colour for 500 km (310 miles) out to sea. Some 2.4 billion tonnes of sediment washes down the Ganges each year – more than any other river in the world.

Its mouth is the largest on Earth. This vast rugged swamp forest, called the Sunderbans, covers an area of 42,000 km² (16,200 sq miles).

THE GREAT RIVERS

NILE
LENGTH: 6,648 km (4,132 miles)
DRAINAGE AREA: 3,349,000 km² (1,293,000 sq miles)
DISCHARGE AT MOUTH: 1,584 m³/sec (56,000 cu ft/sec)
DISTINCTION: Ancient Egyptian agriculture depended on its seasonal flooding
NOTABLE TRIBUTARIES: White Nile, Blue Nile
MAJOR CITIES: Alexandria, Cairo, Luxor, Khartoum

AMAZON
LENGTH: 6,500 km (4,000 miles)
DRAINAGE AREA: 7,050,000 km² (2,722,000 sq miles)
DISCHARGE AT MOUTH: 180,000 m³/sec (6 million cu ft/sec)
DISTINCTION: The world's greatest river flow. The longest navigable river – navigable for 3,200 km (2,000 miles)
NOTABLE TRIBUTARIES: Negro, Tapajós
MAJOR CITIES: Belèm, Manaus

MISSISSIPPI-MISSOURI
LENGTH: 6,019 km (3,870 miles)
DRAINAGE AREA: 3,221,000 km² (1,247,000 sq miles)
DISCHARGE AT MOUTH: 17,545 m³/sec (600,000 cu ft/sec)
DISTINCTION: The longest river flowing southwards
NOTABLE TRIBUTARIES: Yellowstone, Platte, Ohio, Red, Cumberland
MAJOR CITIES: New Orleans, Memphis, St Louis, Minneapolis, Kansas City

CHANG JIANG (YANGTZE)
LENGTH: 5,525 km (3,434 miles)
DRAINAGE AREA: 1,970,000 km² (698,000 sq miles)
DISCHARGE AT MOUTH: 35,000 m³/sec (1 million cu ft/sec)
DISTINCTION: Lifeline of China
MAJOR CITIES: Shanghai, Nanking, Wuhan, Chonquing

CONGO
LENGTH: 4,700 km (2,900 miles)
DRAINAGE AREA: 3,822,000 km² (1,440,000 sq miles)
DISCHARGE AT MOUTH: 42,000 m³/sec (1.5 million cu ft/sec)
DISTINCTION: The longest river flowing westwards
NOTABLE TRIBUTARIES: Kwa, Ubangi, Sangha
MAJOR CITIES: Kinshasa (Leopoldville), Brazzaville, Kisangani (Stanleyville)

MEKONG
LENGTH: 4,500 km (2,500 miles)
DRAINAGE AREA: 795,000 km² (307,000 sq miles)
DISCHARGE AT MOUTH: 15,900 m³/sec (500,000 cu ft/sec)
DISTINCTION: Longest river in South-east Asia
MAJOR CITIES: Phnom Penh, Vientiane

VOLGA
LENGTH: 3,688 km (2,300 miles)
DRAINAGE AREA: 1,380,000 km² (533,000 sq miles)
DISCHARGE AT MOUTH: 8,000 m³/sec (300,000 cu ft/sec)
DISTINCTION: Flows to landlocked Caspian Sea
NOTABLE TRIBUTARIES: Oka, Kama
MAJOR CITIES: Moscow, Gorky

DANUBE
LENGTH: 2,850 km (1,750 miles)
DRAINAGE AREA: 805,000 km² (325,000 sq miles)
DISCHARGE AT MOUTH: 6,450 m³/sec (200,000 cu ft/sec)
DISTINCTION: Europe's only major river to flow eastwards
NOTABLE TRIBUTARIES: Drava, Sava, Tisza, Prutul
MAJOR CITIES: Belgrade, Budapest, Bratislava, Vienna

GANGES
LENGTH: 2,506 km (1,557 miles)
DRAINAGE AREA: 1,073,000 km² (400,000 sq miles)
DISCHARGE AT MOUTH: 15,000 m³/sec (500,000 cu ft/sec)
DISTINCTION: Sacred river of the Hindu religion
MAJOR TRIBUTARIES: Brahmaputra, Jumna
MAJOR CITIES: Allahabad, Chandpur

RHINE
LENGTH: 1,320 km (820 miles)
DRAINAGE AREA: 224,000 km² (86,000 sq miles)
DISCHARGE AT MOUTH: 2,200 m³/sec (77,000 cu ft/sec)
DISTINCTION: One of Europe's principal commercial rivers
MAJOR TRIBUTARIES: Main
MAJOR CITIES: Cologne, Bonn, Strasbourg, Basel, Düsseldorf

THE DEEPEST LAKE ON THE PLANET

Lake Baikal is situated in southern Russia, close to the border with Mongolia, and plunges to 1.6 km (1 mile) at its deepest point. It is also the world's oldest lake, formed more than 20 million years ago as the Earth's crust tore apart to create a rift valley. Rivers gushing down the mountainsides then filled the canyon with water. Today the lake is itself the source of the world's fifth-longest river, the Yenesei, which stretches 5,540 km (3,440 miles) to the Arctic Ocean.

Baikal holds more water than any other lake on Earth: 23,000 km^3 (5,500 cu miles). The lake is still growing larger as the land on either side of it continues to tear apart. In a few million years' time the rift is expected to join up with the Arctic Ocean and cause Asia to split in half.

LAKES VISIBLE FROM THE MOON

The Great Lakes of North America and their connecting channels form the largest freshwater system on the surface of the Earth. Lakes Superior, Michigan, Huron, Erie, and Ontario cover more than 245,000 km^2 (94,000 sq miles) and hold 22,813 km^3 (5,473 cu miles) of water – that is one-fifth of the world's entire freshwater supply. If all this water was emptied over the mainland of the USA it would be 3 m (10 ft) deep.

The Great Lakes are so vast they have a climate of their own, cooling surrounding areas in summer and causing huge amounts of snow to be dumped over shoreline towns in winter when cold Arctic air sweeps across the water and picks up tremendous amounts of moisture.

If you could stand on the Moon, you would be able to see the lakes and even recognize the shape of the biggest, Lake Superior, which has the largest surface area of any freshwater lake in the world. It has nearly 4,500 km (2,800 miles) of shoreline, stretching 560 km (350 miles) at its widest point west to east, and 260 km (160 miles) north to south. The lakes drain into the St Lawrence River, which empties into the North Atlantic Ocean.

THE LAKE CREATED BY AN EXPLODING MOUNTAIN

Crater Lake, the deepest lake in North America, is the result of a gigantic eruption that ripped out the heart of a volcano some 7,700 years ago. The explosion of Mount Mazama in the Cascade Mountains, in what is now Oregon, left behind a bowl-shaped hole about 1,200 m (3,940 ft) deep that eventually filled up with water.

Since that ancient eruption another, smaller volcano has risen from the

A SATELLITE'S VIEW *On the border of the USA and Canada, the five Great Lakes are so huge that they can be seen from space.*

BLUE GEM *A volcanic explosion in prehistoric times gave birth to Crater Lake in the USA. The island is the result of continuing volcanic activity.*

bed of the lake and formed an island. The lake basin is now being mapped with sonar equipment, and it is believed that there are hydrothermal vents (*see page 354*) on the lake bottom, formed from hot water seeping through cracks in the floor. The water is heated by volcanic rocks, which might also be responsible for colouring the lake's water a beautiful crystal blue.

WATER THAT HARBOURS A DEADLY GAS

On August 21, 1986, Lake Nyos in Cameroon, West Africa, released a deadly gas. A cloud of odourless carbon dioxide gas erupted from the lake and rushed down into the valleys. Some 1,700 people perished.

Just two years before, nearby Lake Manoun killed 37 people with the same deadly gas. Both Nyos and Manoun look like ordinary lakes, except that they lie on a volcanic zone where the Earth's crust is very weak and lava easily erupts. Sometimes carbon dioxide gas wells up from deep under the lakes. A small shift in the water from, say, an underwater tremor, releases the gas and the carbon dioxide erupts. Since the 1980s, gas has again built up inside both lakes to dangerous levels, possibly double that of the past eruptions. The fear is that the lakes could explode with so much gas they could endanger the lives of thousands more people and the wildlife that has re-established itself.

Special degassing works are now being built to avert further disasters. These are designed to suck the gas from the bottom of the lakes.

THE TUFA SPIRES OF LAKE MONO

Pillars of calcium carbonate up to 9 m (30 ft) high project from the shores of Lake Mono in California, USA. The amazing formations, known as tufa, are the result of minerals seeping up from underground springs.

The lake is a geologist's paradise. It is 760,000 years old (the oldest lake in North America), and is ringed by volcanoes. Two of its islands are volcanic domes.

The lake was originally fed by ancient glaciers during the Ice Age when it was 60 times larger than today's 170 km² (66 sq miles).

STILL WATERS *It may look empty but Lake Mono is teeming with invertebrate life. It is also a major stopover point for migratory birds.*

UTAH'S STINKING PREHISTORIC LAKE

At the end of the Ice Age, the shrinking glaciers fed a vast lake across the western part of North America. Today that prehistoric lake has shrunk from 52,000 km² (20,000 sq miles) to a tenth of its original size. The Great Salt Lake sits in the dazzling white Bonneville salt flats of Utah. Its water is five times

more saline than sea water and contains 5 billion tonnes of salt, making it easy for a person to float on the surface.

The lake stinks of rotten eggs from the sulphurous gases given off by rotting plants and animal remains in the shallow waters. There are no fish, and the tiny brine shrimp is one of the few animals that can survive in this hostile environment. But the shrimps are a vital part of the lake's ecology, keeping it clean by feeding on algae

RED TIDE *A multitude of red algae on the shoreline of Great Salt Lake, USA. They are among the few living things that can tolerate the salinity.*

and they in turn are food for one of North America's largest avian populations: almost 2 million eared grebes, numerous migrating birds, with ducks, geese, pelicans, gulls, and hundreds of other types of birds living in the marshes and wetlands surrounding the lake.

THE LIMESTONE LAKES OF THE HINDU KUSH

Under the arid foothills of the Hindu Kush mountains of north-eastern Afghanistan lies a dramatic staircase of lakes. Over hundreds and thousand of years, travertine (calcium carbonate) from the mountain springs, which gently seep through the limestone rocks, has built up walls around the Band-e Amir River and dammed it into a series of six lakes, the largest of which is 6.5 km (4 miles) long. Where the water has spilled over the edges of the lakes, it has formed spectacular cascades of travertine that look like waterfalls frozen in suspended animation. The water has then built up in another lake below, and so a 'staircase' of brilliant blue, green, and turquoise lakes has been created.

THE WORLD'S LARGEST INLAND SEA

Covering 370,000 km^2 (143,000 sq miles), the Caspian Sea is the largest body of inland salt water in the world. It is up to 1,025 m (3,363 ft) deep, and lies on the traditional border between Europe and Asia. This saltwater sea is 28 m (92 ft) below sea level with no outlet and is mostly fed by the River Volga.

It is important for fishing, most notably the sturgeon that make the famous Beluga caviar, as well as herring, pike, and carp.

The Caspian is facing serious ecological damage: the sturgeon are being overfished and poisoned by oil spills. Since 1929 the sea has been shrinking and ports at the northern end are being left high and dry.

A WALL OF POWER

Iguazu, on the border of Brazil and Argentina, is made up of 275 waterfalls arranged in a horseshoe-shape 2.7 km (3,000 yd) wide, forming a natural amphitheatre of water power.

Some cascades drop 82 m (269 ft) straight down a rugged precipice into a yawning gorge below; others tumble over a series of short steps.

Eleanor Roosevelt, wife of the former US President Franklin D. Roosevelt, said the Iguazu Falls 'make our Niagara Falls look like a kitchen faucet [tap].'

A permanent cloud of mist hovers 30 m (100 ft) high, usually sparkling with rainbows. The moisture nourishes the surrounding tropical forest and more than 2,000 species of plants, including begonias, orchids, ferns, lianas, and palms.

Toucans, flocks of parrots, and swifts dodge in and out of the falls to feed on insects, and myriad butterflies flit among the greenery.

FALLING WATER

The thundering waters of the world's great waterfalls are awesome. They are powerful enough to wear down and cut through solid rock, yet their fine plumes of mist create a delicate microclimate for a variety of plants and animals.

THE GREAT SMOKE THAT THUNDERS

Africa's Victoria Falls is one of the most spectacular waterfalls in the world. Situated on the Zambezi River at the border of Zimbabwe and Zambia, the falls are about 1.6 km (1 mile) wide and drop 128 m (420 ft) into a narrow chasm. The spray, which forms a rainbow-filled cloud towering 300 m (1,000 ft) above the waterfall, can be seen from 20 km (12 miles) away. Local Kalolo tribespeople called the mist, spray, and noise created by the falls 'The Smoke That Thunders'.

Victoria Falls is actually made up of five separate waterfalls: the Eastern Cataract, Rainbow Falls, Devil's Cataract, Horseshoe Falls, and Main Falls. The British explorer David Livingstone was the first European to discover the falls, in 1855, and named them after Queen Victoria.

PLUNGE POOL *At the height of the rainy season the Zambezi River cascades over Victoria Falls at 1,080 m³ (38,000 cu ft) a second.*

CASCADE DOOMED TO DISAPPEAR

The huge amount of water flowing over North America's Niagara Falls eats away the bedrock at almost 1.2 m (4 ft) a year. At this rate, in about 25,000 years' time the entire Niagara River will be eroded, Lake Erie will merge with Lake Ontario, and Niagara Falls will disappear altogether.

On average 5.7 million litres a second (1.25 million gallons a second) plunge down the falls from Lake Erie, 98 m (320 ft) above Lake Ontario. Already the waterfall, on the eastern side of North America between the borders of Canada and the USA, is some 11 km (7 miles) upstream of its original birthplace at the end of the last Ice Age.

The falls were formed when the glaciers that crushed North America started to melt, and four of the Great Lakes – Huron, Superior, Michigan, and Erie – began to drain into Lake Ontario through a new channel, the Niagara River, cascading over the cliff now known as Niagara Falls.

THE TALLEST WATERFALL IN THE WORLD

The extraordinary Angel Falls in Venezuela are 15 times taller than North America's Niagara Falls (*see page 321*). In October 1937, an American pilot, James Angel, was forced to crash-land his plane on a flat-topped mountain in Venezuela. Once he had struggled out of the cockpit he noticed that across a vast canyon lay another flat mountaintop, Devil's Mountain, and spilling over its edge tumbled a ribbon of water dropping down almost 1 km (3,300 ft) of sheer cliff face. He had discovered the world's tallest waterfall.

The waterfall was named Angel Falls after him. Today the only practical way of seeing the falls is by aeroplane.

SURGING TORRENTS THAT STRANGLE THE MEKONG

Water surges over the Khone Falls in South-east Asia at the incredible rate of some 9.5 million litres (2 million gallons) a second. This is the greatest volume travelling over any waterfall in the world, yet the 10 km (6 mile) long Khone Falls, close to the border of Cambodia and Laos, drops only 22 m (72 ft).

The falls were discovered in the 1870s when a French expedition paddled up the Mekong River in search of an inland route to China. The savage run of cascades forced them to give up their journey. Even today, the waters are so violent that they have strangled development in the region.

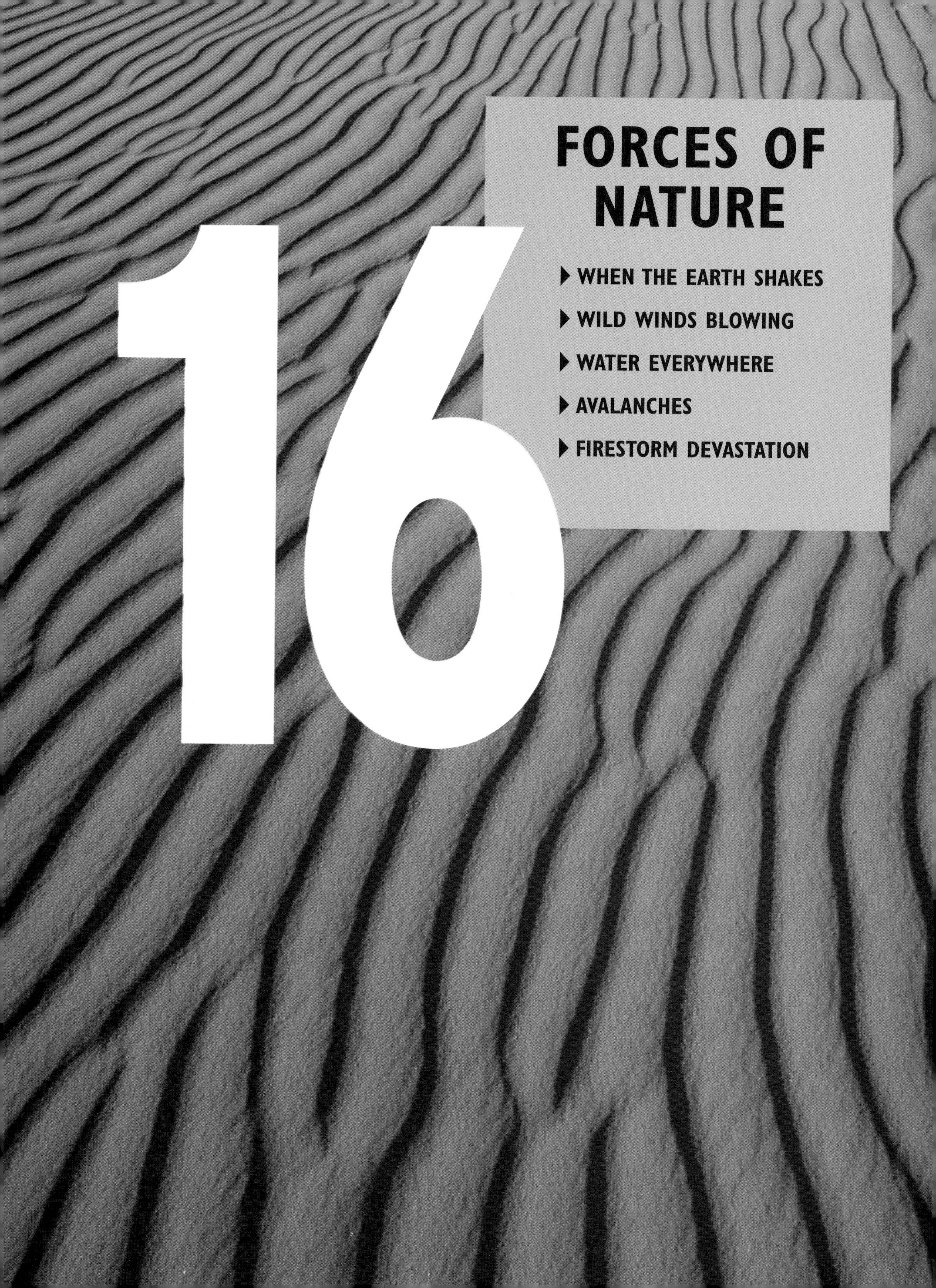

16

FORCES OF NATURE

- WHEN THE EARTH SHAKES
- WILD WINDS BLOWING
- WATER EVERYWHERE
- AVALANCHES
- FIRESTORM DEVASTATION

WHEN THE EARTH SHAKES

Earthquakes are totally unpredictable. In just a few seconds the slow movement of the Earth's crust suddenly accelerates and vast fault lines open up, sending out massive shock waves which can devastate buildings and roads.

THE FORCE THAT ROCKED SAN FRANCISCO

On April 18, 1906, the USA suffered its worst natural disaster when two huge jolts rocked the city of San Francisco. The ground split open for 470 km (292 miles) along the San Andreas Fault, and in some places the land shifted 6 m (20 ft) sideways. The earthquake was estimated at 8.3 on the Richter scale.

Overall the damage from the shock waves was surprisingly slight because the buildings were made of wood and were flexible enough to withstand the shaking. But the damage caused by fire was catastrophic when escaping gas from shattered pipes ignited. The fire brigade was unable to fight the flames because the water pipes had also fractured. The inferno blazed for three days. Some 3,000 people died and 225,000 were left homeless out of a population of 400,000.

QUAKE ZONE *San Francisco has suffered numerous earthquakes. This one in 1989 measured 7.1 on the Richter scale, killed 62 people, and caused damage of around $6 billion.*

THE DAY THAT LISBON TUMBLED

On November 1, 1755, the Portuguese capital of Lisbon was shaken by an earthquake that killed 60,000 people. The city was reduced to rubble. Fires broke out and many of the survivors fled to the harbour. But around 11am a mountain of water reared up out of the sea and drowned most of the unsuspecting onlookers. Several more tsunamis followed (*see page 438*).

The earthquake was caused by the North African plate slipping past the plate holding Europe and Asia. It is estimated to have been a gigantic 8.7 on today's Richter scale.

THE BIGGEST KILLER OF MODERN TIMES

When a fault line slipped under Tangshan, China, 90 per cent of the town's buildings collapsed within seconds. The ground vibrated sideways, then jolted up and down. Some survivors heard a whistling sound, others reported a large bang.

The earthquake, registering 8.3 on the Richter scale, struck this town near Beijing in the early hours of July 28, 1976. Almost the entire population was buried; 242,000 people died and 164,000 were injured. It was the most disastrous earthquake in modern times anywhere in the world.

A BRUTAL WAKE-UP CALL FOR KOBE

The citizens of Kobe in Japan had a tremendous shock on January 17, 1995, when a quake destroyed most of the city. The earthquake,

SWALLOWED UP *During the Kobe earthquake damage to infrastructure was made worse by the region's predominantly sandy soil.*

measuring 7.2 on the Richter scale, struck at 5.46am and many residents died in bed when their homes collapsed. Some 6,310 people died.

Older buildings bore the brunt of the damage. Many of the newer buildings remained upright because they were designed to withstand quakes, but engineers were shocked by the collapse of a large elevated section of the modern Hansin highway that had been specially strengthened.

A SLEEPY TOWN'S BAD DREAM

In 1811, a sleepy little town in the US Midwest turned into some scene from a nightmare: buildings collapsed, trees toppled, and the nearby Mississippi broke its banks.

During the night of December 15, the citizens of New Madrid, Missouri, were woken by violent shaking and a tremendous roar. Land crumpled, and the tremors made church bells ring in Boston, 1,600 km (1,000 miles) away.

Two more quakes followed at what was probably magnitude 8 on the Richter scale and thousands of smaller aftershocks. It was the most powerful series of earthquakes in US history.

ROCK AND ROLL

There are tens of thousands of earthquakes a year, most of them minor tremors. Altogether they release the same amount of energy as 100,000 atom bombs. But what causes them?

The Earth's crust is made of massive plates, around 100 km (60 miles) thick, which are constantly slipping and sliding past each other along fault lines. As the edges of these plates bump and grind against one another they often get snagged. Tremendous pressure builds up until eventually the plates rip past each other, producing an earthquake.

The violence at the centre of the quake is shunted through the surrounding rocks at colossal speeds, spreading the earthquake far and wide. The force can bring down tall structures, fracture roads, and trigger landslides.

Perhaps the best-known fault line is the San Andreas Fault that cuts through California and is an ever-present threat to the cities of San Francisco and Los Angeles *(see page 435)*. Along the Pacific coasts of Japan, Russia, and South America the Earth's plates are sliding under their neighbours, leading to much deeper earthquakes occurring 650 km (400 miles) beneath the surface.

China suffers severe earthquakes where the Earth's crust is being deformed by the impact of the Indian subcontinent hitting Asia and pushing up the Himalayan mountain range.

Other mountain chains produce smaller seismic shakes. Iceland is located on the Mid-Atlantic Ridge, a giant crack in the middle of the Atlantic Ocean which has formed because the two plates are moving apart. As the crust is pulled asunder, liquid rock (magma) seeps up to the surface and creates new mountains.

Plates moving apart have also created the Great Rift Valley (see *pages 390–1*). This 6,400 km (4,000 mile) slash in the Earth's surface from the Dead Sea to Mozambique is a seventh of the Earth's circumference. Magma rises through the cracks, sometimes erupting into active volcanoes. Frequent earthquakes show that the rift is still pulling apart.

SLIP AND SLIDE *As two plates grind past each other they may lock together for a while since their edges are not smooth, but eventually the forces are too great and they will suddenly give way causing an earthquake.*

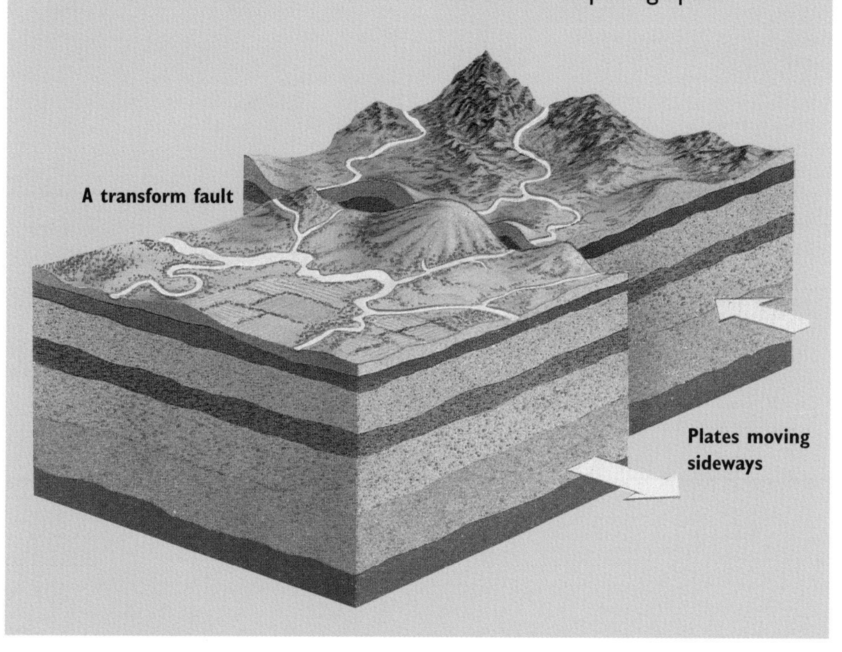

THE SCAR IN THE DESERT

The San Andreas Fault slices through two-thirds of California, a vast fracture in the Earth's surface stretching for more than 1,100 km (680 miles) in a south-east to north-west direction. It is up to tens of kilometres wide, but is mostly hidden underground. In just a few places the fault is revealed as a scar on the landscape.

The San Andreas accounts for more than 10,000 earthquakes (mostly minor tremors) each year, and occasional disasters such as those in San Francisco in 1906 and 1989, and in Los Angeles in 1933, 1971, and 1994.

The fault occurs where the Pacific Ocean plate rubs against the North American plate. The two plates are slipping past each other at about 5 cm (2 in) a year, about the rate of a fingernail growing. This is causing Los Angeles to move north towards San Francisco, until eventually the two cities will meet in 15 million years' time.

WILD WINDS BLOWING

Of all natural disasters, storms are among the biggest killers. Tornadoes can be the fastest winds on Earth, but tropical cyclones, sometimes hundreds of kilometres across, move like swirling torrents and cause immense destruction.

PAYING THE PRICE OF HURRICANE ANDREW

The world's most economically damaging storm took place in the early hours of August 24, 1992, in southern Florida, USA. Hurricane Andrew was rated five out of five on the international scale of storm intensity, with gusts of winds and tornadoes moving at up to 270 km/h (168 mph). More than 1.5 million people were evacuated; 23 people died, and 250,000 were left homeless. In just four hours the damage amounted to $20 billion.

The storm formed off the west coast of Africa, fed on the warm waters of the Atlantic, and turned into a hurricane, causing maximum damage to Florida, before petering out.

WHIRLING WINDS *The world's greatest number of tornadoes strikes Tornado Alley in the Midwest of the USA, with some 100 twisters a year.*

THE FASTEST WINDS ON RECORD

In May 1999 some 71 tornadoes touched down from a massive storm over Oklahoma and Kansas in the USA. It was the biggest tornado outbreak in history. The worst of the tornadoes tore Oklahoma City apart, killing 45 people and leaving hundreds more injured. A billion dollars' worth of damage was caused.

Wind speeds in the tornado's funnel reached 511 km/h (318 mph) – the fastest recorded winds on Earth. The funnel was 1.6 km (1 mile) wide. It tore across the ground for four hours, ripping up almost everything along its 100 km (60 mile) path.

THE TORNADO THAT LEFT DEATH IN ITS WAKE

In March 1925 the largest tornado in history crossed three US states, leaving a trail of destruction 350 km (217 miles) long. In total 689 people were killed, 1,980 injured, four towns destroyed, six severely damaged, and 11,000 people made homeless. The Tristate Tornado, as it was called, began in Missouri, when the people of Annapolis watched a huge, dark funnel-shaped cloud roaring like a giant freight train and sparking with lightning. The tornado headed into Illinois, swelling into a funnel 2 km (1 1/4 miles) wide, and crossed into Indiana before finally dying out.

THE LAND IN THE PATH OF THE CYCLONE

On the night of November 12, 1970, a cyclone swept into Bangladesh on waves surging up to 7 m (23 ft) high. The sea drove deep into the low-lying delta, engulfing everything in its path. Up to a million people died although the devastation was so vast that we will never know the exact number.

Seven of the nine most deadly weather events of the 20th century were cyclones striking Bangladesh.

THE KILLER RAINS OF HURRICANE MITCH

Rain dumped by Hurricane Mitch over Central America in 1998 triggered massive landslides that killed 20,000 people. Hurricanes are deadly not only because of their tremendous winds, but also because they pick up enormous quantities of

WHAT IS A HURRICANE?

The world's greatest tempests are known by different names: hurricanes in the USA, typhoons in South-east Asia, and cyclones in the Indian and Pacific Oceans. But they are all the same type of storm and they blast their way across sea and land with so much power that they could supply the energy needs of the USA for a century.

These tempests are born as tropical winds blowing off warm seas. The warm, wet air rises and as it cools in the cold air above, it condenses into thunderclouds. As the water turns into raindrops it releases heat. That heat rises in the cloud and sucks up air off the sea surface, fuelling the winds further.

The storm grows as it takes in moisture and warmth from the sea. As the winds grow stronger they get sucked into the middle of the storm where the warm air is rising fastest, and eventually the storm spins around a central 'eye'. The tighter the vortex, the faster the storm rotates. The storm eventually dies when it runs out of warm seas from which to draw energy, or when high-altitude winds destroy the storm's structure.

SUPER-SPIRALS *Winds move inside a hurricane in a powerful upward spiral, sucking up anything in their path. The absolute centre or 'eye' of the storm remains calm, however.*

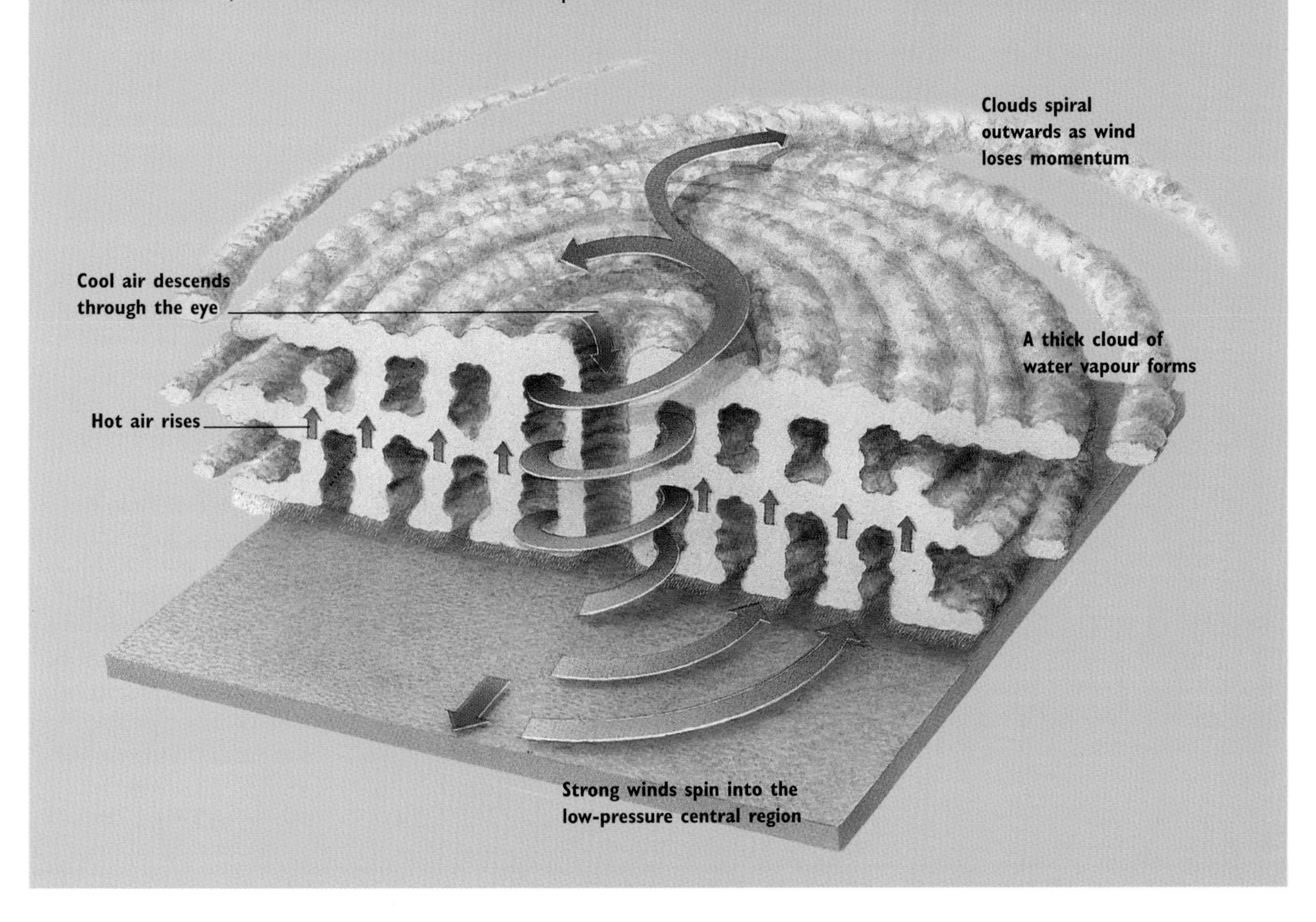

water vapour from the ocean – about 2 billion tonnes per day. When that water falls back down as rain, as it did when Nicaragua, Honduras, Guatemala, and El Salvador were struck by Hurricane Mitch, it can be lethal.

Mitch was the result of a large storm that brewed up in the Atlantic. As it drew close to Nicaragua its winds were howling at 320 km/h (200 mph). Then, as it pushed inland and was cut off from its fuel supply – the warm Caribbean – it released its massive cargo of rain. A million people lost their homes.

SUPER-HURRICANE BLOWS JAMAICA OVER

Hurricane Gilbert, the most intense hurricane of the 20th century, hit Jamaica in 1988. The pressure around the hurricane's eye was a record low of 885 millibars (the lower the pressure, the higher the wind speed), winds gusted up to 330 km/h (205 mph), waves surged up to 6 m (20 ft) high, and 38 cm (15 in) of rain fell in a few hours.

The maelstrom tossed planes into trees and mowed down forests. Once it had savaged Jamaica, the storm then picked up speed as it headed for the Cayman Islands and the Gulf of Mexico. Altogether it left 200 people dead and 800,000 homeless.

WATER EVERYWHERE

When whipped up by the weather or stirred by seismic forces, water can cause untold devastation. Immense waves known as tsunamis, freak floods, and water-sodden landslips are natural events over which we have little control.

A DESTRUCTIVE WALL OF WATER

On July 17, 1998, a tsunami tore a gaping hole through 30 km (18 miles) of the southern coast of Papua New Guinea. Towns and villages in its path were obliterated and 10,000 people left homeless. Witnesses said that the sea suddenly disappeared, followed moments later by a distant roar like a jet engine. Then a towering wall of water appeared from the distance, surging forward so fast that people had little chance to escape. As the tsunami sped inland it scoured the sea floor of sand and literally sandblasted its victims, killing more than 2,000 people.

This tsunami was triggered by a tremor that set off an underwater landslide of sediment. It is thought that a third of tsunamis might be caused by these underwater slumps.

MASSIVE WAVE BREAKS ALL RECORDS

The world's highest recorded tsunami struck Lituya Bay on the Alaskan coast on July 9, 1958. The wave was 530 m (1,740 ft) high, which is ten storeys higher than the Empire State Building in New York. It was caused by a major earthquake that triggered a landslide in the upper part of the bay. An avalanche of rocks cascaded down an almost kilometre-high mountainside, severing the front of the Lituya glacier, which plunged into the bay.

Some 30 million m³ (40 million cu yd) of debris hitting the water caused a massive wave to shoot up a slope on the opposite side of the bay, stripping the mountainside of soil and trees. The two-man crew of a fishing boat perished, though their vessel and another survived.

WHAT IS A TSUNAMI?

A tsunami is an enormous wave or series of waves with enough energy to wipe out a town. Tsunamis happen when a big seismic event shakes the sea floor and causes the sea to ripple. Seismic events include earthquakes, volcanic eruptions, a massive landslide dropping into the sea, a huge landslide under the sea, or even an asteroid, meteor, or comet impact. The vast volume of sea water and tremendous amount of energy create waves that can travel 800 km/h (500 mph) – as fast as a jet airliner.

A tsunami is hard to detect in the deep sea because only the crest of the wave is visible, perhaps 20–60 cm (8–24 in) high, while the bulk is submerged for hundreds of metres below. The length of tsunamis is immense – the distance from crest to crest may be 1,000 km (600 miles) – and these terrible waves can cover vast distances without losing much energy.

It is only when the tsunami arrives at a shallow coastline that it heaves up from the seabed and reveals itself, often arriving as a dark wall of water. Sometimes the sea can be sucked right out of the coast, leaving fish flapping on the exposed sea floor, followed by the arrival of the tsunami. The wave then bulldozes away everything in its path.

MOUNTAIN REGION WAITING FOR THE NEXT DISASTER

The spectacular mountains in the Italian Alps are notorious for flash floods. One such flood in 1994 tore apart the Piedmont region around Genoa, killing 100 people and leaving 10,000 homeless.

The Alps stand like a wall in the way of warm wet air sweeping up from the Mediterranean. The clouds tend to break on the southern flanks of the mountains and the rainwater washes down the steep slopes in torrents, bursting through towns and villages with little warning.

FREAK STORMS ENGULF A CITY IN MUD

Venezuela experienced a series of thunderstorms in late 1999 in which 80 per cent of the average yearly rainfall fell in 16 days. The effect on Caracas, a city set among steep hills, was catastrophic. On December 20 the intense downpours

unleashed deadly mudslides that killed 50,000 people.

It is thought that the tragedy was due to a band of tropical storms that shift north or south of the Equator with the seasons, moving farther north than usual. Although why the storms changed course is not clear.

ERUPTION THAT LEAD TO A WAVE OF TERROR

When Krakatoa erupted on August 27, 1883, it unleashed the deadliest tsunami ever recorded. The massive blast left behind a 290 m (950 ft) deep crater in the ocean floor that created a 40 m (130 ft) wave, which surged 5 km (3 miles) inland over the nearby islands of Java and Sumatra. Thousands of people were drowned and a Dutch warship was tossed 3 km (1 3/4 miles) up a Sumatran valley.

Even 5,000 km (3,100 miles) away in India, the tsunami sank 300 river boats on the Ganges. In total, 36,000 people were killed around the world by the effects of the Krakatoa eruption.

MIGHTY RIVER DROWNED TOWNS AND VILLAGES

The flood that burst out along 1,600 km (1,000 miles) of Mississippi and Missouri riverbanks in July 1993, left 70,000 homeless. For the previous nine months the central region of North America had been continually drenched by intense rains as high-level jet-stream winds steered in endless storms. The ground had become so saturated and the rivers so full that it was only a matter of time before the levees – earth embankment defences – gave way, causing more than $15 billion in damage.

The Mississippi flood was so shattering that the citizens of Valmeyer in Illinois rebuilt their entire town on higher ground.

For most people the chances of surviving a future flood remains questionable. Engineers can build limited flood defences, and weather forecasters can warn of dangerously heavy rain, but the stark truth is that living on the flood plain of a great river is a constant risk.

LIVING IN A FLOOD PLAIN *When the Mississippi burst its banks in 1993, after months of torrential rain, a total of nine states suffered from flood damage.*

AVALANCHES

An avalanche can carry as much as a million tonnes of snow, sending ahead of it a shock wave of air that blasts almost everything in its path. Anyone unfortunate enough to be trapped inside is likely to be crushed or suffocated to death.

AVALANCHE THREAT TO MEN AT WAR

Avalanches in the Tyrol Alps in the First World War took the lives of more than 60,000 men. Italian and Austrian troops fighting in the mountains endured three years of avalanches as well as snow, ice, and lack of supplies. One soldier said: 'The most frightening enemy was nature itself… entire platoons were hit, smothered, buried without a trace.'

Some of the worst incidents struck in December 1916 when 4 m (13 ft) of snow fell in 48 hours, setting off two days of avalanches that killed some 10,000 troops on both sides.

LANDSLIDE THAT KILLED TENS OF THOUSANDS

In Peru an earthquake on May 31, 1979, set off an avalanche, which together with the quake killed 66,000 people. The earthquake, measuring 7.7 on the Richter scale and centred near the industrial port of Chimbote, was the most deadly landslide of the 20th century.

It dislodged a massive slab of rock and ice on Mount Huascarán, which roared down on the town of Ranrahirca, obliterating its 5,000 residents and flattening the mountain resort of Yungay, killing almost 20,000 people.

THE STAGES OF AN AVALANCHE
1 *Loose snow slips on top of layers of harder snow.*
2 *As the mass of snow accelerates it can become suddenly airborne.*
3 *The avalanche gathers momentum, reaching up to 350 km/h (200 mph).*

FIRESTORM DEVASTATION

Natural fires sparked largely by lightning have been sweeping forests and grasslands for millennia, and help to regenerate the vegetation. But once raging they can be difficult to stop and cause destruction to land and life.

NORTH AMERICA SUFFERS A YEAR OF INFERNOS

In 2000 the USA suffered its worst wildfire season, as 96,000 fires raged over an area of 4,300 km^2 (1,660 sq miles). The cost of damage and firefighting totalled $1.6 billion.

The winter of 1999/2000 had been dominated by the La Niña weather phenomenon (*see page 421*) that had brought hot, dry weather to the south and west of the country.

The first fires of the year broke out in February in New Mexico and spread northwards. By May the fires were taking a huge toll. The western parts of North America were ablaze from Canada to Mexico. Dry thunderstorms brought lightning that triggered fresh forest fires but no rain to quench them.

More than 30,000 firefighters, many of them brought in from overseas, battled the flames for months. By September the number of fires had peaked and a month later the biggest fires were under control. But it was not until November that the last fires were extinguished.

CITY CHOKED BY SMOKE FROM FOREST FIRES

Kuala Lumpur, the capital of Malaysia, was plunged into 24-hour darkness by a thick haze of smoke in September 1997. The smoke that descended on the city for seven weeks came from thousands of fires that were raging across Malaysia, Sumatra, and Borneo.

The fires were started deliberately to clear forests so that the land could be used for farming, but they soon burned out of control as hot, dry winds fanned the flames. The normal monsoon rains failed to arrive for two months – a consequence of El Niño (*see page 421*) that had sent the region into its worst drought in 50 years. Altogether, 20,000 km^2 (7,720 sq miles), roughly the size of the US state of New Jersey, were turned into a charred wilderness.

The smoke smothered an area larger than Europe, choking 75 million people and killing more than 4,000, including those aboard a passenger plane which crashed due to the smog over Sumatra.

The economies of the eight affected countries nose-dived as tourists stayed away, crops were ruined, businesses closed down, and damage at an estimated $20 billion was caused. Wildlife also suffered, including the already threatened orang-utans of Borneo, and coral reefs in the surrounding seas which were choked by soil washed down from the burnt forests. It will take many years for them to recover from the incident, now regarded as one of the world's greatest environmental disasters.

FIRESTORM IN THE AUSTRALIAN BUSH

On February 16, 1983, a ferocious firestorm caused the deaths of 75 people in Victoria and South Australia. The Australian bush was like a tinderbox in early 1983: the El Niño weather phenomenon (*see page 421*) had driven temperatures up to a roasting 40°C (104°F) causing widespread chronic drought.

More than 100 fires simultaneously broke out across the south of the country. These built into an inferno that sped through the bush with winds gusting at around 160 km/h (100 mph). Bush fires are common in Australia, with around 15,000 fires sweeping the Australian outback each year. But it was the speed of this firestorm that caught so many people unawares and led to such a high number of fatalities.

OUTBACK INFERNO *The Australian bush burnt with such ferocity in 1983 that trees exploded in the 800°C (1,500°F) heat.*

INDEX

Page numbers in **bold** refer to main entries, those in *italics* to captions and/or illustrations only.

A

B

C

D

E

F

G

T

U

V

ACKNOWLEDGMENTS

Abbreviations:
T=Top; B=Bottom; R=Right; L=Left; M=Centre

BBCNHU - BBC Natural History Unit Picture Library
OSF - Oxford Scientific Films
SPL - Science Photo Library

1 OSF/Brian Kenny. 8-9 Tom Stack & Associates/Dave Fleetham. 12 Tom Stack & Associates/Thomas Kitchin. 13 Ardea/Chris Knights, T; DRK/Fred Bruemmer, B. 14 Ardea/Hans Dossenbach. 14-15 Minden Pictures/Frans Lanting. 15 Auscape/Becca Saunders. 16-17 DRK/Tom Brakefield. 18 Minden Pictures/Frans Lanting. 20 NHPA/Martin Harvey. 21 OSF/Martin Colbeck, T; DRK/Anup Shah, B. 22-23 BBCNHU/Karl Amman. 23 NHPA/Andy Rouse. 24 DRK/Fred Bruemmer. 25 DRK. 26-27 Gallo Images/ Beverly Joubert. 28 OSF/Peter Gathercole. 29 Auscape/Tui De Roy, T; BBCNHU/Niall Benvie, B. 30 DRK/ Dwight Kuhn. 31 Still Pictures/Mark Carwardine, T; DRK/Wayne Lynch, B. 32 OSF/Doug Allan. 33 Minden Pictures/Mitsuaki Iwago, T; Tom Stack & Associates/Jeff Foott, B. 34 Auscape/Tui De Roy. 35 DRK/Joe McDonald. 36 DRK/Fred Bruemmer. 37 OSF/David Thompson. 38 DRK/Joe McDonald. 39 DRK/Dwight Kuhn, TM, TR, BL; DRK/ D. Cavagnaro, BM; DRK/Jeff Foott, BR. 40-41 DRK/Tom Brakefield. 42-43 Ardea/S. Meyers. 44 Minden Pictures/Gerry Ellis. 45 NHPA/Andy Rouse. 46 OSF/Robin Bush. 48 Tom Stack & Associates/Mike Severns. 49 DRK/M.C. Chamberlain, T; Ardea/Wardene Weisser, B. 50 NHPA/A.N.T. 51 Ardea/Hans & Judy Beste. 52 BBCNHU/Pete Oxford, M; DRK/Kim Heacox, B. 53 Tom Stack & Associates/Barbara Gerlach, B; OSF/H.L. Fox, T. 54-55 Tom Stack & Associates/Ken W. Davis. 55 DRK/Wayne Lynch, T; Tom Stack & Associates/John Gerlach, M. 56-57 DRK/Manfred Pfefferle. 58 Ardea/Alan Weaving. 59 Ken Oliver/Wildlife Art Ltd., T; Auscape/Reg Morrison, B. 60 OSF/Scott Camazine, T; DRK/N.H. Cheatham, B. 61 Barry Croucher/Wildlife Art Ltd, L; BBCNHU/Anup Shah, R. 62 OSF/D.H. Thompson, T; BBCNHU/Premaphotos, B. 63 Minden Pictures/Matthias Breitner, L; DRK Wayne Lankinen, M; NHPA/Rich Kirchner, R. 64 Auscape/Jean-Marc La Roque, T; Ian Jackson/Wildlife Art Ltd., B. 66 Ardea/Kenneth W. Fink. 67 BBCNHU/Francois Savigny, T; Peter Scott/Wildlife Art Ltd., B. 68 Ardea/Clem Haagner, T; NHPA/Stephen Dalton, B. 69 Minden Pictures/Fred Bavendam. 70 DRK/Ford Kristo, M; Tom Stack & Associates/Dave Watts, B. 71 Minden Pictures/Flip Nicklin, T; Ken Oliver/Wildlife Art Ltd., B. 72 BBCNHU/Mike Wilkes. 72-73 Sandra Doyle/Wildlife Art Ltd., 74 DRK/James Rowan. 75 OSF/David Haring. 76 DRK/Joe McDonald. 77 Minden Pictures/Mitsuaki Iwago, T; NHPA/N.A. Callow, M; DRK/David D. Dennis, B. 78 BBCNHU/Armin Maywald, L; Peter Scott/Wildlife Art Ltd., R. 79 NHPA, T; Ardea/Chris Harvey, B. 80 Barry Croucher/Wildlife Art Ltd. 81 Minden Pictures/Mitsuaki Iwago. 82 Auscape/Jean-Paul Ferrero. 82-83 DRK/Stephen Krasemann. 83 Auscape/D. Parer & E. Parer-Cook. 84 NHPA/Rich Kirchner. 85 OSF/Alain Christof, T; DRK/Tom & Pat Leeson, B. 86 Ardea/M.D. England, T; Ardea/D. Parer & E Parer-Cook, B. 87 Nature Production/Goichi Wada, T; BBCNHU/Rico & Ruiz, B. 88-89 DRK/Steve Kaufman. 90 Auscape/D. Parer & E. Parer-Cook. 91 NHPA/Daniel Heuclin, T, M; DRK/Michael Fogden, B. 92 Auscape/Jean-Michel Labat, T; NHPA/Stephen Dalton, B. 94 BBCNHU/Steve Packham. 94-95 NHPA/Daniel Heuclin. 95 NHPA/E.A. Janes. 96 Ardea/Valerie Taylor. 97 NHPA/Daniel Heuclin, M; DRK/Michael Fogden, B. 98-99 DRK/Norbert Wu. 100 NHPA/Stephen Dalton. 101 Ken Oliver/Wildlife Art Ltd., L; Minden Pictures/Fred Bavendam, R. 102 Auscape/Becca Saunders. 103 Premaphotos/Ken Preston-Mafham, T; NHPA/Daniel Heuclin, B. 104-5 DRK/Michael Fogden. 106 Auscape/Ferrero-Labat. 107 BBCNHU/Staffan Widstrand. 108 Still Pictures/Norbert Wu. 109 DRK/Stanley Breeden, T; BBCNHU/Magnus Nyman, B. 110 DRK/Stephen Kraseman. 111 Auscape/Bill & Peter Boyle, T; Bruce Coleman/Jane Burton, B. 112 Premaphotos/Ken Preston-Mafham, T; DRK/John Cancalosi, B. 113 Still Pictures/M. & C. Denis-Huot. 114 DRK/John Gerlach, T; DRK/Michael Fogden, B. 115 BBCNHU/Jeff Rotman, T; DRK/Marty Cordano, B. 116 Peter Scott/Wildlife Art Ltd., T; DRK/John Cancalosi, B. 117 NHPA/Stephen Dalton. 118 Minden Pictures/Mitsuaki Iwago. 119 BBCNHU/Peter Blackwell, T, Still Pictures/Roland Seitre, B. 120 Tom Stack & Associates/Tom and Therisa Stack. 121 Premaphotos/Ken Preston-Mafham, T; BBCNHU/Francis Abbott, B. 122 Ian Jackson/Wildlife Art Ltd. 123 Gallo Images/Anthony Bannister, T; NHPA/Daniel Heuclin, B. 124-5 OSF/Michael Fogden. 126 Tom Stack & Associates/Dawn Hire. 127 OSF/Kathie Atkinson. 128-9 DRK/Stephen Krasemann. 129 Auscape/Francois Gohier. 130 NHPA/Daniel Heuclin, L; DRK/Wayne Lynch, M. 132 Ardea/Francois Gohier. 133 DRK/John Gerlach. 134-5 NHPA/Manfred Danegger. 135 Ardea/Ferrero-Labat. 136 Minden Pictures/Mitsuaki Iwago, R; NHPA/Daniel Heuclin, B. 137 Minden Pictures/Mitsuako Iwago, L; Minden Pictures/Frans Lanting, B. 138 OSF/G.I. Bernard, T; BBCNHU/Chris O'Reilly, B. 139 Minden Pictures/Tui de Roy. 140 Auscape/D. Parer & E. Parer Cook. 141 DRK/Michael Fogden. 142 Lee Peters. 142-3 Ardea/Jean-Paul Ferrero. 144-5 DRK/M.C. Chamberlain. 146 Auscape/Ron & Valerie Taylor. 147 OSF/P. & W. Ward. 148 OSF/Michael McKinnon. 149 BBCNHU/Premaphotos. 150 Premaphotos/Ken Preston-Mafham. 152 Auscape/Jean-Paul Ferrero. 153 DRK/Michael Fogden, T; Ron Hayward, B. 154 Minden Pictures/Konrad Wothe. 154-5 Minden Pictures/Tui De Roy. 155 DRK/Barbara Gerlach. 156-7 DRK/David Northcott. 158-9 BBCNHU/Anup Shah, T; NHPA/Stephen Dalton, B. 160-1 NHPA/Martin Harvey. 161 NHPA/Stephen Krasemann. 162 Auscape/Nicholas Birks, T; NHPA/John Shaw, B. 162-3 Ardea/D. Parer & E. Parer-Cook. 164 OSF/Mark Deeble & Victoria Stone. 165 OSF/Doug Allan, T; Lee Peters, B. 166 Lee Peters. 167 Ardea/Kenneth Fink, T; DRK/Tom Brakefield, B. 168-9 DRK/M.C. Chamberlain. 170 NHPA/Stephen Dalton. 171 Lee Peters. 172 NHPA/Stephen Dalton. 173 NHPA/Jany Sauvanet. 174 Image Quest/Peter Parks, M; Lee Peters, B. 175 NHPA/Stephen Dalton. 176 Ardea/Jean-Paul Ferrero. 177 OSF/William Gray. 178 DRK/Peter D. Pickford. 178-9 Ardea/M.T. Bomford. 180 NHPA/Bill Coster. 180-1 DRK/Fred Bruemmer. 182 Ardea/David & Kate Urry. 182-3 DRK/Johnny Johnson. 183 Ardea/Jean-Paul Ferrero. 184 Auscape/Hellio & Van Ingen. 185 OSF/Richard Herrmann. 186 Minden Pictures/Frans Lanting, T; Auscape/Jean-Paul Ferrero, B. 188 OSF/Max Gibbs. 189 BBCNHU/Dietmar Hill. 190 DRK/M.P. Kahl, T; NHPA/Anthony Bannister, M; NHPA/Nigel Dennis, B. 191 OSF/Martyn Colbeck, T; Minden Pictures/Gerry Ellis, B. 192 Minden Pictures/Frans Lanting. 192-3 Tom Stack/Dave Fleetham. 194-5 OSF/Paul Kay. 196-7 NHPA/Stephen Dalton. 197 BBCNHU/Pete Oxford. 198 DRK/Belinda Wright. 199 BBCNHU/Adrian Davies, T; Ardea/Stefan Meyers, B. 200 Gerald Cubitt. 200-1 SPL/Andrew Syred. 202 BBCNHU/Jose Ruiz. 202-3 Minden Pictures/Jim Brandenburg. 203 Ardea/Eric Lindgren. 204 NHPA/Kevin Schafer. 205 Minden Pictures/Shin Yoshino, M; OSF/Manfred Pfefferle, B. 206 BBCNHU/Doc White, T; OSF/Mickey Gibson, B. 207 Minden Pictures/Tim Fitzharris. 208 Auscape/Ferrero-Labat. 209 Auscape/Ferrero-Labat, T; BBCNHU/Tony Heald, B. 210 BBCNHU/Georgette Douwma. 210-11 Gerald Cubitt. 212-13 DRK/Larry Lipsky. 214 Innerspace Visions/Ingrid Visser. 214-15 Minden Pictures/Fred Bavendam. 216 Tom Stack & Associates/Dave Fleetham, C; Lee Peters, B. 217 OSF/David Thompson, R; Ron Hayward, L. 218 Auscape/S. Cordier, T; DRK/Michael Fogden, B. 219 BBCNHU/Anup Shah, R; OSF/Clive Bromhall, T; Minden Pictures/Gerry Ellis, M. 220 Auscape/Jean-Paul Ferrero. 221 OSF/Owen Newman. 222 Premaphotos/Ken-Preston-Mafham. 223 Auscape/Kathie Atkinson. 224 Minden Pictures/Michael Quinton. 225 BBCNHU/ Pete Oxford. 226 Ardea/Chris Knights. 227 DRK/Anup Shah. 228 OSF/Densey Cline/

Mantis Wildlife Films. 230 Woodfall Wild Images/David Woodfall. 231 Tom Stack & Associates/ Jeff Foott. 232-3 Auscape/Jean-Paul Ferrero. 233 Richard Bonson/ Wildlife Art Ltd. 234 Gerald Cubitt. 234-5 BBCNHU/Jeff Rotman. 235 Ardea. 236 Premaphotos/Ken Preston-Mafham. 236-7 Ardea/ Francois Gohier. 238 NHPA/ Stephen Dalton, T; DRK/Lynn M. Stone, B. 239 Minden Pictures/ Frans Lanting. 240 DRK/Tom & Pat Leeson. 240-1 DRK/Gary Gray. 241 Gerald Cubitt. 242-3 NHPA/Martin Harvey. 244 OSF/Tim Shepherd, T; DRK/Stephen Krasemann, B. 245 Auscape/Jamie Plaza Van Roon. 246 Ardea/Eric Lingren. 247 BBCNHU/Duncan McEwan. 248 Tom Stack & Associates/Chip & Jill Isenhart. 248-9 OSF/Alastair Shay. 249 DRK/Steve Kaufman, T; Gerald Cubitt, B. 250 NHPA/ Brian Hawkes. 250-1 NHPA/G.I. Bernard. 251 NHPA/John Shaw. 252 BBCNHU/George McCarthy, T; OSF/Alastair Shay, B. 253 OSF/Michael Brooke, T; Ardea/Ian Beames, B. 254 Tom Stack & Associates/Jeff Foott. 255 NHPA/Laurie Campbell, T; Premaphotos/Rod Preston-Mafham, M. 256 DRK/Stephen Krasemann, T; Auscape/Reg Morrison, B. 257 Minden Pictures/Mark Moffett. 258 DRK/ T.A. Wiewandt. 258-9 BBCNHU/ Brian Lightfoot. 259 Minden Pictures/ Mark Moffett. 260 OSF/ T.C. Middleton, T; NHPA/ Anthony Bannister, B. 262 Still Pictures. 263 Richard Bonson/ Wildlife Art Ltd., T; OSF/Earth Scenes, B. 264 NHPA/Stephen Dalton, T; DRK/T.A. Wiewandt, B. 265 SPL/Andrew Syred/Eye of Science. 266 OSF/David Dennis. 266-7 OSF/David Thompson. 268 Ardea/Bob Gibbons. 269 OSF/ Harold Taylor. 270-1 OSF/Babs & Bert Wells. 272-3 DRK/Michael Fogden. 273 BBCNHU/Adam White. 274 BBCNHU/Jose Ruiz, T; BBCNHU/Duncan McEwan, B. 275 Ardea/John Mason, T; OSF/David & Sue Cayless, M. 276-7 BBCNHU/Adrian Davies. 277 Minden Pictures/Frans Lanting. 278 Gerald Cubitt. 279 DRK/Stephen Krasemann. 280 DRK/Michael & Patricia Fogden, M; BBCNHU/David Kjaer, B. 282-3 SPL/Andrew Syred. 283 NHPA/M.I. Walker, T; SPL/CNRI, B. 284 Premaphotos/ Ken Preston-Mafham, M; SPL/ Andrew Syred, B. 285 SPL/Eye of Science, T; SPL/Dr Jeremy Burgess, B. 286 SPL/Alfred Pasieka. 286-7 SPL/Juergen Berger, Max Planck Institute. 287 SPL/Dr Linda Stannard. 288-9 Woodfall Wild Images/Ted Mead. 289 SPL/A.B. Dowsett. 290 SPL/Andrew Syred. 290-1 Auscape/Jean-Michel Labat. 292 Premaphotos/Ken Preston-Mafham. 293 Ardea/Kenneth W. Fink. 294 Ardea/John Clegg. 295 SPL/Eric Grave, T; Premaphotos/Ken Preston-Mafham, B. 296-7 DRK/M.C. Chamberlain. 298 SPL/Moredun Scientific Ltd. 298-9 Auscape/ Becca Saunders. 299 NHPA/G.J. Cambridge. 300 Ardea, T; OSF/ Clive Bromhall, B. 301 SPL/ Andrew Syred, T; Richard Bonson/Wildlife Art Ltd., B. 302 Premaphotos/Ken Preston-Mafham. 304-5 Minden Pictures/ Tui De Roy. 305 Olive Pearson. 306 Olive Pearson, M. 306-7 Minden Pictures/Frans Lanting. 308-9 Auscape/Jean-Paul Ferrero. 310 DRK/Johnny Johnson, T; BBCNHU/Doug Allan, B. 311 DRK/Norbert Wu, T; Minden Pictures/Michio Hoshino, B. 312-13 DRK/Fred Bruemmer. 312 Ardea/M. Watson. 313 Auscape/Colin Monteath. 314 Olive Pearson, T; BBCNHU/ Mats Forsberg, B. 315 Ardea/M. Watson, T; BBCNHU/Cindy Buxton, B. 316 DRK/Michael Ederegger, T; BBCNHU/Peter Blackwell, B. 317 Auscape/Jean-Paul Ferrero. 318-19 Auscape/Tui De Roy. 319 OSF/Clive Huggins. 320 Olive Pearson. 320-1 Auscape/ Francois Gohier. 321 Still Pictures/ Silva-UNEP. 322 DRK/Joe McDonald. 323 NHPA/Otto Rogge, T; SPL/Dr Morley Read, B. 324 DRK/Belinda Wright. 325 DRK/Joe McDonald. 326 DRK/Stephen Maka, T; SPL/C.K. Lorenz, C. 327 DRK/ Fred Bruemmer. 328 Olive Pearson, M; OSF, B. 329 OSF/ Michael Fogden, T; DRK/Michael Fogden, M. 330 Minden Pictures/ Gerry Ellis. 331 BBCNHU/Nick Gordon, T; Ardea/Bob Gibbons, B. 332 NHPA/Kevin Schafer. 332-3 Minden Pictures/Mark Moffett. 334 Olive Pearson, T; DRK/ Michael Fogden, B. 334-5 NHPA/ Jany Sauvanet. 336 Minden Pictures/Konrad Wothe, L; Ardea/M. Iijima, B. 337 OSF, T; NHPA/Daniel Heuclin, B. 338 DRK/Stanley Breeden. 339 OSF/G.H. Thompson, T; OSF/Colin Milkins, B. 340 Olive Pearson, T; OSF/Pat Caulfield, B. 340-1 OSF/Jack Dermid. 341 DRK/Jeff Foott. 342 SPL/ Eye of Science. 342-3 SPL/Simon Fraser. 343 DRK/Jeff Foott. 344 DRK/Tom and Pat Leeson. 344-5 Minden Pictures/Frans Lanting. 345 NHPA/Hellio & Van Ingen. 346 Olive Pearson, T; BBCNHU/Premaphotos, B. 346-7 OSF/David Cayless. 347 DRK/M.Harvey. 348-9 BBCHNU/David Shale. 349-50 DRK/Norbert Wu. 351 DRK/Norbert Wu, T; OSF/ Peter Parks, B. 352-3 DRK/Steve Wolper. 354 Tom Stack & Associates/OAR/NOAA. 355 Images Unlimited/Al Giddings. 356 BBCNHU/Hanne & Jens Eriksen. 356-7 Tom Stack & Associates/Jeff Foott. 357 BBCNHU/Pete Oxford. 358 NHPA/Laurie Campbell, T; OSF/Ben Osborne, B. 360 SPL/ Luke Dodd. 361 SPL/Gordon Garrado, T; SPL/Jerry Looriguss, B. 362 SPL/Pekka Parvianen, T; Galaxy Picture Library, M. 363 SPL/NASA, T; Galaxy Picture Library/Y. Hirose, B. 364-5 SPL/David Parker. 365 SPL/ D Van Ravenswaay, T. 365 Images of Africa/Vanessa Burger, B. 366 Julian Baker. 367 SPL/NASA, T; Galaxy Picture Library, B. 368 SPL/NASA. 369 Galaxy Picture Library. 370 SPL/ European Space Agency. 370-1 SOHO. 371 Galaxy Picture Library. 372 Novosti, T; SPL/ John Sanford, M; SPL/NASA, B. 373 Richard Bonson/Wildlife Art Ltd. 374 SPL/Peter Ryan. 375 SPL/ David Nunuk, T; Mountain Camera/ John Cleare, B. 376-7 SPL/Frank Zullo. 377 SPL/Jerry Schad. 378-9 SPL/Chris Madeley. 380 Tom Stack & Associates/ Mark Allen Stack. 381 Richard Bonson/Wildlife Art Ltd., T; Fortean Picture Library/Werner Burger, B. 382 Tom Stack & Associates/TSADO/NCDC/ NOAA, T; SPL/Peter Menzel, B. 383 Richard Bonson/Wildlife Art Ltd., 384 OSF/Warren Faidley, T; Robert Harding Picture Library/ Tony Waltham, B. 386 Tom Stack & Associates/Thomas Kitchin, M; Tom Stack & Associates/Sharon Gerig, B. 387 Frank Spooner/ Gamma. 388 Robert Harding Picture Library/Minden Pictures. 389 NHPA/Daniel Heuclin. 390 Corbis/Kaehler. 390-1 Minden Pictures/Gerry Ellis. 392 OSF/ Stephen Downer. 393 OSF/ Martyn Chillmaid. 394-5 DRK/ Jeff Foott. 396 Magnum Photos/ Dennis Stock. 397 Minden Pictures/Carr Clifton. 398-9 DRK/ Barbara Cushman Rowell. 400 Richard Bonson/Wildlife Art Ltd. 400-1 Mountain Camera/ Colin Monteath. 402 NGS Image Collection/Richard Olsenius. 402-3 Minden Pictures/Mitsuaki Iwago. 404 Tom Stack & Associates/Dave Watts. 405 Gerald Cubitt. 406-7 Ardea/Jean-Paul Ferrero. 407 Lee Peters. 408-9 Woodfall Wild Images/ Nigel Hicks. 410 DRK/Stephen Krasemann. 410-2 Ardea/Francois Gohier. 412-3 Geoscience Features Picture Library/Dr B. Booth. 414 NGS Image Collection/ O. Louis Mazzatenta. 416 Ardea/ Edwin Mickleburgh. 417 SPL/ NOAA. 418-19 Minden Pictures/ Michio Hoshino. 419 NHPA/Paal Hermansen. 420 Tom Stack & Associates/Brian Parker. 421 Still Pictures/G. Griffiths-Christian Aid, T; Julian Baker, B. 422 Still Pictures/Michel Roggo. 422-3 NHPA/Nigel Dennis. 424 Galaxy Picture Library, M; SPL/Worldsat International Inc., B. 424-5 Minden Pictures/Gerry Ellis. 426-7 DRK/Barbara Gerlach. 427 DRK/Michael Collier. 428-9 NHPA/Martin Wendler. 430 Gerald Cubitt. 432 Robert Harding Picture Library. 432-3 Magnum Photos/P. Jones-Griffiths. 433 Lee Peters. 434 Tom Stack & Associates/ Merrilee Thomas. 435 Richard Bonson/Wildlife Art Ltd. 436-7 Ardea/Francois Gohier. 438-9 Katz Pictures. 439 SPL/ Earth Satellite Corporation. 440 Still Pictures/Roberta Parkin. 441 NHPA/Ralph & Daphne Keller.

FRONT COVER: Auscape/Yann Arthus-Bertrand, TR; Tom Stack & Associates/Dave Fleetham BR; NHPA/Jany Sauvanet, TL; Ardea/Adrian Warren, M. BACK COVER: Ardea/Adrian Warren

1000 Wonders of Nature was published by The Reader's Digest Association Limited, London

Reprinted 2002

We are committed to both the quality of our products and the service we provide to our customers. We value your comments, so please feel free to contact us on 08705 113366 or by email at: cust_service@readersdigest.co.uk

If you have any comments or suggestions about the content of our books, email us at gbeditorial@readersdigest.co.uk

1000 Wonders of Nature was edited and produced by Toucan Books Ltd, London, for The Reader's Digest Association Limited, London

Contributing Authors
Michael Bright
David Burnie
Tamsin Constable
Paul Simons

Editors
Celia Coyne
Daniel Gilpin
Jane Hutchings

Picture Researcher
Wendy Brown

Proofreader
Roy Butcher

Indexer
Laura Hicks

Design
Bradbury and Williams

Designer
Bob Burroughs

For The Reader's Digest, London

Reader's Digest, General Books, London

Project Editor
Lisa Thomas
Art Editor
Louise Turpin
Proofreader
Barry Gage
Book Production Manager
Fiona McIntosh
Pre-press Manager
Howard Reynolds
Senior Production Controller
Sarah Fox
Pre-press Technical Analyst
Martin Hendrick

Editorial Director
Cortina Butler
Art Director
Nick Clark
Executive Editor
Julian Browne
Publishing Projects Manager
Alastair Holmes
Development Editor
Ruth Binney
Picture Resource Manager
Martin Smith
Style Editor
Ron Pankhurst

Origination
Colour Systems Ltd

Printing and binding
Mateu Cromo, Madrid, Spain

ISBN 0 276 42614 2
BOOK CODE 400-077-02
CONCEPT CODE
GR0693/G-UK